Innovations and Breakthroughs in the Gold and Silver Industries

Vaikuntam Iyer Lakshmanan • Barun Gorain
Editors

Innovations and Breakthroughs in the Gold and Silver Industries

Concepts, Applications and Future Trends

Springer

Editors
Vaikuntam Iyer Lakshmanan
Process Research Ortech Inc.
Mississauga, ON, Canada

Barun Gorain
Hindustan Zinc Limited
Udaipur, Rajasthan, India

ISBN 978-3-030-32548-0 ISBN 978-3-030-32549-7 (eBook)
https://doi.org/10.1007/978-3-030-32549-7

This Springer imprint is published by the registered company Springer Nature Switzerland AG
The registered company address is: Gewerbestrasse 11, 6330 Cham, Switzerland

This book is dedicated to Shelina and Sarada for their infinite patience and lifelong unwavering support.

Preface

Gold and silver have been considered stores of value since the dawn of human civilization. Historically, gold and silver have received love and affection from all, whether poor or rich. Gold and silver have been used as currency in the past but now are traded as commodities. The gold and silver mining industry creates jobs all over the world, in developing and developed countries as well as in and rural and urban areas. It creates well-paying jobs in distant communities that last for many years.

The mining of gold and silver is highly capital-intensive. Many tonnes of feed ore are mined and processed for recovering a few grams of gold. The trend of decreasing ore grades along with increasing ore complexity necessitates more water, energy, and advanced technologies, leading to higher operating costs and lower profit margins for companies. In addition, there are significant environmental and safety challenges. Industry uses reagents that are not environmentally friendly. Recent incidents of tailings dam failures are posing a major threat of "license to operate." Resource nationalism and disputes on royalties are taking a heavy toll on mining companies in some countries. Regulatory guidelines have become stringent across the world.

It is now widely accepted that a holistic approach to problem-solving is the key to resolving many of these challenges. But the mining industry has been slower in adopting an integrated approach involving the value chain, resulting in lost opportunities for the stakeholders. There is a growing awareness that the existing mining paradigm of massive waste generation and tailings storage is unsustainable.

A positive trend, however, is now emerging. The industry is looking at leveraging innovation and digital technologies to address the challenges. Automation along with a digital way of working is now being embraced to drive productivity across the value chain.

This book presents a value-chain perspective on gold and silver processing focusing on the present innovative trends including digital technologies, be it for resource identification, mining, process efficiency, improvement, safety, environment, and corporate social responsibility. Various advancements are also presented with actual case studies from the gold and silver mining industry.

We would like to express our sincere gratitude to all the contributors to this book, which include Dr. Aghil Ojaghi, Dr. Raja Roy, Dr. Luisa Moreno, and Prof. Kumar Murty. We appreciate Mr. Jonathan Chen for his support. Thanks are due to Ms. Savitha Ananth for her administrative support. Special gratitude is expressed to Ms. Brinda Megasyamalan, Springer, for her constant support and encouragement.

Mississauga, ON, Canada Vaikuntam Iyer Lakshmanan
Udaipur, Rajasthan, India Barun Gorain

Contents

About the Editors

V. I. Lakshmanan, Ph.D. is an internationally renowned Teacher, Scientist, and Innovator in the areas of sustainable development and clean technologies. He has more than 40 years of hands-on experience in technology commercialization and skills development initiatives with both private and public sectors including the United Nations. He has successfully guided process technologies from concept to commercialization for resource, energy, and chemical industries. Born and educated in India, he moved to Canada in 1974 after serving as a Lecturer at the University of Birmingham, UK. His corporate experience includes roles in Noranda, Eldorado Nuclear, Ortech Corporation, and Process Research Ortech Inc., the last of which he founded and guided to become a Global Leader in sustainable process technologies development through its laboratory and pilot plant. He is a Fellow of the Canadian Academy of Engineering and an Adjunct Professor at the Department of Materials Science and Engineering, University of Toronto. He is an active Mentor of new entrepreneurs and past chairman of the Canada India Foundation. He received the prestigious Queen Elizabeth II Diamond Jubilee Medal in 2013 for his services to civil society.

B. Gorain, Ph.D. has more than 25 years of experience in precious and base metals plant operations, capital projects, and technology and innovation roles. He is now the Chief Technology and Innovation Officer for Hindustan Zinc, Vedanta Resources. He was previously a Director with Barrick Gold based in Toronto for 12 years focusing on strategic areas, namely, innovative technologies, operations optimization, digitization, and automation. He also worked as Group Leader in Mineral Treatment with Teck (Cominco) for 6 years based in Trail, British Columbia, after working with Mt. Isa Mines (now Glencore) in Australia and Coal India. He has a Ph.D. in Metallurgical Engineering from JKMRC (University of Queensland) in Australia and a B.Tech. degree in Mineral Engineering from Indian School of Mines, Dhanbad.

About the Editors

About the Authors

Luisa Moreno, Ph.D. has more than 15 years of experience in finance, business development, and technical research, with a focus on technology, mining, and metal industries. She is a Physics Engineer (FCT/UNL), with a Ph.D. in Materials Science and Mechanics from Imperial College London in the UK. She held positions as Senior Analyst at Toronto based-investment banks and as an Investment Research Analyst at a global investment research firm. She has authored a number of advanced technical reports on key technology metals and strategic industrial minerals. She has great interest in the equitable, environmental, and sustainable development of mineral resources and advises governments and corporations on topics related to mine development, commodities supply security, supply chains, mineral sector development, and related economic growth. She is a recognized Analyst and common Guest Speaker on television and at international conferences. She is often quoted in national and international newspapers and news (e.g., Reuters, *The Globe and Mail*, Bloomberg, etc.). Currently, she is the Managing Director of Tahuti Global Inc., a company that she founded.

V. Kumar Murty, Ph.D. is Professor of Mathematics at the University of Toronto. His research is in number theory, arithmetic geometry, and their applications to information technology, especially security and privacy. He has authored more than 120 papers and authored or edited eight books in these areas. He also has two patents and, in 2011, was named Inventor of the Year at the University of Toronto. He is a Fellow of the Royal Society of Canada, a Fellow of the Fields Institute, and a Foreign Fellow of the National Academy of Sciences (India).

Aghil Ojaghi, Ph.D. obtained his Ph.D. in Mineral Processing Engineering from the University of Tehran. He is known for simplifying process operations, developing and implementing mineral processing and hydrometallurgical solutions, and applying expertise to ensure accomplishment of goals. He has over 14 years of experience in gold and base metals process development, plant commissioning, and implementation of innovative technologies for complex, refractory, and double refractory gold ores. He is a Metallurgy Manager and has conducted research and development work at Process Research Ortech Inc., Mississauga, Ontario, Canada.

Raja Ram Mohan Roy, Ph.D. obtained his undergraduate degree in Metallurgical Engineering from Indian Institute of Technology, Kanpur, and Ph.D. in Materials Science and Engineering from the Ohio State University. He has conducted research and development work for a number of years at the University of Toronto and Process Research Ortech Inc. He has over 20 years of experience in extractive metallurgy and materials science with emphasis on process flow sheet development. He has supervised and managed a large number of projects involving bench-scale and pilot-scale test work. He received Light Metals Recycling Award from TMS in 1998. He has coauthored 40 technical papers and was Coeditor of Chloride 2011 symposium organized by TMS.

Chapter 1
Key Challenges and Opportunities in the Gold and Silver Industry

B. Gorain and V. I. Lakshmanan

Gold and silver have been known to mankind since the fourth millennium BCE. For a large part of history, gold and silver have been considered stores of value and used as currencies. With the switch to fiat money and the advent of the industrial age, gold and silver have become commodities. The profitability of gold and silver industry is directly dependent on the prices of gold and silver, which is no longer dependent on supply and demand only but many other factors including investor sentiment. To get a complete picture of the price dynamics of gold and silver, it is important to understand the role gold and silver have played in human history.

1.1 Gold and Silver as Currency

Gold has always been considered more valuable than silver. Menes, the founder of the first Egyptian dynasty, has described the value of gold to be two and one half times the value of silver. Gold coins were first produced in Lydia, Western Turkey, in the sixth century BCE from electrum, a natural alloy of gold and silver. The use of gold coins called aureus was prevalent in the Roman Empire. Emperor Julius Caesar (49–44 BCE) set the weight of aureus gold coin to be 1/40 of a Roman pound (328.9 g) of pure gold or about ~8.2 g. The weight of aureus was successively reduced during the reign of Roman emperors (Wikipedia 2018a).

B. Gorain (✉)
Hindustan Zinc Limited, Udaipur, Rajasthan, India
e-mail: Barun.Gorain@vedanta.co.in

V. I. Lakshmanan (✉)
Process Research Ortech Inc., Mississauga, ON, Canada
e-mail: llakshmanan@processortech.com

V. I. Lakshmanan, B. Gorain (eds.), *Innovations and Breakthroughs in the Gold and Silver Industries*, https://doi.org/10.1007/978-3-030-32549-7_1

It fell to 1/45 of a Roman pound (~7.3 g) during the reign of Nero (54–68 CE) and to 1/50 of a Roman pound by the time of Caracalla (211–217 CE). New gold coin named solidus, meaning sold gold, was introduced by Diocletian (284–305 CE) around 301 CE and weighed 1/60 of a Roman pound or about 5.5 g each. The value of one gold coin solidus was set to 1000 silver coins called denarius. Only limited amount of solidus gold coins were struck and aureus also remained in circulation. Emperor Constantine the Great (306–337 CE) reintroduced the solidus gold coins but set its weight at 1/72 of a Roman pound or about 4.5 g each. Through all these changes, the purity of gold coins remained the same at over 99%. However, this was not the case with the silver coin denarius, which contained less and less silver as the time passed.

Around 211 BCE, denarius weighed 1/72 of a Roman pound or around 4.5 grams. During the rule of Augustus (31 BCE–14 CE), the weight of denarius decreased to 1/84 of a Roman pound or around 3.9 g. The silver content of denarius was 95–98% by weight between 211 BCE and 37 CE. The weight of denarius coins decreased to 1/96 of a Roman pound or around 3.4 grams during the time of Nero (37–68 CE) and silver content decreased to 93.5% (Wikipedia 2018b). The weight of denarius coins remained constant at 1/96 of a Roman pound or around 3.4 g after this time, but the silver content kept on decreasing. The denarius contained about 85% silver during the reign of Trajan (98–117 CE), 75% silver during the reign of Marcus Aurelius (161–180 CE), 60% silver during the reign of Septimius Severus (193–211 CE), 50% silver during the reign of Caracalla (198–217 CE), and only 0.5% in 268 CE during a period of intense civil war and foreign invasions (Peden 2017). The silver content of denarius was set to 5% in 274 CE by Aurelian. There was runaway inflation in the Roman Empire during the fourth century, which ultimately led to its disintegration. One Roman pound of gold was worth 50,000 denarius coins in 301 CE, 120,000 denarius coins in 311 CE, 300,000 denarius coins in 324 CE, and 20,000,000 denarius coins in 337 CE in the year of Constantine's death (Peden 2017).

After the fall of Western Roman Empire, Byzantine gold solidus coin commonly called bezant was widely used during middle ages throughout Europe and the Mediterranean. As the influence of Byzantine Empire waned, silver currencies became prevalent in Europe. Silver pennies, Italian denari, French deniers, and Spanish Dineros circulated in Europe. This led to use of a silver standard, which was followed in many countries before being abandoned in the early twentieth century. The ratio of value of gold to silver was 15½:1 from the sixteenth century to nineteenth century. The price of an ounce of gold was set at £0.89 in 1257 CE by Great Britain, which was gradually increased to £4.25 in 1717 CE and then kept at £4.25 per ounce of gold till 1944 Bretton Woods Agreement. In the nineteenth century, most countries printed paper currencies that were backed by gold reserves. It became known as the gold standard. The United States followed the British gold standard prior to 1791 CE. It set the price of gold at $19.49 per ounce in 1791, raised to $20.69 per ounce in 1834, and adjusted to $20.67 per ounce by the Gold Standard Act of 1900. Bretton Woods Agreement was adapted by most developed countries in 1944 to settle their international balances in US dollar effectively making the US dollar the official global currency. Under this agreement, the US dollar was made fully convertible to gold at the exchange rate of $35/ounce. Till 1960, the United States maintained the price of gold close to $35 per ounce by selling gold to the free market whenever the price of gold increased. In 1961, the London Gold Pool was established by pooling of gold reserves from eight central banks in the United States and seven European countries. The London Gold Pool tried to maintain the Bretton Woods System of fixed-rate convertible currencies and defend a gold price of US$35 per troy ounce in the London gold market. This resulted in major losses to the member countries, and the London Gold Pool collapsed in 1968. In 1972, the Gold Standard or the pegging of gold price to US dollar was abandoned, and in 1974 all restrictions on trading of gold were lifted. Gold became a commodity to be traded on stock exchanges.

1.2 Gold and Silver as Commodity

After gold started trading in the open market, the gold price rose sharply. It was trading over $160 per ounce in 1975 and over $600 per ounce in 1980. After that, the gold price dropped slowly to below $260 per ounce in 2001 and then rose again to peak at over $1900 per ounce in 2011. The gold price fell to below $1100 per ounce in 2015 and was trading above $1200 per ounce in October 2018. Figure 1.1 shows the price of gold over the last 6 years.

Silver price was around $1 per ounce in 1961 and rose slowly to around $6 per ounce in the beginning of 1979. The Hunt brothers, Nelson Bunker Hunt and William Herbert Hunt, tried to corner the silver market, and the silver price rose rapidly to a record high of $49.45 per ounce on January 18, 1980. In response to the rising silver price, COMEX (Commodity Exchange, Inc.), a division of the New York Mercantile Exchange (NYMEX), introduced Silver Rule 7 on January 7, 1980, placing heavy restrictions on the purchase of commodities on the margin. The price of silver started to fall and silver market crashed. The price of silver dropped by

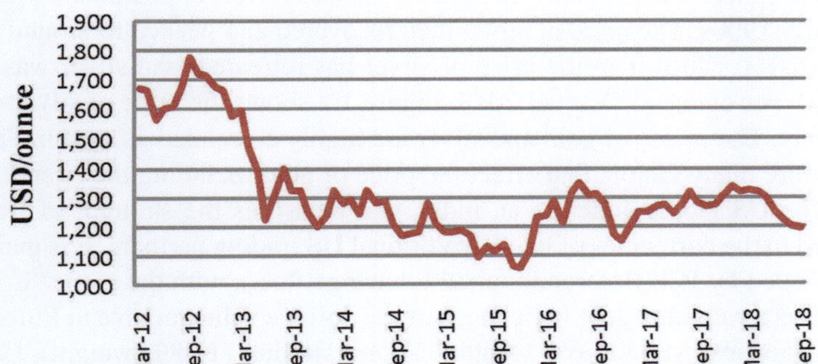

Fig. 1.1 The price of gold between 2012 and 2018 (Data from investing.com)

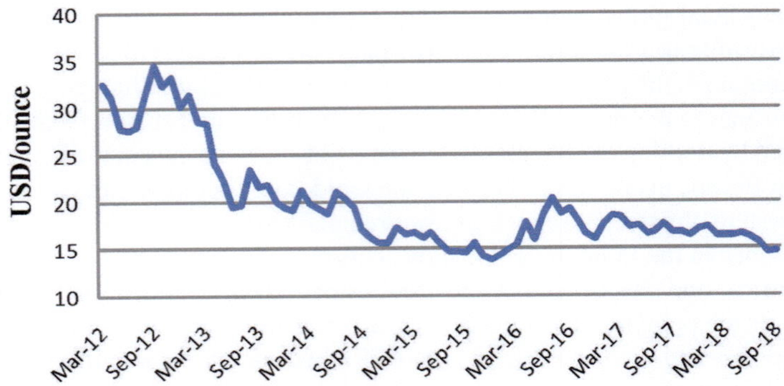

Fig. 1.2 The price of silver between 2012 and 2018 (Data from investing.com)

Fig. 1.3 The comparison of gold (thick line) and silver (thin line) prices between 2012 and 2018 (Data from investing.com)

50% on March 27, 1980, known as Silver Thursday, from \$21.62 per ounce to \$10.80 per ounce. Subsequently, the price of silver slowly fell to under \$4 per ounce in the early 1990s. The price of silver then recovered and peaked to around \$50 per ounce in 2011. Since then the price of silver has retreated, and silver was trading below \$15 per ounce in October 2018. Figure 1.2 shows the price of silver over the last 6 years. The prices of gold and silver are highly correlated as seen in Fig. 1.3.

There are many factors that affect the price of gold including the strength of US dollar. The US Dollar Index is an index that measures the strength of US dollar compared to the currencies of major traditional US trading partners. It is maintained and published by ICE (Intercontinental Exchange, Inc.), with the name "US Dollar Index". It is a weighted geometric mean of the dollar's value relative to Euro (57.6% weight), Japanese yen (13.6% weight), Pound sterling (11.9% weight), Canadian dollar (9.1% weight), Swedish krona (4.2% weight), and Swiss franc (3.6% weight). It was established in March 1973 after the Bretton Woods system was dismantled.

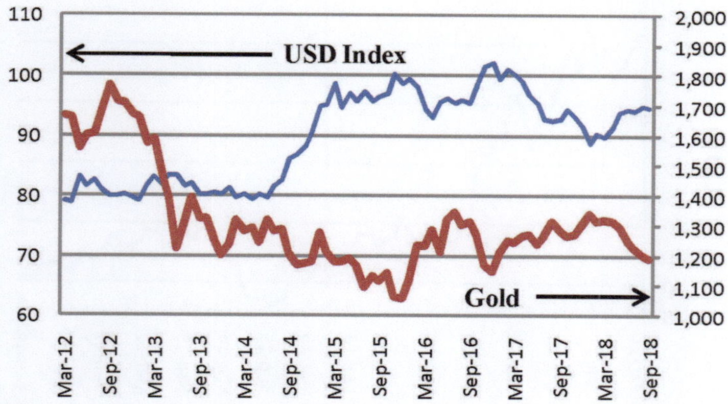

Fig. 1.4 The comparison of the gold price (thick line) and the US Dollar Index (thin line) between 2012 and 2018 (Data from investing.com)

Its value was 100 to begin with. Its highest value was 164.72 in February 1985 and lowest value was 70.70 on March 16, 2008 (Wikipedia 2018c). Figure 1.4 shows the value of the US Dollar Index and the price of gold over the last 6 years. It can be seen that the US Dollar Index and the price of gold have been negatively correlated since 2014.

As US dollar strengthens, gold becomes more expensive in other currencies leading to a fall in demand and a lower price of gold. Conversely, as US dollar weakens, gold becomes less expensive in other currencies leading to a rise in demand and a higher price of gold. As both gold and US dollar are considered safe havens, the price of gold and the value of US Dollar Index can rise together during times of economic uncertainty.

1.3 The Competition for Investment

Since gold is now traded as a commodity in the open market, the price of gold is affected by investor sentiments. Gold has traded lower since it peaked in 2011 at over $1900 per ounce. After making a low below $1100 per ounce in 2015, the price of gold has failed to reach $1400 per ounce. This has resulted in many investors shying away from investing in gold mining companies as the return on investment has been poor or negative. This can be judged from the share price performance of Barrick Gold Corporation, one of the largest gold mining companies in the world as seen in Fig. 1.5.

The share price of Barrick Gold Corporation peaked over $50 in 2011 but then fell below $7 in 2015 as the price of gold fell. Subsequently, the share price rose above $20 in 2016 as gold price rose due to economic uncertainty caused by Brexit. Since then share price of Barrick Gold Corporation has been trending lower and was trading above $11 in September 2018. Due to poor returns, investors have been

Fig. 1.5 The share price of Barrick Gold Corporation (Data from investing.com)

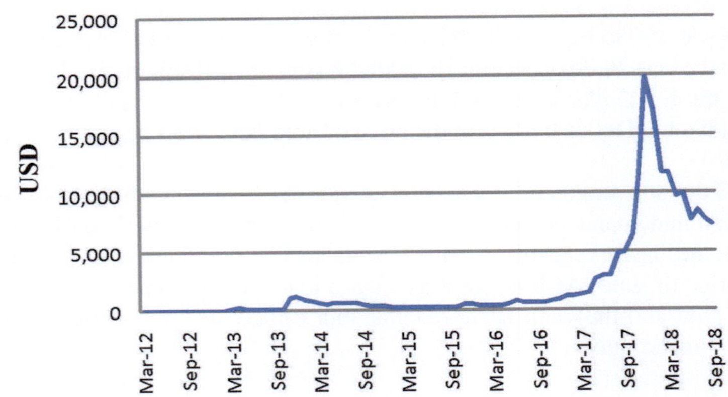

Fig. 1.6 The price of Bitcoin between 2012 and 2018 (Data from investing.com)

flocking to other asset classes in recent years such as cryptocurrencies and marijuana stocks.

A cryptocurrency (or crypto currency) is a digital asset that uses cryptography for secure financial transactions. It uses decentralized control as opposed to centralized banking systems through distributed ledger technology, typically a Blockchain, which serves as a public financial transaction database (Wikipedia 2018d). The most popular cryptocurrency is Bitcoin, first released as open-source software in 2009. Since the release of Bitcoin, over 4000 other cryptocurrencies have been created. Figure 1.6 shows the price of Bitcoin between 2012 and 2018.

The price of Bitcoin rose exponentially in 2017 reaching a peak of $19,891 in December 2017. The price of Bitcoin then fell sharply making a low below $6000 in June 2018. The Bitcoin was trading above $6000 in October 2018. The collective market cap of all cryptocurrencies has fallen from the peak value of over $700 billion

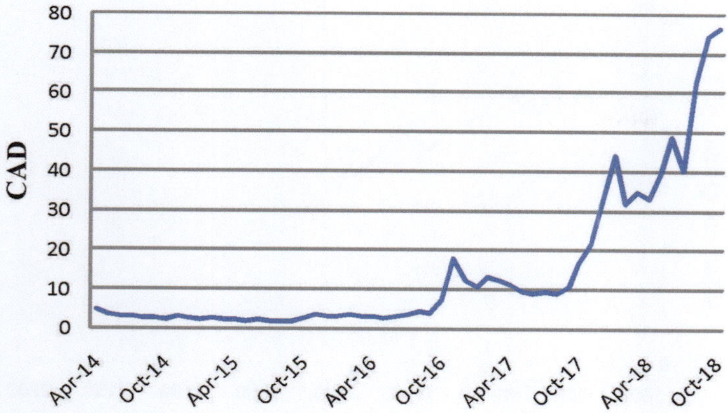

Fig. 1.7 The share price of Canopy Growth Corp (Data from investing.com)

in January 2018 to below \$200 billion in August 2018 and was holding above \$200 billion in October 2018 (Tradingview 2018).

During the last few years, investors were also pumping money in the cannabis sector. Canada is the first country among G7 nations to legalize marijuana. Recreational cannabis was legalized on October 17, 2018, in Canada. In anticipation, many companies were listed on Canadian stock exchanges and attracted significant investment. In May 2018, there were 90 publicly listed cannabis companies in Canada with a market value of about CAD 31 billion (Financial Post 2018b). Figure 1.7 shows the share price performance of Canopy Growth Corp, the biggest company in the world in the cannabis sector. From under 2 Canadian dollar in 2014 and 2015, the stock price of Canopy Growth Corp has rallied to over 75 Canadian dollar in October 2018.

With investors pumping money in cryptocurrencies and cannabis companies, mining companies are struggling to raise money for exploration and capital expenditures. As a result, a number of resource companies have become marijuana companies by reverse takeover and spinoffs. Nearly half of the Canadian marijuana companies, over 40 in March 2018, had started out in the resource sector before converting to marijuana companies (Financial Post 2018a). Reverse takeover of an existing mining or oil and gas "shells" as the listing vehicle is a potentially faster way to list on a stock exchange instead of a traditional initial public offering, which requires a full securities commission review or filing of a prospectus.

1.4 Challenges and Opportunities

The gold and silver mining companies are facing multiple challenges. It is difficult to find a large deposit with good grades. As shown in Fig. 1.8, the average grade of gold ores has been falling. This has resulted in a decreasing yield to spend ratio as

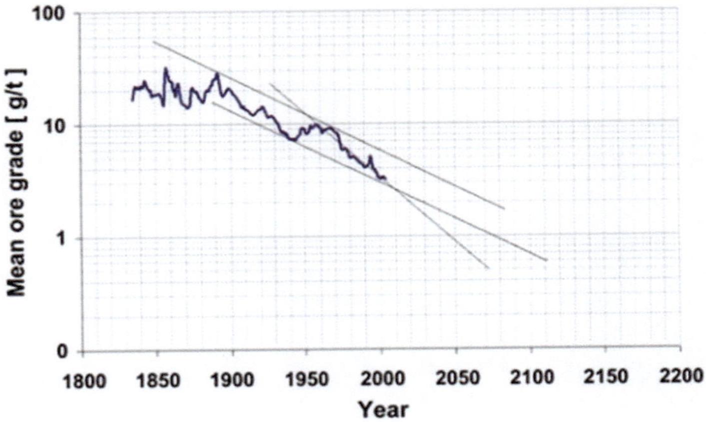

Fig. 1.8 The variation of average gold ore grade with time (Müller and Frimmel 2010)

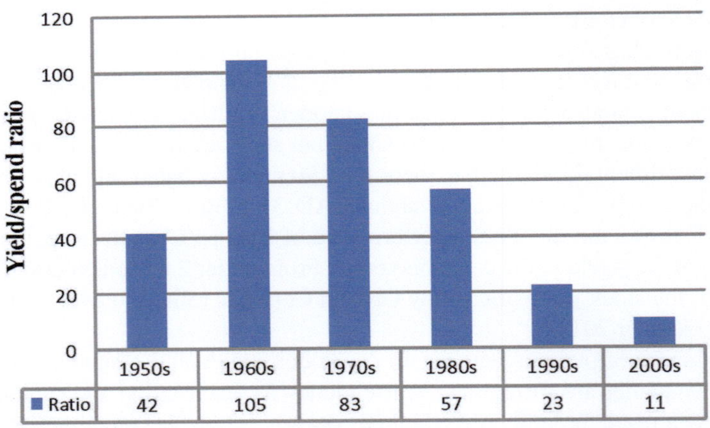

Fig. 1.9 The yield to spend ratio over time (Müller and Frimmel 2010)

shown in Fig. 1.9. In the 1960s, 1 US$ expenditure on exploration yielded 105 US$ of value in the ore discovered on average. This had decreased to 11 US$ of value per 1 US$ expenditure in the 2000s.

The need of the hour is for gold miners to find large economically viable deposits in mining-friendly jurisdictions with no environmental issues. The gold miners are facing the problem of depleting reserves as they are unable to find huge deposits of gold to replace the large gold mines that are close to getting depleted. Many gold miners have also sold some of their mines to reduce their debt. As shown in Fig. 1.10, gold reserves of Barrick Gold Corporation decreased from 140 million ounces at the end of 2012 to 64 million ounces at the end of 2017. This represents a drop in gold reserves of 54% in just 5 years. The amount of gold produced by Barrick Gold Corporation has also been decreasing as shown in Fig. 1.11. The amount of gold

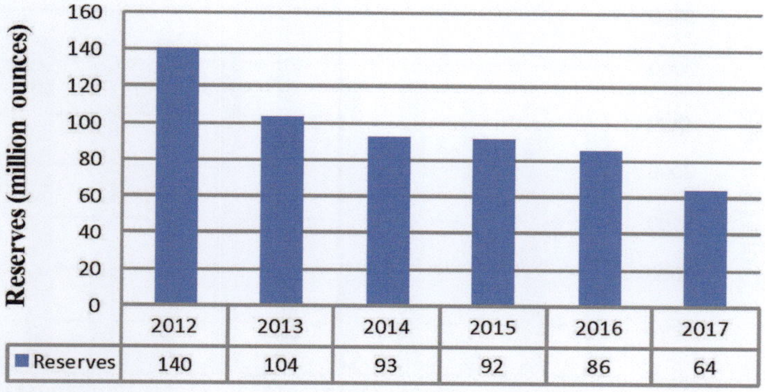

Fig. 1.10 The change in gold reserves of Barrick Gold Corporation

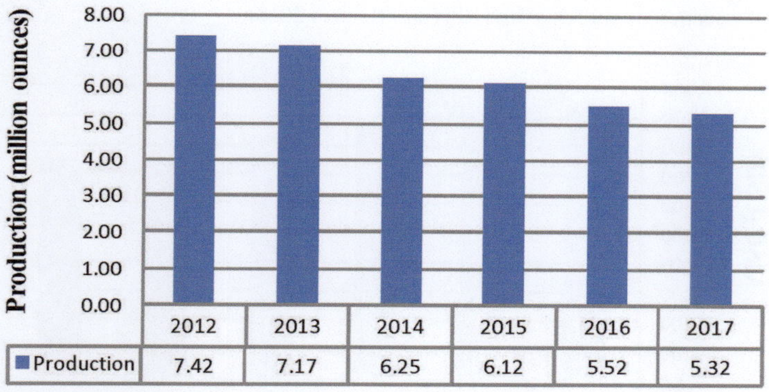

Fig. 1.11 The amount of gold produced by Barrick Gold Corporation (Reuters 2018)

produced by Barrick Gold Corporation decreased from 7.42 million ounces in 2012 to 5.32 million ounces in 2017. This represents a drop in gold production of 28% in 5 years. Other major gold miners have also reported lower gold production. The loss in gold production by major gold mining companies has been compensated by gold production by junior mining companies, and as a result, the world production of gold has reached a plateau. As shown in Fig. 1.12, world gold production has been close to 3250 tons during 2015, 2016, and 2017.

In response to the fall in gold price since 2011, gold companies have tried to be more efficient and reduce the cost of gold production. The gold companies have also become more transparent by developing a measure of what it actually costs them to produce an ounce of gold. This metric called all-in sustaining cost (AISC) was established by World Gold Council in 2013. Figure 1.13 shows the AISC of gold production by Barrick Gold Corporation from 2012 to 2017. Barrick was able to reduce AISC of gold production from $945 per ounce to $730 per ounce between 2012 and 2016. AISC increased slightly to $750 per ounce in 2017. For most gold

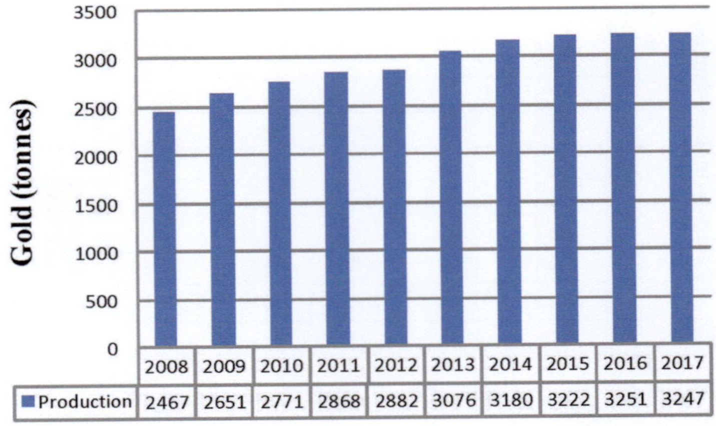

Fig. 1.12 Production of gold over last 10 years (Reuters 2018)

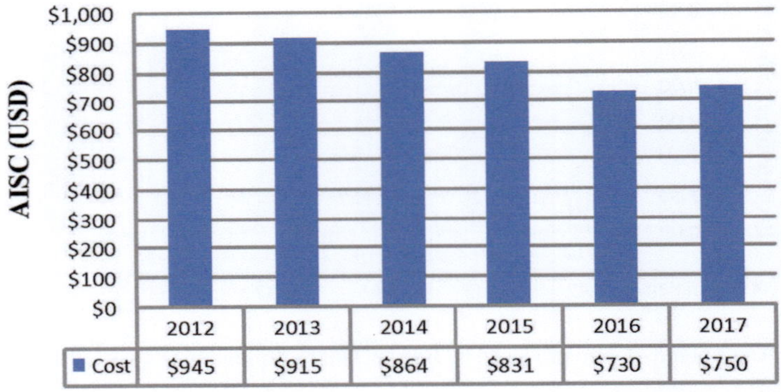

Fig. 1.13 All-in sustaining cost (AISC) of gold production by Barrick Gold Corporation

companies, AISC of gold production is much higher. When gold price fell below $1050 per ounce in December 2015, AISC of many gold miners became higher than the price of gold.

1.5 Embracing Innovation as a Core Value by the Gold and Silver Industry

As the gold mining industry is evolving, so are the challenges associated with mining. Some of the key challenges are low-grade complex ore bodies at greater depths, high capital and operating costs, high energy costs, water scarcity and quality issues, complex environmental issues including tailings management, worldwide constraints on resources availability, stringent regulations impacting the permitting processes, increasing stakeholder expectations, changing demographics of mining operations, and the urgent need for a better assessment of project viability, all these in a very dynamic metals market. This existing mining paradigm is just not sustainable.

The need of the hour for the gold and silver industry is to focus on innovation. There has been no major technology breakthrough in the industry for decades. The last major advancement took place in the 1980s with the emergence of pressure leaching to recover gold from refractory ores. Mining industry needs to invest heavily in research to develop technologies that will drastically lower the cost of finding new ore deposits and processing of ores to produce gold and silver. This is only possible if the fundamental issue of enormous generation of mining waste production, handling, and disposal is addressed as a priority. Selective mining and highly efficient metal extraction right at the mining face or in situ metal recovery is what we need to transform mining, with minimal use of water and energy and zero tailings generation.

The ideal scenario of zero waste mining is not impossible if the industry is determined to embrace innovation as its core value leveraging digital and advanced technologies along with continuously investing in research and development. This is only possible if we engage the young and bright minds to participate in the industry's strategic growth, operational, business, environment, and corporate stewardship activities. The digital era is upon us, and mining needs to gear up to prepare the industry for the next generation of jobs that don't exist today. This is an exciting journey, but the success and investor confidence will only happen if the industry rapidly embraces the digital way of working and innovative technologies.

References

Financial Post. 2018a. Marijuana industry now worth $31 billion looks abroad for growth. *Financial Post*. https://business.financialpost.com/cannabis/spain-colombia-next-frontier-for-canadas-deal-hungry-pot-firms.

Financial Post. 2018b. 'Like a Yukon gold rush': Marijuana is the new gold as former mining companies go to pot. *Financial Post*. https://business.financialpost.com/cannabis/marijuana-is-the-new-gold-as-former-mining-companies-go-to-pot.

Müller, J., and H.E. Frimmel. 2010. Numerical analysis of historic gold production cycles and implications for future sub-cycles. *The Open Geology Journal* 4: 29–34.

Peden, J.R. 2017. Inflation and the fall of the Roman Empire. https://mises.org/library/inflation-and-fall-roman-empire.

Reuters Thomson. 2018. GFMS gold survey report, London.

Tradingview. 2018. Cryptocurrency market. https://www.tradingview.com/markets/cryptocurrencies/global-charts/.

Wikipedia. 2018a. https://en.wikipedia.org/wiki/Aureus.

Wikipedia. 2018b. Denarius. https://en.wikipedia.org/wiki/Denarius.

Wikipedia. 2018c. U.S. Dollar Index. https://en.wikipedia.org/wiki/U.S._Dollar_Index.

Wikipedia. 2018d. Cryptocurrency. https://en.wikipedia.org/wiki/Cryptocurrency.

Chapter 2
Ore Body Knowledge

B. Gorain, V. I. Lakshmanan, and A. Ojaghi

2.1 Geometallurgy

Understanding of the geology and the uncertainties associated with the ore bodies is critical to the success of any mining operation. This is typically carried out using data generated through drill holes, geological mapping, geophysical surveys, and the geologist's interpretation. Ore body modelling and resource estimation are the foundation on which the business case for future mine development and operation is intricately dependent. One major input for resource estimation is quantitative mineralogy information, which is increasingly being recognized as more important than mere elemental assays. Interpretation of mineralogy through assays has been traditionally carried out, but with increasing complexities and uncertainties associated with ore bodies, a direct way of determining mineralogy without the need to make certain assumptions is a positive trend. Having more mineralogy attributes for both valuables and gangue in the resource and reserve models allows for a better integration of geology and metallurgy, which is an emerging area widely known as "geometallurgy" (Lakhsmanan et al. 2016).

Geometallurgy is an interdisciplinary approach which links the geological, geochemical, and mineralogical characteristics to the metallurgical performance of an ore body (Li and Zhou 2017). The extractive metallurgy of gold ores is largely driven by mineralogical factors due to the fact that gold often occurs in at least two forms in an ore. These factors include particle size, association with other minerals, coatings and rimmings, presence of cyanicides, oxygen consumers and preg-robbers, presence of refractory gold minerals, locking of submicroscopic gold in sulfide and sulfarsenide mineral structures, etc.

B. Gorain (✉)
Hindustan Zinc Limited, Udaipur, Rajasthan, India
e-mail: Barun.Gorain@vedanta.co.in

V. I. Lakshmanan (✉) · A. Ojaghi
Process Research Ortech Inc., Mississauga, ON, Canada
e-mail: llakshmanan@processortech.com

V. I. Lakshmanan, B. Gorain (eds.), *Innovations and Breakthroughs in the Gold and Silver Industries*, https://doi.org/10.1007/978-3-030-32549-7_2

Recent advancements in process mineralogy have furthered the cause of integrating mine site geology and process plant to a great extent with a main focus on improving the value of the mine. Geometallurgy relates to the practice of combining geology and geo-statistics with extractive metallurgy to create a geologically based predictive model for mineral processing plants. It is used for risk management and mitigation during plant design and also to assist and enhance mine production planning (Kittler et al. 2011). McCullough et al. have also highlighted the challenges associated with recognizing and embedding the value of geometallurgy in the mining value chain. The main challenge is to permanently change the behavior of people and the processes they follow. This requires executive sponsorship, technology, and data integration along with automated intelligent analysis to realize the full value of geometallurgy (McCullough et al. 2013). Kittler et al. have emphasized the need for proper sample selection with a clear focus on spatial and grade distribution. It is best to avoid blending of geometallurgy samples prematurely to capture ore variability and to retain their spatial provenance. The requirements for operational geometallurgy have been presented by Kittler et al. (2011).

The typical framework, as illustrated in Fig. 2.1, consists of several phases and relates the aspects of geology, metallurgy, and mining.

Bye has documented various industrial case studies that demonstrate strategies for gaining value from geometallurgical studies in 2011. Values include both operational benefits such as proactive fragmentation control and better strategic planning process along with building of geometallurgical domain models (Lakshmanan et al. 2016).

A geometallurgical program usually starts from collection of geological information and analysis, followed by geological modelling. Based on the geological model,

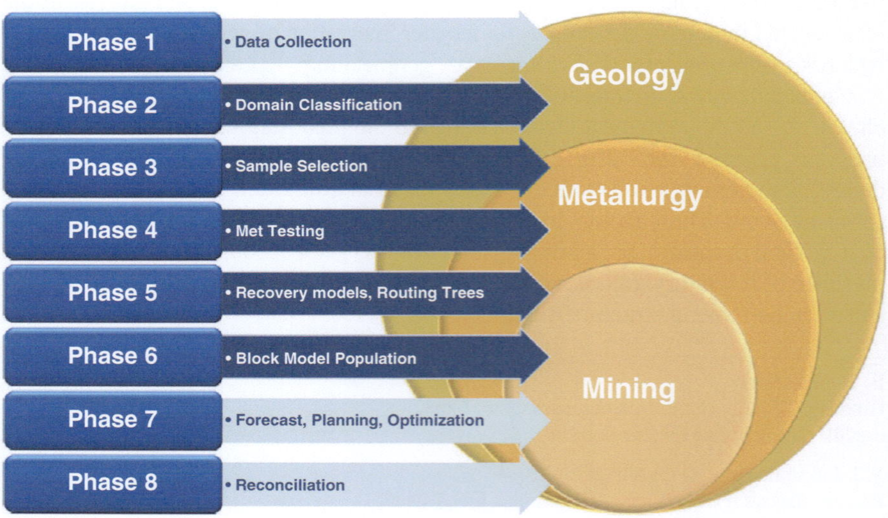

Fig. 2.1 Geometallurgical framework

a geometallurgical matrix is typically developed to guide sampling and compositing for further testing. The most important part of a geometallurgical program is ore characterization, including mineralogical studies and metallurgical testwork. Overall, ore characterization is the quantification of physical data on samples that represent an ore body and provide criteria for process design and flowsheet development (Li and Zhou 2017). Ore bodies are naturally occurring phenomena with little consistency and similarity between different locations and ore types. Even within a single ore body, variations in rock type, ore grade, chemistry, mineralogy, alteration, and deportment of pay metals and deleterious elements exist both vertically and laterally, thus causing issues in ore processing. To understand the ore body variability and to characterize the metallurgical performance, the first stage is to use geological information as a foundation and geo-statistics as a tool to collect samples from representative ore types with various grades, host rocks, and alterations at different locations of the ore body. This is followed by a comprehensive mineralogical and metallurgical testwork program on these samples to generate data which is processed and integrated with other information into the geological model to ascertain the distribution and variation of the mineralogical and metallurgical parameters within the ore body.

In recent years, geometallurgy has advanced rapidly due to the need for developing a number of large deposits (particularly gold ores) and also to the implementation of some sophisticated analytical techniques such as MLA, QEMSCAN, D-SIMS, TIMA, and TOF-SIMS.

Several technologies have been developed to identify changes in the ore being processed and modify operational parameters and geometallurgical models accordingly. At the Phu Kham copper-gold deposit in Laos which has an extremely heterogeneous ore body, a throughput prediction model has been developed which involves integrated site-specific models of drilling, blasting, crushing, and milling operations. These models were developed, calibrated, and validated using SmartTag ore tracking technology which links the ore source (and properties) in the mine with blasting and plant performance, viz., throughput, grade, and recovery in real time. Geometallurgical ore domains were defined along with modelling of blasting and comminution processes using measured plant results, thus enabling short- to long-term strategic planning and optimization (Bennett et al. 2014).

At the Tropicana gold mine in Western Australia, Bond Mill Work Index estimates were derived from X-ray fluorescence and hyperspectral scanning of grade control samples to construct special geometallurgical models. Since inaccuracies in block estimates exist due to limited calibration between grade control derived and laboratory work index values, a method was developed to update block estimates using actual mill performance data. This allowed better refining of work index estimates in the geometallurgical model and a real-time reconciliation of already extracted blocks and a recalibration of future selected blocks (Wambeke et al. 2018).

2.2 Gold Mineralogical Types

Mineralogically, gold associations can be classified into three forms based on its deportment: microscopic gold, submicroscopic gold, and surface gold. Microscopic gold, also known as visible gold, comprises all gold minerals such as gold alloys, gold tellurides, gold sulfides, gold selenides, gold sulfotellurides, and gold sulfoselenides (Zhou et al. 2004). Microscopic gold is found in many gold ores and is the main form of gold in non-refractory gold ores. Gold that is invisible under optical and scanning electron microscopes is referred to as submicroscopic gold and is the major form of gold in refractory gold ores such as Carlin-type gold deposits, some epithermal gold deposits, and volcanogenic massive sulfide (VMS) deposits. Surface gold is gold that was adsorbed onto the surface of other minerals (such as carbonaceous material and iron oxide) during mineralization and subsequent oxidation or metallurgical processing (Li and Zhou 2017).

2.3 Gold Ore Characterization

Overall, ore characterization is the quantification of physical and chemical properties of ores in an ore body. This information forms the basis for any geometallurgical approach. Table 2.1 lists the possible ore characterization investigations for a geometallurgical program.

Table 2.1 Ore characterization test programs for quantifying various parameters

Discipline	Parameter	Tests
Geology	Field relationships	Mapping, drilling, decline
Geochemistry	Grade of valuable and deleterious elements; whole rock analysis	Assays
Mineralogy	Zonation, deportment of valuable metals and deleterious elements	Mineral identification, association, grain size, textural, and liberation data all available via AMICS, QEMSCAN, MLA, D-SIMS, TOF-SIMS, and other instruments
Metallurgical response	Recovery, concentrate grade, reagent dosage	Flotation kinetics, locked-cycle tests, GRG, bottle rolls
Physical properties	Hardness—Grinding	Bond work indices, JK drop-weight test, SPI index, MacPherson 18″ mill test
Geotechnical measures	Site preparation, environmental review	Soil density, ground water flow, slope stability

2.4 Impact of Mineralogy on Gold Ore Processing

The extractive metallurgy of gold ores is largely driven by mineralogical factors due to the fact that gold often occurs in at least two forms in an ore. These factors include particle size, liberation, association with other minerals, coatings and rimmings, presence of cyanicides, oxygen consumers and preg-robbers, presence of refractory gold minerals, locking of submicroscopic gold in sulfide and sulfarsenide mineral structures, etc.

Among the factors mentioned above, liberation, grain size, and association are regarded as three important factors that are seen in all gold ores and must be characterized in a geometallurgy program. Presence of refractory gold minerals such as gold tellurides, aurostibite, and maldonite often causes lower gold recoveries due to their slow leaching kinetics. Submicroscopic gold and carbonaceous matters are the major troublemakers in carbonaceous sulfide ore processing (Li and Zhou 2017).

2.5 Gold Deportment Study

Gold deportment study is a method designed to investigate the occurrence of gold in an ore or mill products and to predict the response of a gold ore to various processes. It is critical to any gold project, and the information generated from such a study can be used for assisting in process selection and flowsheet development. To properly and precisely determine the deportment of gold in an ore or mill product, some comprehensive approaches involving advanced and conventional techniques have been developed and used for gold deportment study. Compared to other procedures using a single technique and a small quantity of material, the advantages of using these comprehensive approaches are that a large sample will be studied to acquire better statistics and each issue will be addressed properly using specific techniques.

For project geometallurgy of gold ores which is often conducted during the scope to feasibility study stages, gold deportment study should be conducted on medium-sized samples (at least 10–15 kg each) based on different lithology, grade, alteration, texture, and mineralization to acquire the results for assisting in process selection and flowsheet development. Other key process mineralogical factors for the ore variability should be investigated as well in this stage. Failure to estimate these key factors will increase the difficulty in building a proper geometallurgical model, hence reducing the value of the model for subsequent mining operation.

For the operational geometallurgy which is conducted for the mine planning or plant optimization based on the results generated from project geometallurgy, quick mineralogical studies can be conducted on small samples (2–3 kg) collected from different blocks of the deposit to determine gravity recoverable gold and/or cyanidation recoverable gold. The information acquired can be used for predicting the metallurgical performance and economical evaluation of the blocks, with reconciliation by metallurgical testing. The mineralogical approaches for project geometallurgy and operational geometallurgy must be quantitative, efficient, and reproducible (Zhou and Gu 2016; Williams 2013).

As an important part of gold mineralogy, deportment study is an important approach used widely by the minerals industry and research institutes in various stages of feasibility study and during plant operation. It is a powerful tool for both project geometallurgy and operational geometallurgy and should be considered for all gold geometallurgy programs (Li and Zhou 2017).

2.5.1 Quantitative Gold Deportment for Complex Ore Bodies

With increasing complexities of gold deposits, it is becoming vitally important to have a deeper understanding of ore gold mineralogy and deportment behavior to identify the best processing route for maximizing the value from a deposit. Mineralogical characterization and metallurgical testing, which leads to the understanding of the ore body, is the core of a geometallurgical program. It aims to correlate geology and mineralogy with data from metallurgical testwork and develop a model to predict variability. Advances in mineralogy technologies, which are well established as key tools in geometallurgical assessments, have made it possible to accurately quantify gold deportment.

An important development in gold mineralogy is the ability to carry out quantitative gold deportment for refractory and double refractory ores. The characterization of these ores (e.g., Carlin-type deposits) is challenging due to low gold ore grades, presence of variable proportion of preg-robbing total carbonaceous matters (TCM) with sulfide inclusions, and visible and invisible gold in different sulfide phases (e.g., pyrite, chalcopyrite, chalcocite, and bornite) and also in iron oxide phases (e.g., magnetite, hematite, and goethite). This section will focus on Carlin-type deposits to demonstrate the value of carrying out detailed gold deportment studies.

The gold ore body in north-eastern Nevada, USA, is commonly known as Carlin-type deposits. Mineralization covers an extensive area, and numerous ore bodies have been localized by the complex structural and lithological controls. Most of the economic gold mineralization is hosted in limy to dolomitic mudstones. The characteristics of the host rocks that are believed to enhance their favorability to gold deposition are the presence of reactive carbonate, porosity, permeability, and the presence of iron, which can be sulfidized to form auriferous pyrite (Bettles 2002). Most of the gold at Goldstrike is located in arsenian pyrite overgrowths or pre-ore pyrite (Arehart and Chakurian 2003).

Naturally occurring organic carbon or TCM is found in many of these ores within Carlin trend and has the ability to adsorb the gold cyanide from leach solution, commonly known as preg-robbing (Stenebraten et al. 2000; Helm, et al. 2009). The proportion of TCM in these ores ranges from 0 to 8.5% with varying preg-robbing activities from high to low which correspond to gold recoveries from ~20 to ~90% (Stenebraten et al. 2000).

Quantitative gold deportment of these complex ores requires an integrated approach to process mineralogy involving various tools such as optical microscopy, mineral analyzer (MLA, QEMSCAN), X-ray diffraction (XRD), dynamic secondary ion mass spectrometry (D-SIMS), Tescan integrated mineral analyzer (TIMA), Time-of-Flight Secondary Iion Mass Spectrometry (ToF-SIMS), X-ray photoelectron spectroscopy (XPS), and laser ablation microprobe-inductively coupled plasma mass spectrometry (LAM-ICPMS). Details of these quantitative gold deportment techniques are presented elsewhere (Chattopadhyay and Gorain 2012, 2014; Chryssoulis and McMullen 2005; Chattopadhyay et al. 2017). These measurements are immensely useful but could be expensive and time-consuming, and therefore it is important to be prudent on representative sample selection with a clear purpose. Improper use of these tools could result in poor diagnosis and wrong interpretation.

Figure 2.2 shows morphology of different total carbonaceous matters (TCM) in Carlin-type deposits. These TCMs are typically fine-grained irregular veins, interstitial stringers, and composite types, suggesting the complication of separating them from pyrite through flotation, for example. Figure 2.3 shows different types of pyrite morphology for Carlin-type ores.

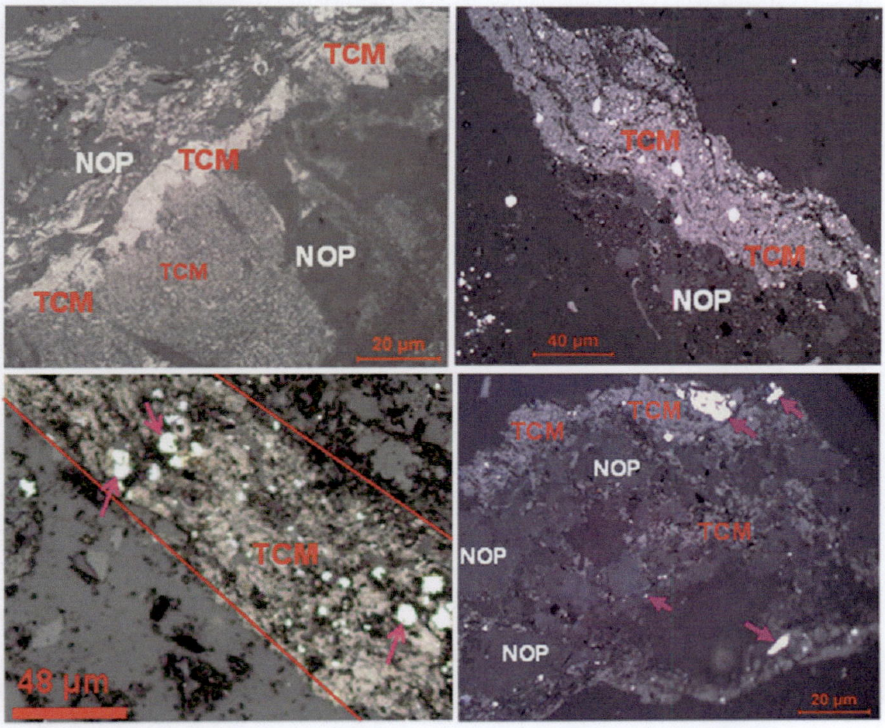

Fig. 2.2 Microphotographs of different types of total carbonaceous matter (TCM) in Carlin-type deposits (Lakhsmanan et al. 2016)

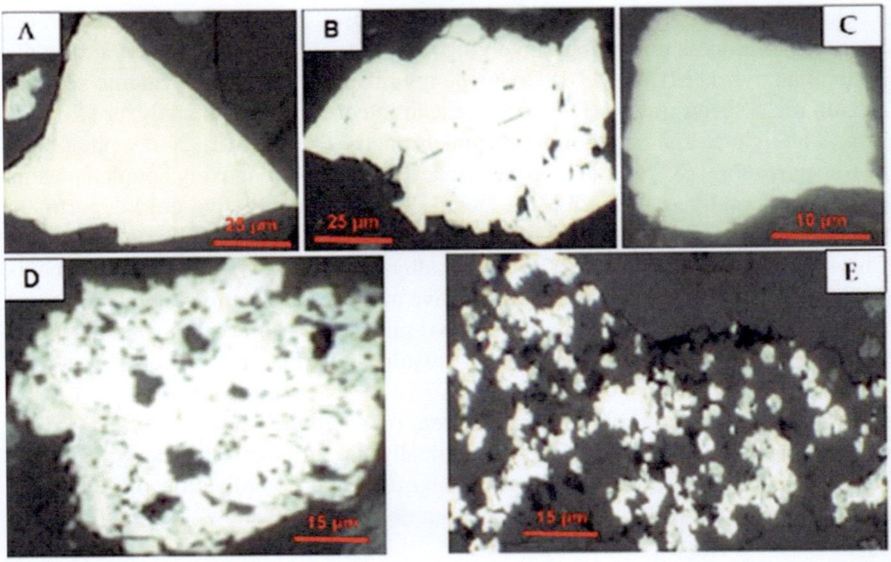

Fig. 2.3 Different types of pyrite identified in Carlin-type deposits (Lakshmanan et al. 2016)

Figure 2.4 shows quantitative gold deportment for one of the Carlin-type deposits in Nevada. Information obtained in Fig. 2.4 is based on a novel gold deportment technique through extensive studies (Chattopadhyay 2016; Lakhsmanan et al. 2016). This provides an in-depth understanding of gold deportment in various ore types and process streams, which is vital for problem diagnosis and allows well-informed decision-making such as defining strategies to develop innovative process solutions for metallurgical issues.

2.5.2 Deliverables

Gold deportment study is an important part of geometallurgy program. It is becoming a powerful tool in predicting the metallurgical performance of a new ore and in troubleshooting both gold and silver losses in an operating plant. If the testwork is well designed and properly executed, a gold and silver deportment study will provide very useful information on process selection, flowsheet development, recovery improvement, and reagent consumption optimization. The results acquired from such a study program should be able to reflect future metallurgical performance of a new ore or identify the cause(s) for gold and silver losses. To ensure that a gold-silver deportment study will provide correct and accurate information as required, selection of mineralogical techniques is most important for any gold-silver project. Deliverables of a typical gold deportment study include the following information.

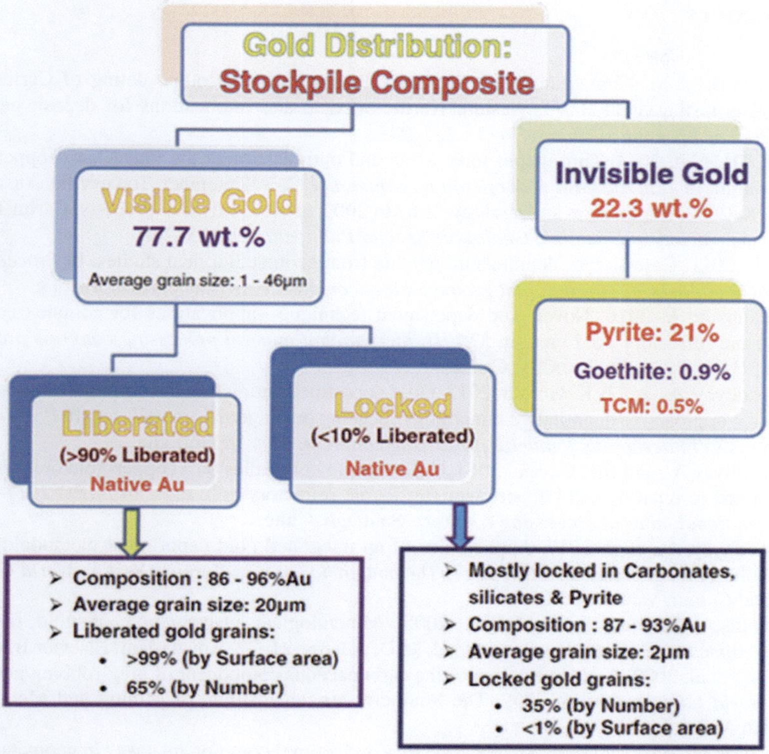

Fig. 2.4 Quantitative gold deportment for one of the Carlin-type deposits in Nevada (Lakhsmanan et al. 2016)

- Quantity and size distribution of liberated gold. This part of gold is recoverable by gravity, flotation, cyanidation, or a combination of these processes.
- Quantity of gold associated with sulfide minerals and the size of gold and gold-bearing sulfides. This part of gold is recoverable by flotation with/without cyanidation or non-cyanide leaching process.
- Quantity and size distribution of gold associated with non-sulfide minerals. This part of gold is potentially recoverable by leaching. Fine-grinding extraction may be considered to liberate or expose the locked gold particles.
- Preg-robbing capacity of carbonaceous matter, if present. This will help to evaluate the necessity and scale of pretreatment by roasting, autoclaving, bio-oxidation, or other techniques.
- Mineralogical factors that may affect or have affected gold.

References

Arehart, G.B., A.M. Chakurian, et al. 2003. Evaluation of radioisotope dating of Carlin-type deposits in the Great Basin, Western North America, and implications for deposit genesis. *Journal of Economic Geology* 98 (2): 235–248.

Bennett, D., et al. 2014. Throughput forecasting and optimisation at the Phu Kham copper-gold operation. In *12th AusIMM Mill operators' conference*, 2–3 September, Townsville, Australia.

Bettles, K. 2002. Exploration and geology, 1962 to 2002, at the Goldstrike Property, Carlin Trend, Nevada. *Society of Economic Geologists Special Publication* 9: 275–298.

Bye, A.R. 2011. Case studies demonstrating value from geometallurgical studies. In *Proceedings of the first AusIMM international geometallurgy conference*, Brisbane, Australia.

Chattopadhyay, A. 2016. Novel gold deportment technique—applications for complex copper-gold and refractory gold ores. In *XXVIII international mineral processing congress proceedings, IMPC 2016*, Québec City, Canada

Chattopadhyay, A., and B.K. Gorain. 2012. Gold deportment studies on a copper-gold ore—a systematic approach to quantitative mineralogy focusing on diagnostic metallurgy. In *Proceedings of the XXVI international mineral processing congress,* New Delhi, India.

Chattopadhyay A., and B.K.Gorain. 2014. Gold deportment studies on a copper-gold ore—a novel approach to quantify gold deportment for double refractory gold ores. In *IMPC 2014—27th international mineral processing congress*, Santiago, Chile.

Chattopadhyay, A., et al. 2017. Applications of an integrated gold deportment methodology for metallurgy diagnostics. In *COM 2017, The conference of metallurgists hosting World Gold & Nickel Cobalt*, Vancouver, Canada.

Chryssoulis, S.L., and J. McMullen. 2005. Mineralogical investigation of gold ores. In *Developments in mineral processing*, ed. M.D. Adams, 21–72. Amsterdam: Elsevier B.V.

Helm, M., et al. 2009. An investigation of the carbonaceous component of preg-robbing gold ore. In *World gold conference 2009*. The Southern African Institute of Mining and Metallurgy, South Africa.

Kittler, P., et al. 2011. 'It seemed like a good idea at the time' common mistakes in geometallurgy. In *First international AusIMM geometallurgical conference,* Brisbane, Australia.

Lakshmanan, V.I., et al. 2016. In *Innovative process development in metallurgical industry concept to commission*, ed. V.I. Lakshmanan, R. Roy, and V. Ramchandran. Cham: Springer.

Li, J., and J. Zhou. 2017. Geometallurgical and mineralogical characterization of gold ores. In *COM 2017 the conference of metallurgists hosting World Gold & Nickel Cobalt*. Canadian Institute of Mining, Metallurgy and Petroleum, Vancouver, Canada.

McCullough, G., B. Mutsaers, and J. Jessop. 2013. Obstacles to embedding geometallurgy decision-making into operations and how to increase maturity of this critical capability. In *2nd AusIMM international geometallurgy conference,* Brisbane, Australia.

Stenebraten, J.F., W.P. Johnson, and J. McMullen. 2000. Characterisation of Goldstrike ore carbonaceous material, Part 2: physical characteristics. *Mineral and Metallurgical Processing* 17 (1): 7–15.

Wambeke, T., et al. 2018. Real-time reconciliation of a geometallurgical model based on ball mill performance measurements—a pilot study at the Tropicana gold mine. *Journal of Mining Technology* 127: 115–130.

Williams, S.R. 2013. A historical perspective of the application and success of geometallurgical methodologies. In *Second international geometallurgy conference,* Brisbane, Australia.

Zhou, J., and Y. Gu. 2016. Geometallurgical characterization and automated mineralogy of gold ores. In *Gold ore processing project development and operations*, ed. M.D. Adams, 2nd ed. Amsterdam: Elsevier B.V.

Zhou, J., B. Jago, and C. Martin. 2004. Establishing the process mineralogy of gold ores. In *36th annual Canadian mineral processors conference CIM*, Vancouver, Canada.

Chapter 3
Emerging Trends in Mining

V. I. Lakshmanan, A. Ojaghi, and B. Gorain

Despite significant advances over decades, gold and silver mining still involves the old paradigm with a significant amount of waste management during haulage, processing, and disposal:

- The first step in a mining operation typically involves removal of overburden to access the ore body for an open-pit mine. This overburden is then either stockpiled as a waste rock or to be re-used at closure.
- After drilling and blasting, haulage of blasted ore which is basically mineralized rock above a certain cut-off grade inclusive of waste dilution material cannot be segregated further during the blasting and mucking activities. The ore is then carried out to the primary crusher or process plant. An ore after mining typically contains mostly gangue minerals. The amount of non-valuable gangue minerals could be more than 99.0% for gold and silver ores.
- The next step is to crush and grind the ore in the process plant mainly to liberate the valuable from non-valuables. Comminution in the mining industry is intrinsically very energy intensive using 2–3% of world's energy. Energy consumption could range from 10 to 25 kWh/ton treated. When treating 100,000 tons/day, this equates to one to two GW/day treated by a comminution circuit drawing 50–100 MW power (Powell 2013).
- Intensive mineral processing steps are then carried out on the ore such as pre-concentration, beneficiation, or concentration to separate valuables from gangue.

V. I. Lakshmanan (✉) · A. Ojaghi
Process Research Ortech Inc., Mississauga, ON, Canada
e-mail: llakshmanan@processortech.com

B. Gorain (✉)
Hindustan Zinc Limited, Udaipur, Rajasthan, India
e-mail: Barun.Gorain@vedanta.co.in

V. I. Lakshmanan, B. Gorain (eds.), *Innovations and Breakthroughs in the Gold and Silver Industries*, https://doi.org/10.1007/978-3-030-32549-7_3

- Once the valuables are recovered in the process plant, tailings management is a major step involving storage of tailings, water reclaim, and detoxification of the mill and/or final site effluent prior to any environmental discharge, when applicable.

It is estimated that the overburden waste produced globally is about 10,000 million tons (Mt) per year, assuming a mining strip ratio of 2.5. This is a significant amount of waste generated by the mining industry. Generally, the lower the head grade, the larger the volumes of ore and waste that are produced. In the Kennecott mine in Utah alone, almost 100 Mt of ore and additional 200 Mt of waste are handled every year (Ericsson 2012). To meet the needs, bigger trucks and shovels are utilized, and the payload trucks have increased from around 200 tons in 1990 to almost 350 tons in 2012. At the same time, the installed power has increased by a similar factor. In just the past 5 years, the bucket volume of wheel excavators has increased from 25 to 40 m^3. The processing plants have followed suit with large 42 ft diameter mills, 500–700 m^3 flotation cells along with larger dewatering and tailings pond facilities.

The trend of increasing throughput and larger equipment is continuing despite all associated challenges along with an ongoing trend of extremely high waste generation. The mining industry is still using the old paradigm. A radical shift is needed, starting with seeking alternatives to handling and processing large amount of waste.

3.1 The New Mining Paradigm

The basic extraction paradigm in the mining industry is "drill, blast, load, haul, dump, crush, grind, separate, leach, dewater, and tailings disposal." There are many variations, but fundamentally, the paradigm has not changed since ancient times. Almost all the innovations so far have made operations in this paradigm safer, more efficient, automated, and even autonomous (Dunbar 2014). What's needed is a major shift in this paradigm for a major breakthrough.

Since both mining and processing operations handle a significant amount of waste, the first consideration for any innovation should be to cut down on waste as early as possible in the mining value chain. Some of the major opportunities are:

1. **Minimal removal of overburden to access ore body**: Use of small-diameter drill holes to access the ore body along with use of novel biotechnologies to recover metals are currently in use (Dunbar 2014). In situ recovery of metals such as uranium and copper oxides using solution mining is presently being pursued by the industry, which is a step in the right direction though new technologies and novel ideas are necessary to make this practical for most other ore types. It appears that companies like Rio Tinto are looking at opportunities with in situ recovery as key to the "mine of the future" (Batterham 2003).
2. **Highly selective liberation of ores early in the mining process**: It is important to ask ourselves "Why create waste in the first place?" Comminution begins with mining, and a new generation of drilling and blasting for selective mining of ores

is critical to avoid or reduce waste removal early on in the mining process. This concept is referred to as "grade engineering" and is becoming a major focus of research and development (CRC-ORE 2014).

3. **Minimal haulage of waste**: This focuses on innovations that will allow processing to be closer to the mine site. There is an increasing interest in pre-concentration technologies such as sensor-based mass sorting (ROM shovels/trucks), classification using screens, stream-based ore sorting, gravity, and dense media separation, which is definitely a positive trend.
4. **Efficient comminution and processing**: These focus on highly selective separation of valuables utilizing efficient technologies resulting in minimal capital and lower operating costs along with significantly smaller footprint.
5. **Minimal generation of tailings and high reuse potential**: The important focus here is on the recovery of by-products and non-conventional value-added products such as building materials with a generation of benign tailings for backfilling underground and open-pit mines.

Reflecting on these five opportunities suggests that the ideal scenario will be a "zero waste mining." Is this really possible? History tells us that the capability and ingenuity of mining and processing professionals have been phenomenal (Lynch et al. 2010). If there is a dire need with a strong vision and adequate resources, there is no reason why "zero waste mining" cannot happen. Various innovative technologies that have been developed by the mining industry definitely provide us the confidence. The area of "zero waste mining" is not the focus of this chapter, but all trends in innovations must be seen in light of this ultimate goal.

3.2 A Drive Toward Selective, Continuous, and Automated Mining

Hard rock mining is essentially a batch process involving drill and blast, followed by load, haul, and dump. The challenge with drill and blast is that, despite the best efforts, there is a limit to how much one can control overbreak and external dilution resulting in issues with selectivity and grade control for both underground and open-pit operations. Also drill and blast introduces cycle into mining, and the mining process must fit into the defined time between blasts. Recent emergence of continuous rock cutters in hard rock mining such as surface miners for open-pit and continuous rock cutters for underground has the potential to transform gold and silver mining.

Continuous rock cutters are typically faster than drill and blast methods, allowing faster access to the ore body with significant addition to NPV of a mine (Vogt 2016). Rock cutting generates significantly lower vibrations and noise levels than the use of explosives, and thus has significant advantages where local communities live closer to the mine. The cutting process affects the rock surrounding the excavation less than the explosives; thus, the rock is stronger and safer or easier to support.

Cutting can produce excavations very close to the desired shape and size of tunnels, so that the cost of extra lining due to overbreak is reduced. Also the number of people exposed to dangerous underground conditions is lowered due to elimination of the need to store and handle explosives. For narrow reefs, cutting offers the opportunity to greatly reduce waste dilution in a process that has similar advantages to ore sorting. If the mine can be created with smaller excavations with cutting, the seismicity and ventilation requirements will both be reduced significantly.

Barrick Gold already has now extensive experience using Sandvik roadheaders, with the first MH620 unit successfully tested at Cortez gold mine in Nevada, USA. This roadheader weighs 125 tons and is driven by a 300 kW cutting motor. This rock cutter at Cortez was used at the Range Front mine decline, which is one of the world's largest roadheaders. According to Barrick, this technology has allowed to improve safety, increase throughput, and reduce mining cost per tonne. This has led Barrick Gold to acquire a second rock cutting machine Sandvik MR361 roadheader in 2017 (Fig. 3.1) for their Turquoise Ridge operation in Nevada to help with their mechanization, automation, and innovation drive (International Mining Journal 2019).

Continuous mining has been practiced over a century now for coal and soft rock mining; it is only now that slowly technology is evolving after decades of research and development. Caterpillar commercially launched the Cat Rock Straight System during the MINExpo 2016, a fully mechanized long wall system for continuous hard rock mining that features the cutting-edge HRM220 hard rock miner. Atlas Copco is planning a commercial launch of its continuous cutter after various efforts

Fig. 3.1 Sandvik cutting machine, Roadheader MR341 roadheader for mining (Sandvik Mining and Construction 2007)

involving collaboration of Rio Tinto during 2009 and also with Anglo American on a project named Rapid Mine Development System (RMDS), since 2012. Joy Global, now Komatsu Mining, has been testing prototype for its DynaMiner, which features DynaCut hard rock cutting technology, since September 2016 with promising results appearing to be demonstrated at Newcrest's Cadia operation. Sandvik is also developing a continuous hard rock miner called the MX650 (CIM Magazine 2017). It seems that by 2020, there should be several launches of hard rock cutting systems by various OEMs.

3.3 A Vision for Continuous Mining Integrated with Processing

A vision for continuous mining integrated with processing was presented at the International Mineral Processing Congress at Quebec City (Gorain 2016). A schematic is shown in Fig. 3.2.

This vision entails continuous rock cutting for mine development for access to ore body in underground operations. Also rock cutting for selective mining of ore to produce fragmentation is carried out that does not require any primary crushing. The mined ore will undergo a secondary crushing step followed by ore sorting to remove any undesired waste rock or dilution material. The rejects from ore sorting will be utilized for backfilling underground. The concentrate from ore sorting will

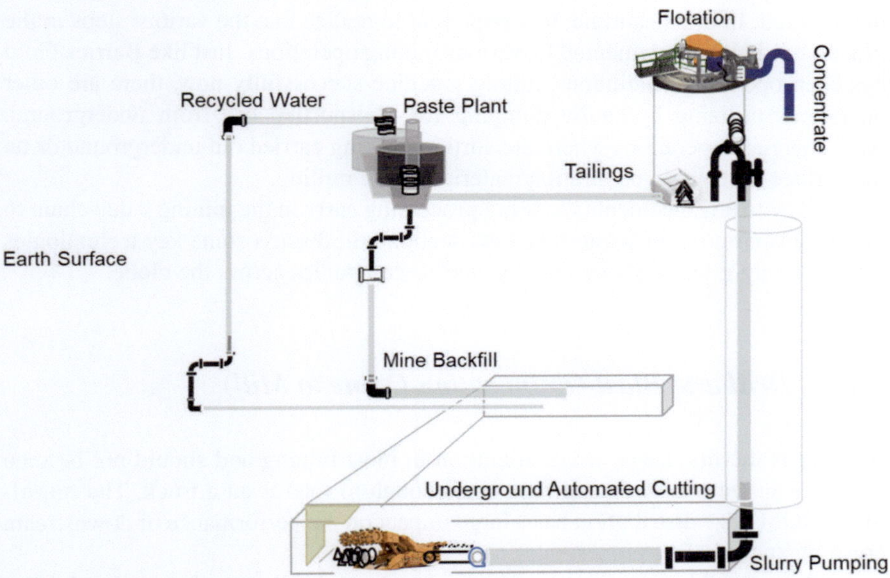

Fig. 3.2 A vision for a continuous integrated mining and processing (Gorain 2016)

be ground to a maximum particle size of 6 mm or below using a small milling circuit underground. The ground ore is then pumped to the surface using a hydraulic system or a suitable positive displacement pumps. The time required for pumping could be utilized for reagent conditioning of the ore for any concentration or hydrometallurgical step on the surface.

On the surface, a coarse flotation step is involved to reject any gangue minerals to reduce comminution energy costs and to further upgrade the ore. A finer grinding stage is then carried out on the concentrate from coarse flotation followed by fine particle flotation in a normal milling circuit. The concentrates produced from fine flotation are then sent for further processing. All flotation tailings on the surface are then collected for making paste fill, which is then sent underground for paste or backfilling underground.

A major advantage of this integrated mining and processing is that the footprint required is significantly lower than a conventional mining operation, with minimal or no need for tailings storage on the surface. This is a continuous process, and hence the whole chain could be automated with minimal need for manual intervention underground. In addition, this system has the potential to be a 100% electric mine with minimal ventilation requirements. This will be a safer, low-cost, and highly productive mine with much lower environmental challenges.

3.4 Early Integration with Processing

The vision presented above will need some time as the technology is slowly becoming matured. In the meantime, it is important to realize that the various steps in the process are being implemented by various mining operations. Just like Barrick Gold has been operating continuous cutting machine successfully now, there are other operations utilizing hydraulic pumping for transporting ore from underground. Also, there are operations where ore sorting is being carried out underground or on the surface to reject mine dilution material before milling.

The key to advancement is to bring processing early in the mining value chain to avoid transportation of waste. The next section will discuss some key technologies that are being tested and operated by various companies across the globe.

3.4.1 Drill and Blast Optimization (Mine to Mill)

Blasting is the first stage of comminution in most mining and should not be seen solely as a means of reducing rock size enough to load it on a truck. The run-of-mine (ROM) size distribution has a large impact on the performance of downstream crushing and grinding processes.

Drilling and blasting is an established technology that evolved over the last century. This is a batch process and often closely embedded in the macho culture of a traditional tough and rough miner (Ericsson 2012). Hard rock cutting, a viable alternative to enable continuous mining, is gaining wider acceptance in metal mining since its success with softer and non-abrasive bedrock and minerals such as potash and coal. This innovation seems to be attractive for narrow veins and reef mining allowing a more selective mining with less waste rock and dilution with positive impact in processing along with the economic benefits associated with continuous mining.

Mine to mill optimization in various operations over the years has shown significant benefits such as high mill throughput rates from reduced top size from mining through increased powder factor or blast energies (Dunbar 2014). There are however some safety and environmental areas where special care and attention are needed to prevent any high wall damage, fly-rocks, noise, and vibrations from using higher blast energies which can become challenges. The introduction of electronic detonators and other techniques however allows mitigating some of these risks. New blasting technologies are emerging that have the potential to significantly increase powder factor through innovative blasting practices such as dual blast layers within a single blast event (Brent et al. 2013).

3.4.2 Ore Tracking

All stages of a mining operation are influenced by ore characteristics. The SmartTag™ system tracks ore from the mine through the process, allowing ore characteristics to be correlated with important operating parameters in the mine and processing plant. These parameters can then be adjusted and optimized for different ore types to increase profitability.

SmartTag™ can be used to track ore from the mine into the plant using the system shown in Figs. 3.3, 3.4, and 3.5. When used this way, we can correlate ore characteristics with important operating parameters in the mine and processing plant, such as ore dilution, ore losses, fragmentation, throughput, and energy consumption. This provides an understanding of how different ores affect mine and plant operation and the final product. Operating parameters and control strategies in the mine and processing plant can be adjusted and optimized for different ore types, thereby reducing costs and increasing profitability.

Metso patented SmartTag™ system is used to track parcels of ore from the mine into the plant and beyond. SmartTag™ is an innovative system which uses hardened Radio-Frequency Identification (RFID) tags, detectors, and custom software to track ore through the mining process. SmartTag is a radio-frequency identify cation (RFID)-based technology designed to allow tracking of ore from its source through blasting, run-of-mine (ROM) pads, crushers, and intermediate stockpiles and finally into the concentrator (Wortley et al. 2011).

The SmartTag™ system is also used to automate geometallurgical modelling enabling throughput forecasting and life-of-mine optimization.

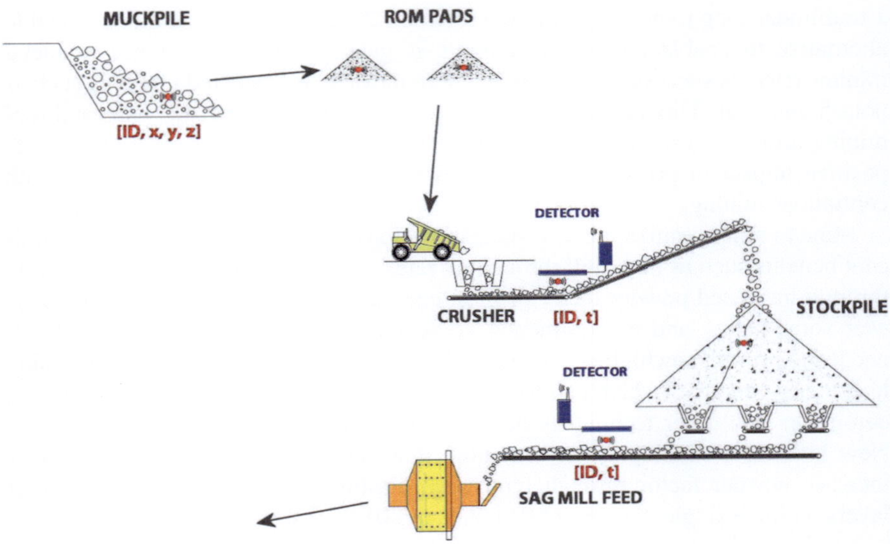

Fig. 3.3 Schematic of RFID tag-based material tracking system (Isokangas 2012)

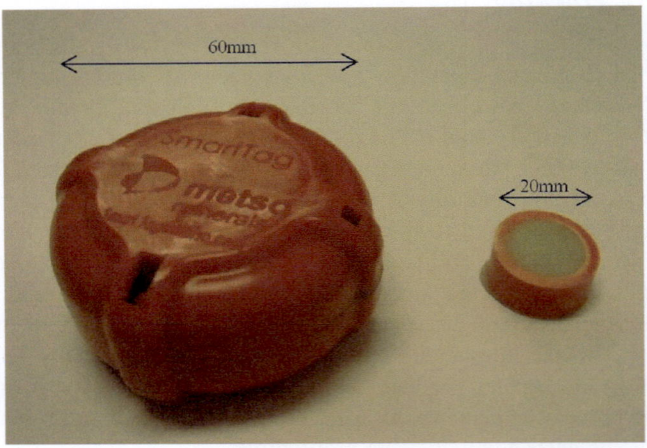

Fig. 3.4 Normal and mini SmartTag (Wortley et al. 2011)

SmartTag™ work principles

- SmartTag™ are placed with source ore in the mine (blast holes, muck piles, etc.), and the starting location of each unique tag is recorded using a ruggedized hand-held computer.
- Tags survive the blast and travel with the ore.
- They are detected when they pass antennas at critical points before milling.

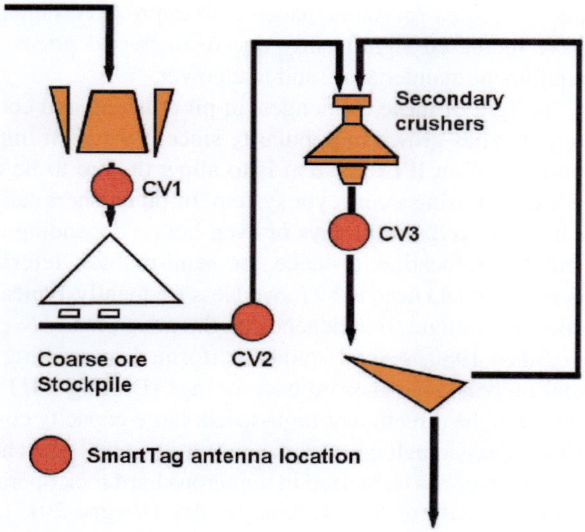

Fig. 3.5 Secondary crushing circuit (Wortley et al. 2011)

- With no internal power source, they can remain in stockpiles for extended periods.
- The physical ore properties in the mine can then be linked with time-based performance data from the plant.

SmartTag™ benefits

- Inexpensive and versatile material tracking from mine to the plant and beyond.
- Valuable information on material movements.
- Link the physical properties of the ore in the mine to the time-based performance data of the plant to optimize the process as a whole.
- Track material through the delivery chain to monitor and optimize product supply management, transport logistics, and margin management.

3.4.3 *In-Pit Crushing and Conveying System (IPCC)*

Transporting ore and waste rock is one of the most crucial elements of an open-pit mining operation. A successful haulage system can make the difference between a mine that is consistently profitable and one that is struggling to meet its marginal costs.

Haul trucks are generally favored based on their flexibility of operation compared to conveyors or trains, which were the preferred method until the 1960s. Since then, technical developments have gradually led to much larger-capacity vehicles capable of hauling more ore in fewer cycles. More recently, diesel emissions regulations have resulted in haul trucks that emit less pollutant while retaining performance and productivity. Telematics and GPS technology now allow haul trucks to be tracked and scheduled for maintenance with improved efficiency. Mining operational costs have increased significantly due to increased prices for diesel, large truck tires, equipment maintenance, and manpower.

In light of these challenges, in-pit crushing and conveying (IPCC) is a technology that has grown in popularity since it was first implemented in the 1950s. The purpose of an IPCC system is to allow the ore to be crushed in the pit and transported out using a conveyor system. In-pit crushers can be mobile which means they can be moved within days or even hours, depending on their size and complexity and the relocation distance, or semi-mobile, referring to units that are more permanent and need to be moved less frequently, typically everyone to 10 years. The specific configuration depends on the mine plan.

Since 1980, several studies performed on existing in-pit crushing installations and models have showed cost savings (Darling 2011). As a result, there is a tendency in the direction of high-speed, large-capacity conveyors due to better productivity. Classic belt conveyors can transport materials at angles up to 37°. This type of conveyor has been used in numerous hard rock operations in Canada with a maximum capacity of 57,500 tons per day (Vergne 2017). High-angle belt conveyors, such as the sandwich and the pocket wall conveyors, can transport material at high angles up to 90° while keeping the positive features of conventional conveyors (Duncan and Levitt 1990). The Conveyor Equipment Manufacturers Association (CEMA) recommends designing belt conveyors with a maximum width of 3.2 m and belt speeds of 8 m/s for rough crushed material.

Capital investment for IPCC system is greater than a truck-shovel (TS) system. When the mine goes deeper, the IPCC operational costs are considerably cheaper compared with TS, and the total NPV increases throughout the project life. An analysis showed that both systems were sensitive to production rate (Andres 2017) (Fig. 3.6).

Fig. 3.6 Comparison slope of ramps in conventional and IPCC system (Szalanski Scot 2009)

3.4.3.1 IPCC Advantages

There are several advantages of an IPCC over a haul truck system. The first and most obvious is a marked reduction in costs due to less need for road and truck maintenance, along with significantly less fuel use and labor costs (a small number of haul trucks are retained in an IPCC operation). IPCC systems are usually reliable, requiring less maintenance and service employees than trucks, and are not impaired by bad weather.

IPCC systems cut truck haulage to a minimum and build operational resilience; they substantially reduce the operational expenditures and hold other environmental and safety benefits too.

The capital and operational cost directly depend on the material transport system. Conventional truck haulage as today's predominant means of material transport in surface mines are well established and provide excellent flexibility; however, they contribute up to 60% of the overall mining cost. IPCC represents a viable, safer, and less fossil fuel-dependent alternative. They comprise of fully mobile, semi-mobile, or fixed in-pit crusher stations connected to conveyors and spreaders (for waste) or stackers (for ore) to transport material out of the mine.

Besides the potential operational expenditure reduction, IPCC systems offer a number of other benefits to mining operations (Radlowski 1988; Darling 2011; Johnson Marc 2014; Andres 2017):

- Reduced manning requirements of between 40 and 60% due to higher automation
- Reduced spear part requirements
- Possible electrical power generation at downhill conveyors
- Reduced greenhouse gas emissions of up to 25%
- Noise level reduction of up to 35%
- Dust emission reduction of up to 40%
- Reduced bad weather downtimes
- Improved safety due to fewer mobile vehicle usage

IPCC system is looking at savings of 20–40% over a truck haulage system, depending on productivity rates and whether the trucks are hauling aggregates or mineral ore.

IPCCs work well where there is horizontal movement of the mine faces, but not so great in deep mines with a lot of benches.

3.4.3.2 IPCC Types

There are a number of different system configurations, largely depending upon whether truck haulage forms any part of the of the material movement chain or not. The three major configurations are:

1. **Fixed or semi-fixed systems**: These are where trucks dump directly into a fixed crusher location. The fixed crusher is not able to be relocated within a reasonable period of time (not be for less than 10 years) (Figs. 3.7 and 3.8).

Fig. 3.7 Fixed IPCC system (Szalanski Scot 2009)

Fig. 3.8 Semi-mobile IPCC system (Cooper Allen 2010)

2. **Semi-mobile systems**: This unit works close to the mine face but is moved less frequently than a mobile crusher. Semi-mobile systems are suited to harder rocks and higher capacities. In this method, trucks are used to transport material from the mine face to the in-pit crusher, often moving between levels (Fig. 3.9).
3. **Fully mobile systems**: These types of crushers work at the mine face; are loaded directly by a shovel, front end loader, or excavator; and move in unison with the excavator on their own transport mechanism as mining progresses (or sizer) (Figs. 3.10, 3.11, and 3.12).

Fig. 3.9 Semi-mobile IPCC system (Courtesy of Sandvik) (Johnson Marc 2014)

Fig. 3.10 Fully mobile IPCC (Koehler 2010)

Fig. 3.11 Fully mobile sizer (Crusher) (Courtesy of MMD) (Johnson Marc 2014)

Fig. 3.12 Fully mobile IPCC system

3.4.4 Underground Preprocessing

Underground mineral processing would reduce the costs of bringing ore to the plant and returning backfill waste to the mine workings. As mines become deeper, cost savings from underground processing become more significant. Underground pre-concentration also allows for increased underground mining rates for a fixed shaft haulage capacity. This translates into increased tons of metal transported to the surface and higher grades feeding the surface processing facility. The objective of pre-concentration is to reject barren waste at as coarse a particle size as possible. Therefore, the ability to apply pre-concentration is determined by the liberation characteristics of the ore. Liberation was achievable at sizes that could be produced by crushing. The main technology used for pre-concentration is dense media separation which has been applied to the separation of metal-bearing sulfides from siliceous gangue. Over the last 40 years, a number of coarse particle processing technologies have been developed that may now be ready to apply industrially for applications such as pre-concentration (Klein et al. 2002).

The dense media separation process is characterized by a high processing capacity and the ability to make sharp separations at coarse particle sizes. There are two basic classes of dense media separators referred to as static separators and dynamic separators. Static separators, such as drums, vessels, and baths, are suited for particle sizes ranging from approximately 2 to 500 mm and have a unit capacity of up to 300 tons/h. Dynamic separators, such as the DSM Cyclone, Dyna-Whirlpool, and Tri-Flo, utilize centrifugal acceleration to achieve separation. Large-diameter separators (up to 1 m) can process particles up to 100 mm at a rate of up to 400 tons/h; the bottom size limit is typically close to 0.5 mm. An underground dense media processing facility would include crushing, screening, dense media separation, and a dense media recovery circuit. The products are essentially dry (Klein et al. 2002).

A Chamber of Mines of South Africa research project developed and piloted underground processing of gold ores from the Witwatersrand. The process design was based on consideration of underground space limitations for a processing facility, mineralogical/liberation properties of the ore, and the capacity to return waste backfill to excavations. The raw ore is ground in a centrifugal mill and screened at 3 mm. The minus 3 mm fraction is floated in specially designed flotation cells for use with very coarse particles to recover gold and gold-bearing sulfide minerals. The flotation tailings are classified using hydrocyclones, which return coarse gangue, moderately coarse valuable minerals, and locked valuable minerals to the mill. The cyclone overflow is a waste product which is thickened for backfill. The estimated gold recovery is 98% while rejecting 60% of the raw feed. In this case, pumping would be required to transport a bulk concentrate to the surface and to return the tailings to the underground excavations. Bulk sulfide flotation followed by selective flotation is a common practice for surface processing plants (Klein et al. 2002).

Ore sorting process is another popular technique which has been used for underground preprocessing. Several coarse particle sorting technologies can be considered for pre-concentration. Sorting machines must be capable of effectively eliminating waste rock at coarse particle sizes and at high throughput rates. The suitability of technologies will depend on the principle of separation, the capacity and applicable particle size range, and factors relating to infrastructure requirements, space, and portability. Processes that do not require water and are relatively small and mobile, which allows them to be mobile for operation close to the mining face, are favored over processes that do not meet these requirements (Klein et al. 2002).

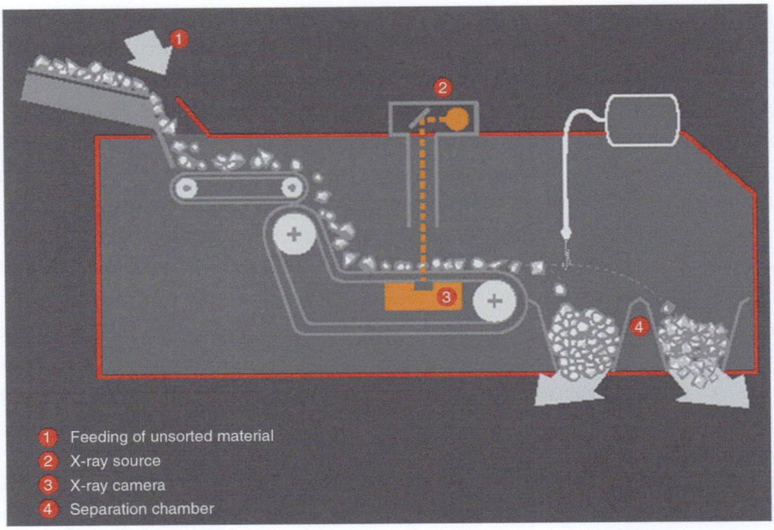

Fig. 3.13 Diagrammatic representation of mechanized ore sorting at underground preprocessing (Lynch and White 2013)

Optical sorters consist of an image scanner and processor to distinguish between particles of different colors, brightness, sizes, and shapes. Optical sorting has been applied to ore processing and recycling for particles up to 350 mm and rates up to 300 mtph (Klein et al. 2002).

Ore sorting testing has been undertaken by Citigold at both laboratory scale and also preliminary bulk sample trials using mechanized ore sorting technology (Fig. 3.13) to reduce the level of waste in the ore to be hoisted. Initial trials by Citigold in 2003 and 2005 have demonstrated the potential for success with ore sorting using photometric (PM) (He-Ne laser light), electromagnetic (EM), and radiometric (RM) detection systems. Responses were only achieved with PM. Samples of +20 to −45 mm material were successfully processed within excess of 85% recovery of ore from waste. Processing of the screened −20 mm material was not successful, with significant misplacing of high-grade dark sulfide ore material into the waste stream. However, advances in technology since that date have indicated greater promise for gold ore using color, X-ray transmission (XRT), and near-infrared (NIR) sensors. Repeat trials will be carried out shortly using a range of techniques. Larger-size particles (50–100 mm) will be used, and the philosophy will be to sort waste from the ore to avoid grade loss. The particles will be washed and screened, with finer material (undersize) reporting directly to the "concentrate" stream. As with the geophysics, the target will be the sulfide proxies for the gold as they represent a more substantial target and this should be a potentially more reliable approach. Selection of waste, rather than ore, means that if there is any doubt about a particular particle, it will report to ore, thus obviating loss of values due to miss-sorting. It is currently anticipated that 50% of the waste component of the run-of-mine ore should be removed, to provide a head grade to the mill of approaching mineral resource grade. The retrieved waste will be used as fill material for disposal in the mined out stope voids, thus further reducing hoisting and disposal costs (Lynch and White 2013).

3.4.5 Bulk Ore Sorting

Conducting bulk ore sorting at the mining face, either in-pit or underground, allows ore and waste to be directed to the appropriate destination (waste dump or process) immediately. In open-pit mines, bulk ore sorting could potentially fit well with in-pit crushing and conveying systems.

Because the natural heterogeneity of the ore makes sorting more effective, a bulk ore sorter should be placed as early as possible in the process. This also avoids extra costs and the lost capacity associated with treating non-economic material. In fact, the benefits of discarding barren material early on are carried through all the downstream process steps.

In almost all cases, bulk ore sorting would need to be implemented after primary crushing to present material at a size that can be handled by the sorter. Each additional crushing or material transfer stage incorporated prior to sorting increases mixing, costs, and unnecessary energy consumption, so these should be avoided if possible.

Conducting bulk ore sorting at the mining face, either in-pit or underground, allows ore and waste to be directed to the appropriate destination (waste dump or process) immediately. The feasibility of this approach will depend on having suitable space available for the sorter as well as on the mining method employed and potential impact on mining productivity.

Underground pre-concentration, prior to haulage or hoisting, produces solid waste underground, which can be combined with tailings and cement from the surface as required and disposed of as fill. In open-pit mines, bulk ore sorting could potentially fit well with in-pit crushing and conveying systems. The sensor(s) would be located on a conveyor leaving the pit and would use a flop gate to separate waste and ore onto their respective conveyors. Furthermore, a system equipped with multiple flop gates could be used to separate different ore types onto separate stockpiles, which could be either blended according to the downstream process requirements or sent to different process routes (Metso 2018).

3.5 Automation and Digitization

Mineral processing operations present many challenges due to variations in unmeasured ore properties, material transport delays, and non-linear response characteristics. Advanced process control technologies such as expert systems and multi-variable model predictive controllers are often used to improve the control of these processes to increase productivity and recovery efficiency. Typically these systems are applied after the mine has been in operation for a period of time (Carr et al. 2017).

3.5.1 *Model Predictive Control*

David Carr et al. evaluated the application of BrainWave Model Predictive Controller (MPC) and Advanced Control Expert (ACE) Advanced Supervisory Control System on the SAG mill, ball mill, and rougher flotation cells for a new gold processing line at the OceanaGold (Carr et al. 2017).

The advanced control was tested against the IDEAS™ dynamic simulator and commissioned during the startup of the processing plant to help achieve target production as quickly as possible. Operators were able to run the equipment in a consistent and optimized manner from an early stage of the plant operation, increasing efficiency and profitability. MPC is able to reduce process variability beyond the best performance that can be obtained with conventional proportional-integral-derivative (PID) control methods. MPC is able to optimize the control of processes that exhibit an integrating type response in combination with transport delays or variable interaction, which are characteristic of many gold processes including SAG mill weight control and flotation cell froth level control (Carr et al. 2017).

Model predictive control provides an additional tool to improve the control of critical processes where PID- or rule-based expert control is not well suited to the application. In particular, MPC is able to optimize the control of processes that exhibit an integrating type response in combination with transport delays or variable interaction. This type of response is particularly difficult to control, and it is common in mineral processing for many different processes including level control of flotation cells and crushers and SAG mill weight control.

MPC has been successfully applied to many of the critical processes at a mineral processing plant including SAG mill control, ball mill control, and rougher flotation control. MPC has consistently demonstrated the ability to reduce process variability and increase stability compared to PID- or expert system-based control (Carr et al. 2017).

3.5.1.1 SAG Mill

An MPC was installed to control the SAG mill weight, the feed rate in the ore feed system, and the SAG mill sound emissions as part of the ACE solution. A diagram of the control strategy is shown in Fig. 3.14. The ACE control system manages the SAG weight using both the ore feed rate and the SAG mill rotation speed (RPM). This approach ensures that the maximum possible production rate is achieved at all times while automatically reducing the RPM when required to reduce energy consumption and retain an adequate minimum mill fill weight to reduce liner wear or when excessive sound emissions are detected.

3.5.1.2 Ball Mill

The ANDRITZ Ball Mill ACE control strategy adjusts the pump speed to maintain stable cyclone pressure. The ball mill sump level is allowed to vary over a specified range so that minor changes in production rate are managed in the sump. This

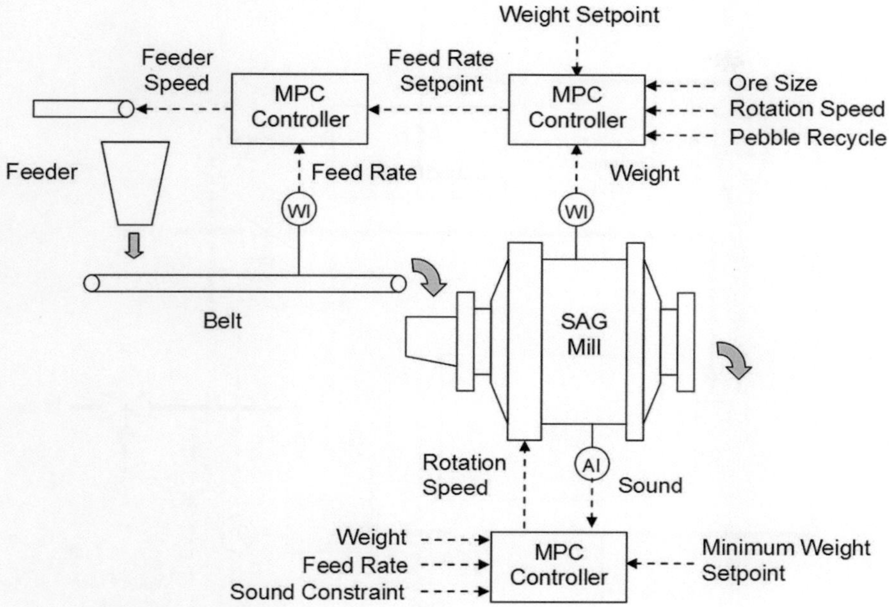

Fig. 3.14 SAG mill MPC control strategy (Carr et al. 2017)

approach avoids unnecessary disturbances to the cyclone pressure and the number of cyclones in service, resulting in increased stability of the cyclone pressure and the P80 to the flotation cells. Dilution water to the sump is adjusted to maintain infeed density to the cyclones, which is also an important factor in stabilizing cyclone performance. The control strategy is shown in Fig. 3.15.

The MPC is used to control the cyclone feed pressure and cyclone feed density by adjusting sump pump speed and dilution water flow. The ACE system is used together with the MPC to control the sump level by adjusting the number of cyclones in service. The ACE system also adjusts the cyclone pressure setpoint over a small specified range to assist with the sump level control for the purpose of minimizing the number of cyclone openings or closings.

Optimal performance of the flotation process requires stable control of the froth depth in each flotation cell. If the froth level can be maintained, a stable froth layer is formed without excessive bubble breakage, and there is a steady overflow of froth over the lip of the flotation cells. A well-chosen froth depth allows time for some "drainage" of entrained water and gangue particles before the froth overflows the cell; however, greater froth depths can result in lower recovery and increased use of expensive frother chemicals. The target froth level for the cell is often determined by the operator, but can be automated using a camera-based supervisory system.

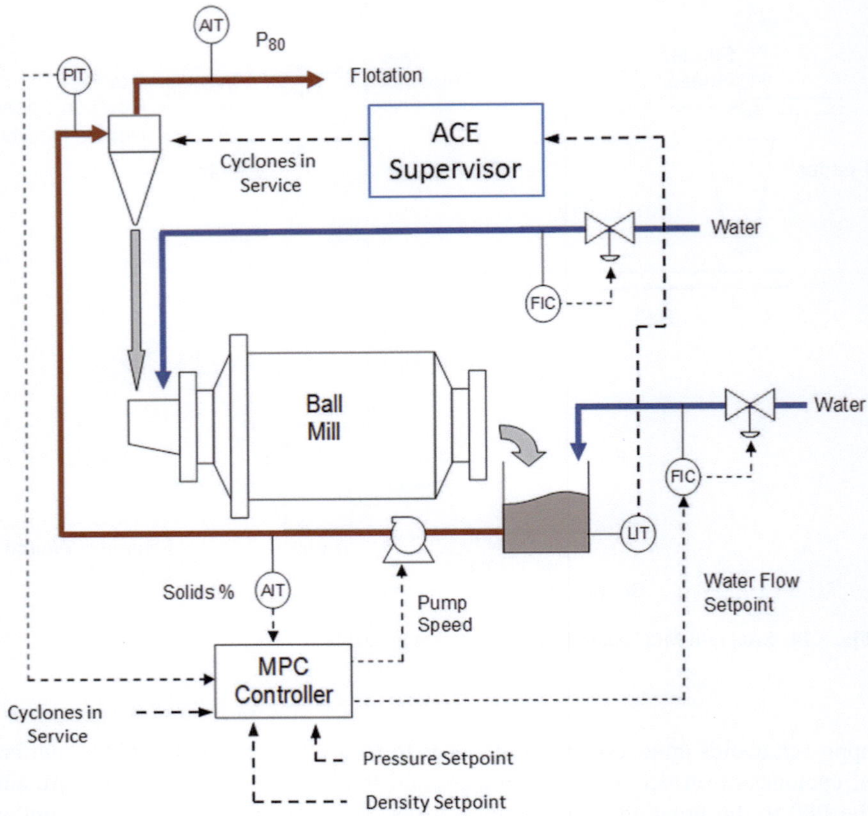

Fig. 3.15 Ball mill control strategy diagram (Carr et al. 2017)

3.5.1.3 Advanced Control of Flotation Cell Level

Optimal performance of the flotation process requires stable control of the froth depth in each flotation cell. If the froth level can be maintained, a stable froth layer is formed without excessive bubble breakage, and there is a steady overflow of froth over the lip of the flotation cells. A well-chosen froth depth allows time for some "drainage" of entrained water and gangue particles before the froth overflows the cell; however, greater froth depths can result in lower recovery and increased use of expensive frother chemicals. The target froth level for the cell is often determined by the operator, but can be automated using a camera based supervisory system.

The MPC was installed on the flash cell and two rougher flotation cells. The control strategy uses the cell levels and cell valve positions to estimate the flow between the cells. These low estimates are used as feed-forward inputs to the MPC to improve the rejection of level disturbances caused by the actions of the upstream controllers.

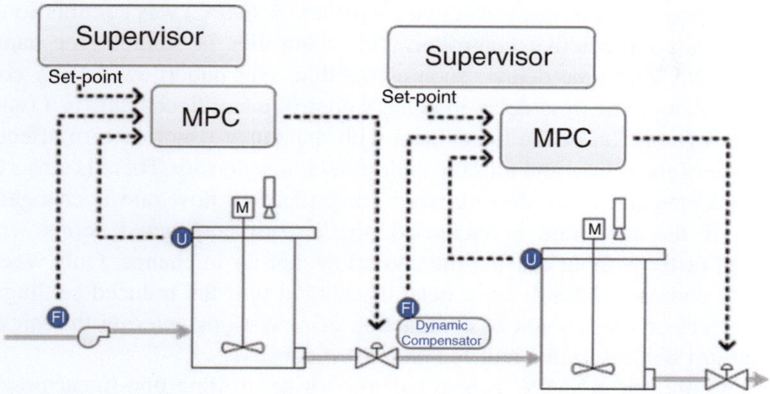

Fig. 3.16 Flotation cell level control strategy (Carr et al. 2017)

Separate MPC are used for each flotation cell instead of a single multi-variable MPC as the interaction of the actuators is coupled only in the direction from the upstream cell to the downstream cell. In addition, as the flow change from a given actuator change is a function of the relative cell levels, a dynamic compensator is used to convert the valve movements into estimated flows as this is a better feed forward to the downstream MPC than the actual valve position at the outlet of the upstream cell. A diagram of the control strategy is shown in Fig. 3.16.

3.5.2 Robust Non-linear Model Predictive Controller (RNMPC)

Reagent consumption in gold processing plants significantly contributes to the total operational costs. In a typical carbon-in-pulp (CIP) circuit, the reagent consumption associated with leaching, adsorption, elution, and regeneration contributes to approximately 15% of the total operational costs (Stange 1999). At Loulo Gold Mine, the reagent consumption associated with leaching, adsorption, elution, and regeneration contributes to 6% of the total operational costs. As such, a reduction in reagent consumption will present a significant opportunity for savings on operating expenditure. Variability in feed ore characteristics results in continuous changes in steady-state operating regions, which presents process control challenges as a result of the long residence times in some of the process units associated with gold processing plants. Overdosing often occurs in the presence of process variability to ensure safe operation and to maintain recovery targets. As a means of reducing reagent consumption and accounting for significant changes in the feed to gold plants, some of the plants have resorted to improving their process control strategies around their circuits.

A robust non-linear model predictive controller (RNMPC) was commissioned as the model-based predictive controller. This controller is suitable for handling processes with long time delays, such as the thickener underflow density control using flocculants. The time delay in density control using flocculants is a function of the rate of reaction of the flocculants with the slurry which in turn affects the settling rate of the solids, and thus the underflow slurry density. There is also a delay in the thickener underflow density when the underflow flow rate is changed, for example, if the flow rate is increased, the already-compacted solids will be discharged faster without causing the underflow density to change. Only when the previously compacted solids have been discharged will the reduced settling time have an effect on the amount of compaction of new solids entering the thickener, which in turn will result in a reduced underflow density.

Furthermore, the RNMPC is also suitable for controlling non-linear processes such as pH control which is highly non-linear due to the non-linear characteristics resulting from the feed and the ion concentration in the controlled system (Lazar et al. 2007).

The main objectives of the controller were to stabilize the feed to the CIL process by controlling the thickener underflow density to setpoint and also to control the pH of a CIL tank to setpoint, taking into account the long residence times (Makhado and Sagara 2017).

Figure 3.17 illustrates the RNMPC configuration of the thickener underflow density and pH (CIL Tank 1) control commissioned at a gold mine in Mali. The aim of the RNMPC is to control the thickener underflow density feeding CIL Tank 1 to setpoint, without violating the thickener torque and slurry bed level constraints, and control the pH in CIL Tank 1 to setpoint.

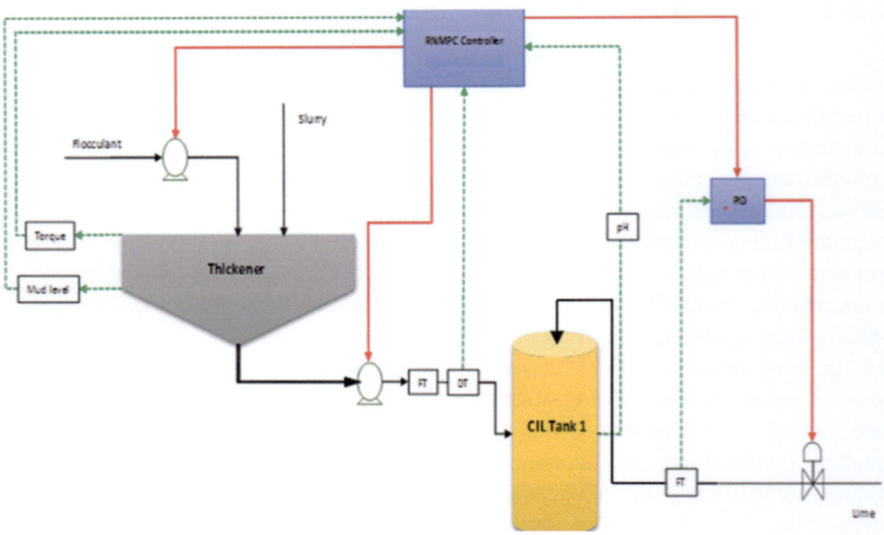

Fig. 3.17 Process configuration and commissioned control loops (Makhado and Sagara 2017)

3.5.3 Era of Automation and Digital Mine

Until few years ago, the mining industry has been fairly conservative with implementing new technologies in comparison to other industries. Australian mining is entering a new era of automated operations. The two imperatives of reducing operating costs and improving safety are the key drivers. It is interesting that many mineral processing plants are state of the art and fully automated, yet surface and underground mining operations have evolved at a much slower pace.

Processing plants, being one integrated unit, were easier to automate. Mining operations, the excavation and moving of rock, have traditionally been a sequence of disconnected steps, but these can now be digitally integrated using technologies that have evolved over the last 5–10 years. With automation comes big gains in productivity and machinery output, which leads to cost reductions. Safety is substantially improved because operators are removed from the more hazardous areas. Also operator error is reduced by eliminating what are often boring repetitive tasks that can now be done by algorithm-guided computers (Lynch and White 2013).

Digitization of mining operations is now strategic to the growth and sustainability of today's mining companies. Recent evolution of sensor and network technologies along with easier access to computing power and subject matter expertise from almost anywhere in the world has created an ecosystem which has the potential to transform mining. Real-time monitoring and predictive analytics along with the ability to control an operation remotely have allowed mining companies to redefine their business models. Autonomous machines are now increasingly being adapted to many operations across the globe. Companies are starting to realize the benefits of digitization through better safety and higher productivity along with lower maintenance costs.

A conventional underground mine is almost like a black box with no visibility of actual progress in underground activities until the end of the shift at best. This lack of visibility and communication leads to poor productivity. To bridge this gap, mining companies are investing to establish high bandwidth network to have a two-way communication, track activities in real time and manage the data centrally. This enables us to provide the right information at the right time and in the right format to the right person to help in data-driven decision-making.

A digital mine essentially consists of the following modules that enhance interoperability across the operations:

- Reliable high bandwidth communication network
- Two-way voice and data communication with workforce and equipment
- Real-time tracking of assets—man, machine, and material
- Real-time status of the mining processes
- Obtaining equipment health and productivity data
- Integrating auxiliary services, viz., dewatering, ventilation, electric substations, seismic monitoring system, CCTV, ambient air quality system, and other data through smart sensors
- Automated traffic control system

- Fleet management system
- Three-dimensional visualization of complete mine for real-time monitoring of operations
- Digital twin with predictive models and prescriptive action items

Details of the different aspects of mine digitization will be presented in Chap. 11.

References

Andres, Jimenez Builes Carlos. 2017. A mixed-integer programming model for an in-pit crusher conveyor location problem. University of Montreal.

Batterham, R.B. 2003. The mine of the future, will it be visible? XXII In *International mineral processing congress*, Cape Town, South Africa.

Brent, G.F., P. Dare-Bryan, S. Hawke, and M. Rothery. 2013. Ultra-high-intensity blasting—a new paradigm in mining. In *Proceedings of world gold conference*, Brisbane, Australia.

Carr, David, Johnson Quenton, and W.A. Gough. 2017. In *Optimization of the Haile Gold mine grinding and flotation line using model predictive control, COM 2017,* Vancouver, Canada.

CIMMagazine.2017.http://magazine.cim.org/en/technology/hard-rock-revolution-on-the-horizon.

Cooper Allen. 2010. In-pit crushing and conveying (IPCC).

CRC-ORE. 2014. Transforming mining productivity with grade engineering. www.crcore.org.au

Darling, P. 2011. SME mining engineering handbook, 3rd ed. https://books.google.ru/books?id=5uq-kdfHLWUC

Dunbar, W.S. 2014. Will the mine of the future be a mine at all? In *Globe and Mail*.

Duncan, L.D., and B.J. Levitt. 1990. Belt conveyor. In *Surface mining*, ed. B.A. Kennedy, 2nd ed., 692–705. Littleton: SME.

Ericsson, M. 2012. Mining technology—trends and development, Polinares consortium journal, working paper # 29.

Gorain B.K. 2016. A vision for continuous mining integrated with processing. In 28th International Mineral Processing Congress 2016 in Québec City.

International Mining Journal. 2019. https://im-mining.com/tag/cortez.

Isokangas, E. 2012. Using smarttag to track ore in process integration and optimization: some case studies in a variety of applications. The Southern African Institute of Mining and Metallurgy Platinum.

Johnson Marc. 2014. Impact of in-pit crushing and conveying on pit shell optimization. D CONVEYING Deswik.

Klein, B., W.S. Dunbar, and M. Scoble. 2002. Integrating mining and mineral processing for advanced mining systems. *CIM Bulletin* 95 (1057): 63–68.

Koehler, F. 2010. In-pit crushing looms the way into Australia. In *Mining magazine congress*.

Lazar, C., R. Pintea, and R. de Keyser. 2007. Nonlinear predictive control of a pH process. In *17th European symposium on computer aided process engineering*, Vol. 24, 829–834.

Lynch, M., and B. White. 2013. Brave new world autonomous mining systems. In *Proceedings world gold 2013*, 157–166. Brisbane: The Australasian Institute of Mining and Metallurgy.

Lynch, A.J., G.J. Harbort, and M.G. Nelson. 2010. *History of flotation*. Melbourne: AusIMM.

Makhado, V.G., and D. Sagara. 2017. Implementation of an advanced thickener underflow density control at LOULO gold mine, MALI.

Metso. 2018. www.metso.com.

Powell, M. 2013. Utilising orebody knowledge to improve comminution circuit design and energy utilization. In 2nd AusIMM International Geometallurgical Conference, Brisbane, 30 September to 2nd October.

Radlowski, J.K. 1988. In-pit crushing and conveying as an alternative to an all truck system in open pit mines. The University of British Columbia.
Sandvik Mining and Construction. 2007. Bulk materials handling in mining [Report].
Stange, W. 1999. The process design of gold leaching and carbon-in-pulp circuits. *Journal of the South* 99: 13–25.
Szalanski Scot. 2009. Optimizing in-pit crusher conveyor performance: P&H Mining Equipment.
Vergne, J.N. de la. 2017. Hard rock Miner's handbook, ed. S.L. Anderse. Alberta: Stantec Consulting.
Vogt, D. 2016. A review of rock cutting for underground mining: past, present and future. *The Journal of South African Institute for Mining and Metallurgy* 116: 1011–1026.
Wortley, M., E. Nozawa, and K.J. Riihioja. 2011. Metso SmartTag the next generation and beyond. In *35th APCOM symposium,* Wollongong, NSW.

Chapter 4
Beneficiation of Gold and Silver Ores

V. I. Lakshmanan, A. Ojaghi, and B. Gorain

This chapter provides an overview of various innovation and technology developments in mineral processing that have shaped the current gold and silver industry. A glimpse of the present and future challenges in mining is also presented. A holistic approach to problem solving involving various stakeholders is gaining momentum, and this has been reflected in this chapter. Various developments that are being pursed in mineral processing focusing on a change in the existing mining paradigm are also presented in this chapter.

4.1 Introduction

Traditionally, a gold and silver operation is segmented into distinct core disciplines, viz., geology, mining, and processing from a technical perspective with a focus on extracting an ore body to produce marketable concentrates or metal. There are various support roles that are critical to the success of these technical disciplines such as administration, maintenance, IT, supply chain, safety, environment, and community relations.

Mining companies have started to look at their mining business holistically and also from different points of views. An integrated approach to mining is gaining prominence not just from an economic perspective but also allows them to take into account the needs of various stakeholders. Mining companies are slowly tapping

V. I. Lakshmanan (✉) · A. Ojaghi
Process Research Ortech Inc., Mississauga, ON, Canada
e-mail: llakshmanan@processortech.com

B. Gorain
Hindustan Zinc Limited, Udaipur, Rajasthan, India
e-mail: Barun.Gorain@vedanta.co.in

V. I. Lakshmanan, B. Gorain (eds.), *Innovations and Breakthroughs in the Gold and Silver Industries*, https://doi.org/10.1007/978-3-030-32549-7_4

into some successful concepts that have transformed other industries such as manufacturing, chemical, and pharmaceuticals along with oil and gas.

Some mining companies are already looking beyond the traditional disciplines of geology, mining, and processing to focus on the key value drivers that tend to integrate these disciplines and their interfaces. This is allowing mining companies to innovate, and the focus is on step change in their cost structures and productivity.

Though this chapter is about innovation in mineral processing, the authors strongly feel that this has to be seen in the light of other disciplines and the mining value chain as a whole. Any innovation focusing on mineral processing alone will only provide part of the solution and will miss big opportunities. The question then is "What are these big opportunities that will allow a step change in profitability of a mining operation in a safe and sustainable manner?"

This chapter will emphasize on the following three key focusing questions related to the innovation in mineral processing in the present mining context:

1. What innovations are required to address the existing mining challenges?
2. What are the various mineral processing innovations that have already shaped and advanced the mining industry?
3. What can be done to bridge any innovation gaps for realizing the full potential?

4.1.1 Innovations in Mineral Processing

To understand the various innovations in mineral processing, we will need to understand the various challenges the mining industry has handled so far. In the early days, high-grade ores and ore bodies were exposed on the surface. To extract these ores required minimal generation and treatment of waste. One could say, this was close to the ideal situation of "zero waste mining" scenario.

As head grades started to deteriorate with ore bodies located at a considerable depth, the need for overburden removal and therefore the generation of mining waste started to increase. With industrialization and the growth in demand of metals, the urgent need to treat low-grade ores at much greater depth intensified. Innovation focused on how to mine higher throughputs and then process them mainly to meet the immediate needs. This paved the way for mass open pit mining and larger mining machineries. Open pit mines and unfavorable terrain resulted in building concentrators that are far away from mining. Also, the consideration for a large tailings facility added to the complexity, resulting in a large mine footprint which increased in proportion with mill throughput.

In the early days when the environmental regulations were not as stringent and the access to power or energy sources was relatively easier, the main focus of innovation activities was to build larger and more efficient equipment to treat the large amount of mined product (containing mostly waste) to meet the needs of metal production targets. Hence, there was not much incentive to challenge the existing mining paradigm of mass scale waste generation, processing, and storage. This situation is still ongoing, although there is now a strong realization that this cannot continue for very long due to mounting energy costs, stringent environmental regulations, and lack of water and other resources along with an ever-increasing opposition by local communities.

The mining industry is now facing a dilemma, and there is a strong push by companies, some governments, and stakeholders to focus on innovation to address the challenges. The following are the main areas that have attracted serious attention for innovation in the area of mineral processing:

- Energy consumption and GHG reduction
- Economy of scale (capital cost)
- Efficient use of consumables (media, liner, reagents)
- Instrumentation, control, and automation
- Use of alternative water sources
- Environmental management of tailings and waste
- Modeling and simulation tools (to reduce risk in design and operations optimization)
- Seamless integration with value chain
- Continuous improvement along with training and education to improve productivity
- Breakthrough technologies (focus on step change such as dry processing, in situ mining, etc.)

4.2 Comminution

This chapter will focus on the key mineral processing innovations that have already resulted or have the potential to make a significant contribution to the mining industry. Principles of operation and design of unit operations will not be the main focus of this chapter as they are readily available for reference in various publications.

Comminution costs are typically one of the largest cost items in a gold milling operation. Since the advent of Autogenous Grinding (AG) and Semi-Autogenous Grinding (SAG) milling technologies in the late 1950s, they have established themselves as the present standard and are commonly used in the gold industry now. These technologies have replaced the previous "conventional" comminution circuits involving crushing-ball mill or rod mill-ball mill circuits. It is estimated that over 90% of the gold ounces are produced by milling operations, and the majority of these ounces come from high aspect ratio SAG circuits with significant contributions from low aspect SAG circuit such as in South Africa (Mosher 2005). The main reasons for the attraction for SAG milling are circuit simplicity and typically lower capital and operating costs to meet the needs for operations with higher throughput. SAG mills work well for handling clay rich sticky ores, which is a challenge for multistage crushing circuits requiring washing plants.

4.2.1 Crushing

Crushing is an important step in comminution that prepares the ROM ore for primary grinding and other downstream processes such as heap leach. The selection of the right crushers and crusher flow sheet is dependent on the upstream mining method and blasting characteristics and the downstream process requirements. As process plant throughputs are increasing, the need to reduce operating costs and capital cost per ton of ore processed is critical. Both crusher and circuit design have evolved with larger crushers employing more horsepower and speed to treat higher throughputs at a reduced cost.

With the advent of SAG milling, cone crushers are now more or less eliminated from comminution flow sheets. Instead, the cone crushers are employed in an SABC circuit to crush the recirculating pebbles. These pebbles tend to be more resistant to impact breakage in the SAG mill, and crusher creates ore surfaces more conducive to breakage in the grinding mills (Major 2002).

During the last 20 years, the size of gyratory crushers has not changed significantly, though their installed horsepower has increased allowing these crushers to treat higher throughputs. An example of the largest gyratory crusher is FLSmidth's 60′ × 113″ UD design with an installed power of 1 MW. Another feature of the new design of gyratory crushers is the development of top service gyratory crusher which appears to enhance safety, reduce maintenance downtime, and lower cost of crusher installation (Erickson 2014).

4.2.2 Mine to Mill

Mine to mill optimization in various operations over the years has shown significant benefits such as high mill throughput rates from reduced top-size from mining through increased powder factor or blast energies (Gillot 2006; Kanchibotla 2014). New blasting technologies are emerging that have the potential to significantly increase powder factor through innovative blasting practices such as dual blast layers within a single blast event (Brent et al. 2013).

Differential blasting for grade (DBG) involves conditioning of sub-volumes of material at bench or stope scale using customized blast designs that generate imposed size distributions with higher grade, concentrated in the finer fractions. This is referred to as differential blasting as it leverages the application of differential charge energies to different blast holes to achieve the desired result (Rutter 2017)

4.2.3 Classification of Comminution Method

4.2.3.1 Existing Technologies

- HPGR
- AG/SAG mill
- Ball mill
- Vertimill™
- IsaMill™

4.2.3.2 Emerging Technologies

- Stirred Media Detritor (SMD)
- HIGmill™
- Superfine IMPTEC crusher
- Horomill®
- Vibrocone™

4.2.3.3 New Technologies

- EDS multishaft mill
- Conjugate anvil-hammer mill (CAHM)
- HPGR—in closed-circuit with air classifiers
- HPGR-stirred mill combined
- Multi-pass HPGR circuit
- Jet micronizer (air jet mills)
- Rotary collider mill (RCM)—DevourX.
- Hicom mill
- Vertical roller mills
- Selfrag

4.2.4 HPGR

The use of HPGR technology has been extended significantly over the past 10 years. When initially introduced to the industry, HPGRs were considered to be a high capital cost and applicable to projects with high throughputs and harder ores. The benefits of the technology are now established, and the impacts on recovery and longer-term operating costs are well proven. HPGR technology is now used in a range of duties, including crushing operations in heap-leach circuits, as an alternative to SAG mills and for recycle crushing. In SAG, mills, however, are less energy efficient in handling harder and abrasive ores. As the ore bodies are steadily becoming more competent along with increasing energy costs, the industry has been looking for alternatives to SAG. It is interesting to note that the traditional crusher and ball milling technology has returned with a more innovative approach to crush rocks using high-pressure grinding rolls (HPGR).

HPGR application continues to expand. AngloGold Ashanti's Tropicana Project commissioned in 2013 treats 5.8 Mtpa of moderately hard free-milling gold ore. The comminution circuit is identical to that of Boddington, with HPGRs utilized as tertiary crushers to produce a P80 2-mm ball-mill feed. A combination of HPGR technology coupled with high-intensity cyanidation (Pyke et al. 2006) was used to recover coarse gold from the quartzitic ore processed by Bendigo Mining. The flow sheet also entailed flotation of auriferous sulfides that were subjected to high-intensity cyanidation. The tailings from the two recovery steps were not subjected to cyanidation.

HPGR application continues to expand. AngloGold Ashanti's Tropicana Project commissioned in 2013 treats 5.8 Mtpa of moderately hard free-milling gold ore. The comminution circuit is identical to that of Boddington, with HPGRs utilized as tertiary crushers to produce a P80 2-mm ball-mill feed. A combination of HPGR technology coupled with high-intensity cyanidation (Pyke et al. 2006) was used to recover coarse gold from the quartzitic ore processed by Bendigo Mining. The flow sheet also entailed flotation of auriferous sulfides that were subjected to high-intensity cyanidation. The tailings from the two recovery steps were not subjected to cyanidation.

Recent successes of HPGR in hard rock application such as at Freeport's Cerro Verde in Peru (Vanderbeek et al. 2006; Koski et al. 2011) and Newmont's Boddington in Western Australia (Dunne et al. 2007; Hart et al. 2011) with a key target of significant grinding energy savings of 15–20% are reported. Both these operations have two stages of coarse crushing using gyratory and cone crushers followed by the third stage of crushing involving HPGR and a single-stage ball milling circuit.

4.2.4.1 HPGR: In Closed-Circuit with Air Classifiers

The application of HPGR technology in comminution circuits is already well established for the processing of hard ores in precious and base metal operations. Of the many possible flow sheets that have been proposed for HPGRs, those using HPGRs as tertiary crushers, in closed-circuit with wet screens (apertures in the range of 6–10 mm), and feeding ball mills seem to have become the standard for these applications.

Although these circuits provide reduced operating costs with a relative increase in energy efficiency and the elimination of steel grinding media when compared to SAG-based circuits, the overall energy reduction of the circuit is compromised by the requirement of complex material handling systems. In addition, the total capital cost is usually higher than for a traditional SABC circuit.

Several different circuits have been proposed to better utilize the HPGR by applying it for higher reduction ratio and therefore the elimination of the less efficient ball mill, e.g., the multistage HPGR. However, the one circuit/application that seems to be the most promising in terms of energy reduction, allied with relatively low capital cost requirements, is one with a single-stage HPGR in closed-circuit with air classification. This dry circuit has already been used in the cement industry for a reliable pre-mill grind or final product grind, reducing power and steel consumption cost and increasing throughput capacity while operating to make a high-quality product as fine as 25 μm (Jankovic et al. 2015).

Weir (KHD HPGR) is advocating this new circuit and has reported on recent project studies, and a few operations (iron ore and limestone processing), with final product sizes of 75 μm or as low as 15 μm, are feasible and enable significant energy reduction (more than 50% overall if compared to traditional circuits with tumbling mills).

This dry HPGR circuit usually is designed with a combination of static and dynamic air classifiers, and a schematic of its flow sheet is shown in Fig. 4.1. It is also interesting to note that "dry ground feed for flotation shows superior performance in pilot testing to wet milled product" (Aidan Giblett—Newmont TS Processing).

4.2.5 Selfrag

Selfrag is an emerging technology showing some promise in pre-weakening of mineral ores by high-voltage pulses. This was pioneered in the 1950s, but it is attracting some serious attention during the last few years. The key benefits of this technology

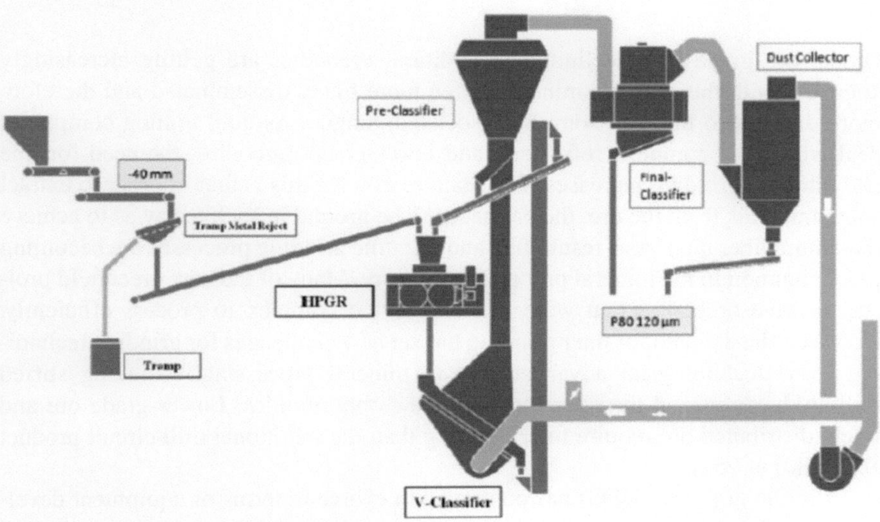

Fig. 4.1 HPGR in closed-circuit with air classification (Van der Meer and Strasser 2012)

compared to conventional comminution machines are that this is highly selective and its ability to weaken the ore with a potential for significant energy savings in further comminution (Shi et al. 2003). Studies on a Newcrest's Cadia East Cu-Au ore samples by JKMRC showed an energy reduction of 5 kWh/ton, along with a potential to remove two MP1000 pebble crushers and one 10 MW ball mill. Simulation studies showed that Selfrag has the potential to reduce the operating cost by $19 M (Shi et al. 2014). Though this study showed some potential, a detailed capital cost estimate is needed to better evaluate the potential of this technology.

4.2.6 IMP Superfine Crusher Technology

The "IMP" superfine crusher technology is presently evolving and has the potential for developing a broad range of application options as the technology matures. This "IMP" superfine crushing concept evolved from the simple premise that the probability of particle breakage increases as the number and intensity of forces simultaneously contacting the particles increase. This is quite different from conventional comminution machines where breakage rate decreases with increased energy inefficiencies as feeds become finer and harder. The IMP superfine crusher is designed to provide a mechanism that could effectively deal with fine hard feeds by maintaining high breakage rates as hard particle became finer and single-particle population rapidly expands (Kelsey and Kelly 2014).

4.2.7 Ultrafine Grinding

Global ore grades are declining. In addition, orebodies are getting increasingly complex, with the valuable minerals often more finely disseminated and therefore more difficult to liberate from the associated gangue. As gold mining companies deal with more complex, refractory, and lower-grade orebodies, the need for fine and ultrafine grinding increases. The main reason for this is that in order to extract value minerals from the ore, the ore needs to be ground in such a way as to achieve maximum liberation. As a result, fine and ultrafine grinding processes are becoming more common in the mineral processing industry. Many of the new greenfield projects involve orebodies that were historically too complex to process efficiently. However, the demand for finer grinding has set new challenges for grinding technology; new technological advancements in mineral processing, including stirred milling, have opened the door for many new opportunities. Lower-grade ore and finely distributed ore require finer grinding than the traditional mill circuit product size (P80) of 75 μm.

Ultrafine grinding (UFG) has continued to evolve in terms of equipment development. A number of specialist machines are commercially available including

Xstrata's IsaMill, Metso's Vertimill and Stirred Media Detritor (SMD), Outotec's High-Intensity Grinding (HIG)_mill, and the Metprotech mill. UFG equipment has been developed with installed powers of up to 5 MW. Compared with conventional ball or pebble milling, the specialist machines are significantly more energy efficient and can economically grind to 10 mm or lower, whereas the economical limit on conventional regrind mills was generally considered to be around 30 mm. Coupled with improvements in downstream flotation and oxidation processes, the rise of UFG has enabled treatment of more finely grained refractory ores due to a higher degree of liberation.

The application of ultrafine grinding to refractory gold ores has proven to offer a viable alternative to conventional oxidative processes in particular cases. Its application to a hydrometallurgical extraction route for other metals beyond gold is open and likely to lead to alternative process routes for these metals also.

As an example, Kalgoorlie Consolidated Gold Mines (KCGM) has successfully operated an ultrafine grinding (UFG) circuit to supplement its roaster capacity for the treatment of the refractory gold flotation concentrate. The capacity of UFG circuit for grinding to 11–12 μm is over 20 tph and achieving over 90% gold recovery (Ellis and Gao 2002).

4.2.8 IsaMill

The IsaMill™, as patented and distributed by Glencore technologies, is a high-intensity, horizontally mounted stirred mill that uses fine (2–6 mm) ceramic media for grinding in fine and ultrafine applications.

The IsaMill™ is a closed mill. Slurry travels through the mill in a "plug" -ow pattern through consecutive grinding disks. Media recirculates between the rotating disks distributing grinding action throughout the IsaMill™. Grinding is by attrition and abrasion of the particles in contact with the high-speed, small, circulating media.

At the discharge end of the IsaMill™, slurry and media reach the patented product separator. Media is centrifuged out to the shell and is pumped back with some of the slurry to the feed end of the mill. This action retains media in the IsaMill™ without the need for fine screens, while the ground discharge slurry exits the center of the product separator. The high efficiency of the IsaMill™ comes from its ability to use small media, with high surface area and high media/particle collision frequency. Small media can still grind coarse feed because of its high speed up to 22 m/s. Figure 4.2 shows a detail schematic of a large IsaMill. The high-power intensity results in a small mill with short residence times less than a minute which avoids overgrinding. This, together with the eight consecutive stages of grinding and product separator, produces a sharp size distribution without external classification (Lakshmanan et al. 2016; Gleencore 2015) (Fig. 4.3).

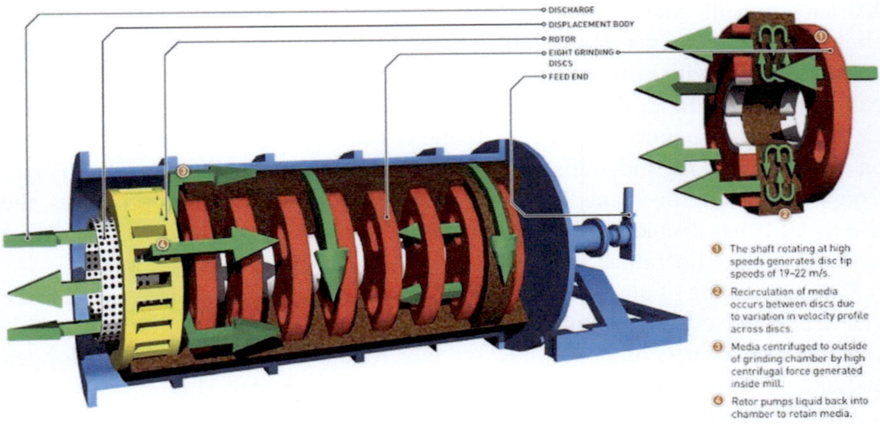

Fig. 4.2 IsaMill schematic (Gleencore 2015; Lakshmanan et al. 2016)

4.2.9 VertiMills

Vertical mills were first designed in the 1950s for applications in fine and ultrafine grinding in Japan. Lately, the vertical mill is making progress toward primary grind application and shown promise in energy reduction relative to ball mills, typical to what is normally seen in regrinding applications (Mazzinghy et al. 2014). This vertical mill would be in series with either an AG or a SAG mill grinding circuit. The vertical mill, or Vertimill™ by Metso® Minerals, has since seen many installations worldwide in regrind applications, noting a significant decrease in energy consumption compared to typical tumbling mill like a ball mill (Palaniandy et al. 2014) (Figs. 4.4 and 4.5).

4.2.10 Stirred Media Detritor (SMD)

The Stirred Media Detritor (SMD) is a fluidized, vertical stirred mill designed for optimum grinding efficiency for fine and ultrafine grinding products. The SMD utilizes the rotational energy of the impeller arms to impart a high-energy motion to the media/slurry mixture inside the mill. This results in particle-to-particle shear and compressive forces which produce the desired grinding mechanism for fine grinding. The vertical arrangement allows the drive train to be entirely supported by the mill body which leads to a small foot print and simple foundation. Also, the vertical arrangement does not require any slurry seals or inlet feed pressure. The SMD is a fluidized media mill, which means the stirrer speed is high enough to distribute the media throughout the slurry regardless of media density, forcing particle and media contact. The SMD power intensity is optimized to achieve efficient

Fig. 4.3 Industrial IsaMill (Gleencore 2015)

grinding, limit wear, and allow for heat dissipation in the case of a high-energy grind. The power intensity (kW/m^3) is relatively high compared with the other mill types but is required to generate a vortex of the media and slurry during operation and to bring the particles in contact with one another for efficient grinding. However, the power intensity is not so high that a cooling system would be required to dissipate the heat generated during a high-energy grind. Also, limiting power intensity limits the sheer force of the media/slurry on the liners and impellers and improves wear life (Fig. 4.6).

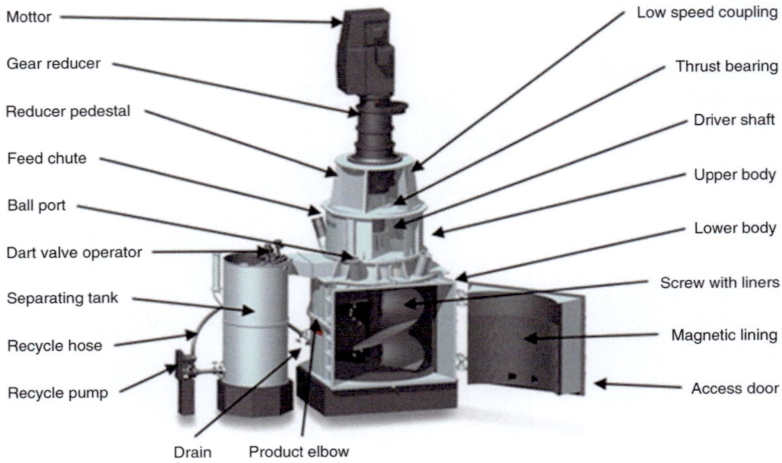

Fig. 4.4 Metso® Vertimill™ schematic (Allen 2013)

Fig. 4.5 Industrial Vertimill (Allen 2013; Lakshmanan et al. 2016)

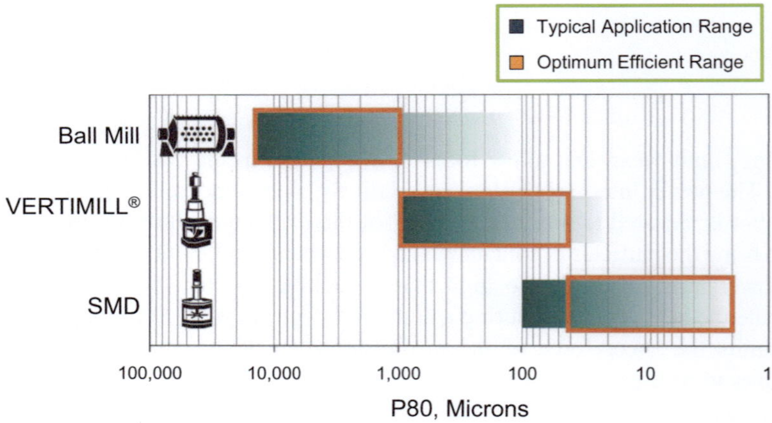

Fig. 4.6 Typical grinding ranges of ball mill, Vertimill, and SMD (Metso Minerals Industries Inc. 2016)

4.2.11 Tower Mill

The tower mill is a vertical stirred grinding mill which can be used in both dry and wet grinding applications. It has provided a means of size reduction in an area where conventional tumbling mills become inefficient (Morrell et al. 1993). The major advantage of the tower mills over the ball mills in regrinding and fine grinding operations is an efficient utilization of the fine grinding media. The normal top-size of media used is 10–25 mm, but for very fine grinding even smaller media can be used. Tower mills were first introduced in the 1950s as an alternative to the ball mills in a regrinding application. Since then, they have been used to grind a range of materials from limestone to coal to sulfide concentrates (AMIRA 1993).

The principal method of grinding in tower mill is attrition, with the possibility of some shearing and compression breakage taking place due to the motion of the media charge created by the rotation of helical screw enhancing the media pressure. The tower mill is filled to a level with the grinding media until developed power equals 80% of motor name plate rating. The tower mill will show a power saving of up to 50% over the conventional tumbling mill. The tower mill has a definite use in comminution and it can also be used for:

- Secondary grind following by ball or rod mill
- Regrind circuit.
- Grinding and leaching, as in gold cyanidation circuits
- Ultrafine grinding
- Given the proper feed and the requirement for a fine grind (Table 4.1)

4.2.12 HIGmill (High-Intensity Grinding Mill)

With installed power up to 5000 kW, the HIGmill™ is the largest fine grinding unit in the marketplace to date. This technology comes in various drum diameters, and the mill heights can be varied to optimize the media load and power input for specific applications. Chamber volumes range from 400 to 27,500 L with the corresponding drives from 132 to 5000 kW. The HIGmill is the only ultrafine grinding technology in commercial use that has the capability to use small-size high-density grinding media in mill sizes above 3000 kW. Over 260 HIGmills have been commissioned (Figs. 4.7 and 4.8).

Table 4.1 Comparing power intensities at different mills

	Typical lower grind size P80 (μm)	Power intensity (kW/m^3)
Ball mill	75	20
Tower mill	20	40
UFG mill	5	280

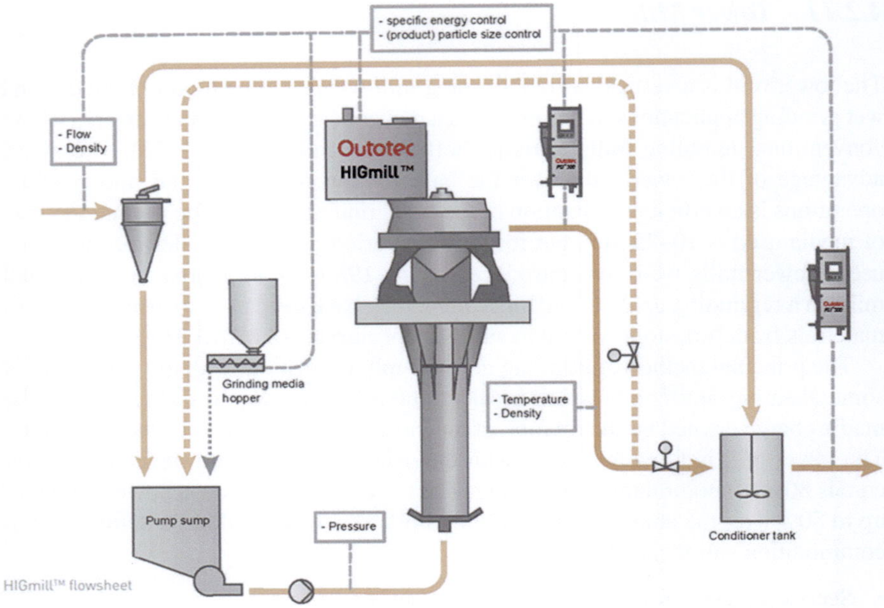

Fig. 4.7 HIGmill flow sheet (Letho et al. 2013)

4.2.13 Comminution Circuit in Large Gold Operations

The most common gold production plants comminution circuit are described in Table 4.2 (Fig. 4.9).

Crush-for-leach includes any number of stages of crushing. *SS-A* single-stage AG mill, *SS-S* single-stage SAG mill, *SS-AC* single-stage AG mill (with pebble crushing), *SS-SAC* single-stage SAG mill (with pebble crushing), *AB* AG mill-ball mill, *SAB* SAG mill-ball mill, *ABC* AG mill (with pebble crushing)-ball mill circuit, *SABC* SAG (with pebble crushing)-ball mill circuit

4.3 Classification

Classification is an important aspect of any comminution circuits. It is well known that a closed-circuit grinding circuit with a classification device results in significant benefits to the comminution process. In the early days, mechanical rake and spiral classifiers were used in closed-circuit grinding circuits. Due to capacity limitations and high operational costs because of wear and tear, these technologies are not commonly used in the present comminution circuits. Hydrocyclones became popular since the early 1950s as they are high-capacity devices that allowed the industry to

Fig. 4.8 Outotec industrial HIGmill (Letho et al. 2013)

meet the demands of increasing throughputs to treat low-grade ores. Since hydrocyclone separation is a function of both size and density, a sharp size separation is not always possible for many ore containing high specific gravity minerals and metals such as gold, PGMs, and lead. This results in building of these small heavies, which are mostly liberated, in the recirculation stream in cyclone underflow leading to unnecessary grinding with poor gravity and flotation responses. The inefficiency of the hydrocyclone separation requires the use of high circulating loads to minimize the mill residence time. High circulating loads increase the power requirements per ton of ore (Albuquerque et al. 2008).

4.3.1 Derrick Stack Sizer™

The key for major improvements in capacity and in energy consumption in closed-circuit grinding is improved sharpness of classification (Hukki and Allenius 1968). This is where screen separation has a significant advantage because of its sharp

Table 4.2 Largest gold-producing operations in 2014 (Adams 2016)

	Operation	Country	Comminution circuit(s)	2014 Production (tons of Au)
1	Muruntau	Uzbekistan	SS-S Crush-for-leach	61
2	Grasberg	Indonesia	SABC Crush-HPGR/grind	35
3	Pueblo Viejo	Dom. Rep.	SABC	35
4	Yanacocha	Peru	Crush-for-leach single-stage SAG	30
5	Carlin	USA	Crush-for-leach SAB Dry crush/grind	28
6	Cortez	USA	SAB Run-of-mine leach	28
7	Goldstrike	USA	SABC Dry crush/grind	28
8	Olimpiada	Russia	SAB	23
9	Veladero	Argentina	Crush-for-leach	23
10	Boddington	Australia	Crusher/HPGR-ball mill	22
11	Kupol	Russia	SAB	21
12	Lihir	PNG	SABC	23
13	Kalgoorlie (KCGM)	Australia	SABC	20
14	Cadia Valley	Australia	SABC	19
15	Oyu Tolgoi	Mongolia	SABC	18
16	Lagunas Norte	Peru	Crush-for-leach	18
17	Driefontein	South Africa	SAB, SS-S, crush-for-leach, crush-rod mill-pebble mill, SS-A/S	18
18	Penasquito	Mexico	SABC (with HGPR)	18
19	Kumtor	Kyrgyzstan	SAB	18
20	Tarkwa	Ghana	SABC	17

separation. Derrick Stack Sizer™ is a recent innovation in fine screening which allows for high separation efficiency and high tonnage capacity on a much smaller footprint that is possible using conventional screens (Lakshmanan et al., 2016; Clark 2007).

4.3.2 *Cavex Recyclone™*

Cavex Recyclone™ is a double classification unit in one stage, which seems to increase the sharpness of separation and reduces the bypass of fines to the underflow. This technology uses wash water injection mainly to rupture the viscous layer to release the trapped fines for proper classification. Recent trials and installations in grinding circuits have shown to significantly reduce fines misreporting to mill

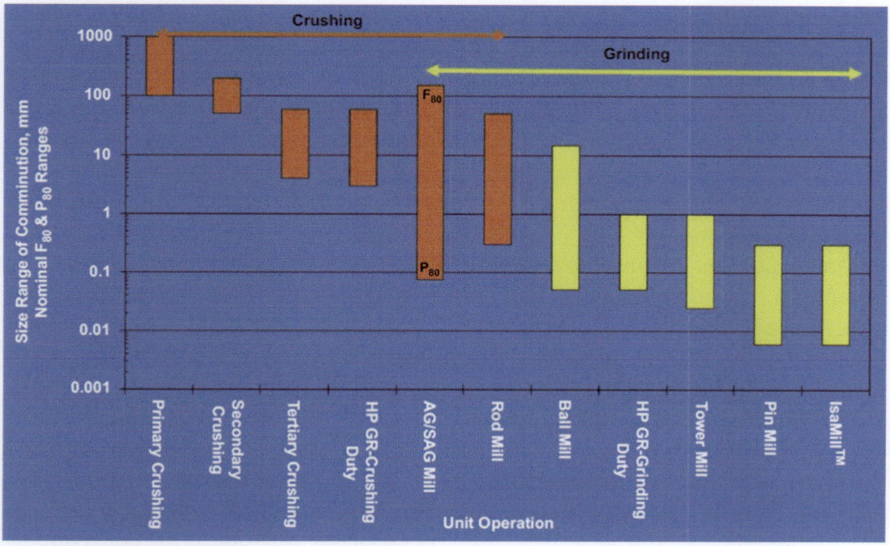

Fig. 4.9 Applicable size range for comminution unit operations (Adams 2016)

from 30 to 15% resulting in reduction in circulating load by around 50–60% (Castro et al., 2009).

4.3.3 Grade Engineering®

Grade Engineering® is an integrated approach to coarse rejection that matches a suite of separation technologies to ore-specific characteristics and compares the net value of rejecting low-value components in current feed streams to existing mine plans as part of a system view (Rutter 2017).

Grade Engineering is focused on extracting metal more efficiently by separating ore from waste before it enters comminution. Early physical rejection of non-valuable material through pre-concentration techniques before processing decreases processing costs and importantly can significantly increase the life of a mine.

Grade Engineering® is the first large-scale initiative to focus on integrated methodologies to deliver maximum operational value (Pease et al. 2015).

4.4 Pre-concentration

Removal of some waste as early as possible after blasting has the potential to reduce haulage costs to the mill in many situations. In addition, this provides an opportunity to upgrade the ore resulting in reduced energy and operating costs per unit of metal input to the plant. A higher head grade to the process plant typically results in better concentration ratio and better unit recoveries as well as the process now has to handle lower gangue content. This concept of pre-concentration is not new with application such as Dense Media separation and sorting on the surface since the 1930s (Munro et al. 1982), also applied underground (Lloyd 1979). It is worthwhile to note that not all ores are amenable to pre-concentration especially when the minerals are finely disseminated in the ore body, resulting in high losses of valuables in the waste product.

Ores types that are amenable to pre-concentration can add significant economic benefit, such as at the Kroondal platinum mines in South Africa, where about 50% of bulk mining feed of UG2 ores to the mill is rejected with a PGM recovery of 95% (Holloway et al. 2009). The value of pre-concentration has now been demonstrated for a wide range of ore types at a coarse-size range (Mohanty et al. 2000; McCullough et al. 1999; Mohanty et al. 2000; Schena et al. 1990). The value of pre-concentration should be looked in a holistic way involving integrated mining, processing, and waste disposal. Bamber has concluded based on his studies that the exploitation of a deposit with ore pre-concentration and waste disposal technologies integrated into the mining process prior to beneficiation on surface is superior to the conventional approach (Bamber 2008).

The following pre-concentration technologies appear to have a significant potential and are slowly finding applications in different mining applications:

- Size classification
- Ore sorting
- Dense media separation
- Coarse particle flotation

4.4.1 Size Classification

Concentration by comminution and size classification alone has been found to be effective for some ore types (Burns and Grimes 1986; Sivamohan and Forssberg 1991; Logan and Krishnan 2012). Size classification presents a low-cost option for rejection of waste, with maximum economic benefit at coarse particle sizes either at naturally arising ROM particle size distribution or at a coarse crush size, as close as possible to the mining face (Klein et al. 2002).

4.4.2 Ore Sorting

Ore sorting in the mining industry has evolved from manual sorting in the early days to basic optical sorting first used in mining in the 1970s using camera technology and digital image processing for industrial minerals. Arvidson and Wotruba (2014) have provided a review of the various applications of ore sorting technologies (Arvidson and Norrgran 2014). The importance of ore sorting in improving economics of marginal deposits is increasing, being realized by the mining industry (Lessard et al. 2014; Foggiatto et al. 2014).

At present, majority of automated ore sorters, outside of the diamond industry, are color or conductivity sorters (Bartram and Kowalczyk 2009). This is a significant innovation in mineral processing with an ability of pre-concentration with significantly improved economics. The optical sensors have quite a few applications, but as the optical properties differ relatively little, laser-based sensors could be useful. If the primary surface properties are distinct, NIR sensors are normally used. Novel sensor technologies are now necessary to extent the application to ore types beyond optical sensors that exploit material properties such as electrical conductivity, magnetization, molecular structure, and thermal conductivity. Combination of sensors also have potential such as optical/NIR, optical/inductive, or XRT/inductive (Arvidson and Norrgran 2014).

The recent development of LIBS (laser-induced breakdown spectroscopy), focused on pattern recognition, appears to have the potential for mass ore sorting. SonicSampDrill and IHC Mining have been working on the development of the LIBS Ore Sorter (Sonic Samp Drill 2014). Though there is potential, much work is required to make this technology commercial for large throughputs.

Normally 10–90% of the mass can be rejected in an early stage of the process at the primary, secondary, or tertiary crushing stage. Also, low-grade waste rocks don't need to be transported, crushed, milled, or further treated.

Ore sorting is applicable for precious metals, gold, base metals, diamonds, coal, ferrous metals, copper, platinum, slag, industrial minerals, and gemstones. The method which has used for ore sorting was shown is Tables 4.3 and 4.4.

4.4.3 Dense Media Separation (DMS)

Dense media separation is widely used in coal, diamond, chromite, and iron ore industries since its introduction by Dutch Sate Mines (DSM) in 1947. Though used occasionally in other industries, its application in precious and base metals, viz., PGMs, gold, copper, and zinc, is slowly gaining some momentum.

Napier-Munn et al. (2014) have provided a review of the various DMS technologies, with a focus on technologies that have been implemented by the industry. Tati nickel plant in Botswana uses DMS to reject 70% of mass with recovery of ~2/3 of the contained nickel and has allowed the mine to target a zero cutoff grade, thus allowing bulk mining and simplifying the mine plan with increased resources. Another impressive application of DMS has been in recovery of fine-grained gold from waste rock dumps at the Witwatersrand gold fields in South Africa. Despite a low feed grade (0.2–0.3 g/ton), DMS was successful in recovering 70% of the gold into 30% of the mass (Napier-Munn et al. 2014).

Table 4.3 Ore sorting methods and mineral application

Industrial mineral	Gems	Ferrous metals	Nonferrous metals	Fuel	Slag
COLOR, XRT, NIR	COLOR, XRT, XRL, NIR	XRT, EM, NIR	COLOR, XRT, NIR	XRT, RM	XRT, EM
Lime stone Dolomite Quartz Rock salt Talc Calcite Feldspar Magnesite	Diamond Emeralds Rubies Sapphires Tanzanite	Iron Manganese Chromite	Gold Silver Zinc Copper Nickel Tungsten PGMs	Coal Oil shale Uranium	Stainless steel slag Ferro silica slag Ferro chrome slag Nonferrous slag

Table 4.4 Sensor ore sorting segmentation

Sensor/technology	Material property	Segment
RM (radiometric)	Natural gamma radiation	Fuel, precious metals
XRF/XRL	Visible and X-ray fluorescence	Diamonds, base metals
XRT (X-ray transmission)	Atomic density	Base metals, precious metals, industrial minerals, fuel, diamonds
COLOR (CCD color camera)	Reflection, absorption, transmission	Base metals, precious metals, industrial minerals, diamonds
PM (photometric)	Monochromatic reflection/absorption	Base metals, industrial minerals
NIR (near-infrared spectrometry)	Reflection, absorption	Base metals, industrial minerals
IR (infrared cam)	Heat conductivity, heat dissipation	Base metals, industrial minerals
(electro-magnetic sensor)	Conductive magnetic	Base metals

Typically, DMS processes are compact processes with high throughput of 300–1000 ton/h. Holloway et al. have concluded that DMS seems to be well-positioned with tremendous growth potential to address the need for reducing energy costs, improving mine-mill integration, and simplifying tailings handling through coarse ore pre-concentration (Holloway et al. 2009). Bamber has carried out various studies and has highlighted the significant potential of DMS in underground applications (Bamber 2008) (Fig. 4.10 and Table 4.5).

4.4.4 *Coarse Particle Flotation*

Flotation of ore at a top-size of 3 mm, followed by gravity concentration of the flotation tailings, has been applied previously in South Africa to treat Witwatersrand gold ores resulting in the production of a 40% mass pull at an overall gold recovery

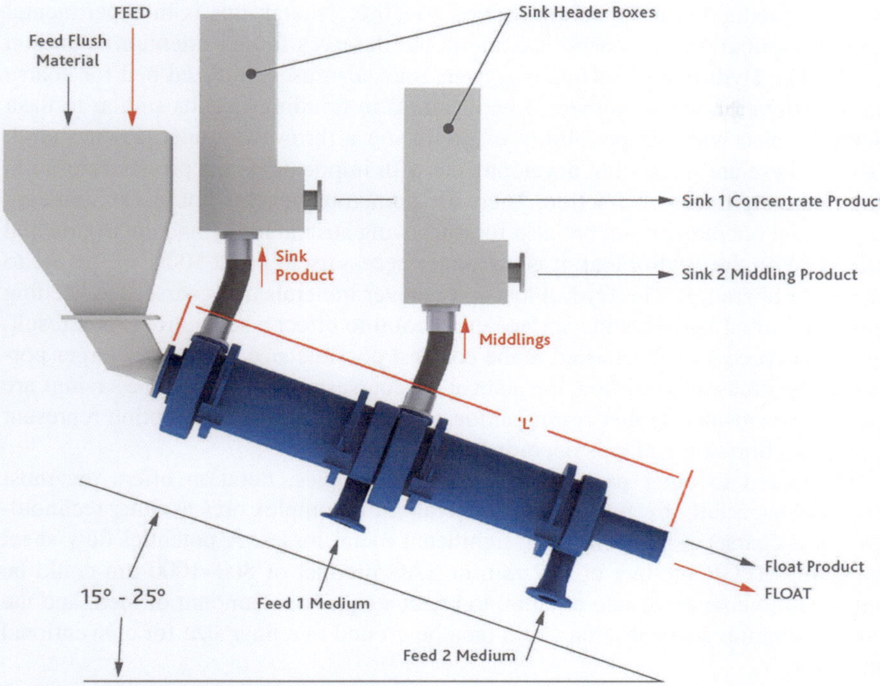

Fig. 4.10 Dense media separator schematic (Sepro Mineral Systems 2019)

Table 4.5 Application of DMS in a variety of minerals

Metals, oxides, and sulfides	Industrial minerals and silicates
Gold	Barytes
Antimony	Coal
Bauxite	Diamond
Chromite	Feldspar
Copper	Fluorspar
Iron	Limestone
Lead	Phosphate
Manganese	Potash
Nickel	Quartz
Tin	Petalite
Zinc	Spodumene

of 98% (Lloyd, 1979). Coarse particle flotation of particle ranging from 3 to 5 mm is common in the potash industry. For many base metals and precious metals applications, coarse flotation is challenging in conventional and flash flotation machines for pre-concentration purposes.

Jameson (2014) has developed a fluidized bed flotation technology for coarse flotation focusing on pre-concentration at a coarse size (600–800 μm) with an estimated reduction in operating cost of 10–20%. Though this is in experimental stage, nevertheless the concept has merit and deserves further attention (Jameson 2014). The HydroFloat® technology from Eriez also uses fluidized bed for coarse particle flotation with commercial applications in grinding circuits similar to flash flotation, also with the possibility of producing a throwaway tails (Franco et al. 2015). These are interesting developments with implications for pre-concentration at a relatively finer-size fraction. There are numerous benefits not just in reducing capital and operating costs but also for improving sustainability and environmental efforts. With the HydroFloat, it is no longer necessary to grind 100% of the ore to complete liberation. The HydroFloat can recover minerals at a coarse size needing only a minimal hydrophobic surface expression to effect a separation. As a result, maximum recovery is achieved at the coarsest possible size such that a larger portion of the ore is rejected from the plant at the coarsest possible size. The savings are enormous considering that comminution energy and media consumption represent a large portion of the plant's operating cost.

Compared to other pre-concentration technologies, flotation offers the most potential especially for fine-grained mineralogical complex ores as other technologies at a coarser grind will incur significant metal losses. A potential flow sheet involving HPGR product of 1–2 mm or SAG product of 800–1000 μm could be subjected to coarse particle flotation to reject a significant amount of feed, and the coarse flotation concentration could then be ground to a finer size for conventional flotation.

4.5 Flotation

Flotation has been used as a means to pre-concentrate refractory ores even before the 1930s in Canada, Australia, and Korea using oils as flotation collectors to produce bulk low-grade gold concentrates (Richart 1912; Taggart 1925). Until the late 1960s, most of the flotation activity was centered in Canada with gold production coming from flotation treatment of refractory, copper-gold and other complex ore types (Dunne 2005). During the late 1960s flotation of pyrite was a key focus in many South African gold operations to meet the needs of sulfuric acid market from the booming uranium industry. The gold boom in the 1980s and 1990s led to the processing of several refractory gold and copper-gold deposits in Australasia, Africa, and the Americas using flotation to produce concentrates with further treatment using bacterial and pressure leaching.

A review of the reagents used in gold flotation is presented elsewhere (Kappes et al. 2011). Flotation using xanthates as the primary collector is common for liberated, placer gold and even for many refractory ores. Dithiophosphates are widely used as secondary collectors along with xanthates in various operations. Monothiophosphates are normally more stable and selective for gold minerals with a high silver content. In acidic flotation circuits, mercaptobenzothiazole (MBT) is often preferred for gold-bearing partially or sometimes fully oxidized pyrite.

Modified thionocarbamates are used in copper-gold ores, which allow some selectively against iron sulfides even at a pH value less than 10. Many gold plants use a blended collector or separate collectors as needed based on mineralogy and ore chemistry.

Copper sulfate is commonly used as an activator of gold-bearing iron sulfides in flotation plants. The dosage, the addition point, and the slurry pH range are critical parameters for maximizing the benefits from copper sulfate. Addition of copper sulfate before collector appears to be important to improve kinetics of iron sulfides. Addition of copper sulfate at pH values ranging from 7 to 10 may slow down floatability of iron sulfides. Excessive addition of copper sulfide may result in lower recovery due to oxidation of xanthate into dixanthogen in solution, thus reducing the effectiveness of collection. Copper sulfate addition also appears to impact froth stability. There is an optimum dosage below which results in high slime recovery with a tenacious froth, whereas a higher dosage leads to a fragile froth behavior.

Sulfidization is also used for naturally oxidized or tarnished iron oxides due to unfavorable milling environment. Solution potential control at −450 mV using sodium sulfide or sodium hydrogen sulfide is typically used prior to flotation. Use of NaHS and silver ions has been reported to improve gold recovery at the Los Pelambres mine in Chile (Chryssoulis 2001). Sulfide ions appear to act as flotation activators at low concentrations less than 10^{-5} M, whereas at concentration above 10^{-5} M, it acts as a depressant (Aksoy and Yarar 1989).

Flotation of Carlin-type double-refractory sulfide ores is considered as challenging (Bulatovic 1997; Kappes et al. 2010). Refractory gold ores commonly contain free gold, submicroscopic gold, carbonaceous material, base metals, pyrite, marcasite, arsenopyrite, and pyrrhotite, with clays and graphitic carbon known to inhibit flotation behavior (Swash 1925). The Carlin refractory gold ores are mainly associated with arsenian pyrite, which is known to oxidize quickly resulting in reduced flotation kinetics. Newmont has developed the N_2TEC process to provide a reducing environment during grinding and flotation mainly to prevent oxidation of arsenian pyrite (Simmons 1997).

Mineralogy and gold deportment studies clearly suggest that an ultrafine grind (sub 10 μm) is necessary to maximize recovery of gold-bearing pyrite. A finer grind, however, results in slower flotation kinetics and poor selectivity against carbonaceous matter resulting in a very high concentrate mass pull. Novel technologies are needed to address this challenge. Newmont has developed the N_2TEC process to float arsenian pyrite in a reducing environment to minimize oxidation of sulfides, and this technology is presently used in (Simmons 1997).

The deleterious preg-robbing organic carbon can also be removed by flotation and either discarded, if the gold content is negligible (Taggart 1925), or separately treated with kerosene prior to cyanide leaching (Beer 1994).

With the trend of declining head grades requiring high throughputs to improve economics, the need for water has increased significantly over years. In temperate and equatorial latitudes, there is no shortage of water in general, although the disposal of used water is a significant issue. In other arid parts of the world such as in Chile, Western Australia, and Southwest USA, water is scarce or unavailable requiring the need to use the sea, hypersaline, or brackish water for mining.

Desalination of seawater is expensive as new desalination plants and associated pumping and pipeline systems can easily have a capital cost exceeding $500 million along with high operating costs (Blin and Dion Ortega 2013). Barrick Gold has developed a new process, air-metabisulfite treatment (AMBS), which allows use of seawater or brackish water for pyrite depression in copper flotation with minimal metallurgical impact compared to that with the conventional lime-based process (Gorain 2012). The AMBS treatment has also resolved the issue of molybdenum flotation in copper-molybdenum ores using seawater.

A holistic approach to reagent selection is becoming more important (Nagaraj and Farinato 2014) due to increased ore complexities and also the advent of quantitative mineralogy tools and better understanding of flotation cell hydrodynamics. In addition, due to stringent environmental regulations, there is a focus on developing greener flotation reagents to meet sustainability goals. More recent application of electrochemical and surface analytical techniques has allowed better understanding of the fundamentals of flotation mechanism and chemistry (Smart et al. 2007). This is an important development that is already helping in better understanding of flotation behavior of complex ores and in developing solutions to complex flotation problems (Ralston et al. 2007; Fuerstenau 2007; Gorain 2013).

4.6 Other Physical Separation Processes

4.6.1 Magnetic Separation

There are tens of thousands of low-intensity magnetic separation (LIMS) and thousands of high-intensity magnetic separations (HIMS) used in the minerals industry today. A significant magnetic separation development that had made a major impact in the mineral industry was the drum separator for magnetite ores, mainly to treat fine-grained and low-grade magnetite ores for producing high-grade concentrates for sinter feed, pellets for blast furnace feed, and also for direct reduction iron processes (Arvidson and Norrgran 2014). The Wetherill separator type is still in use today in various forms for separating precious metals, typically after calcination of ore feed. Use of magnetic separator to recover gold-bearing iron oxides from roaster leach tails is also practiced such as at Barrick's Goldstrike operation (Douglas and Semenyna 2013).

Due to declining grades, the need for sophisticated high-capacity magnetic separators is becoming important. Drums for low-intensity magnetic separators in iron ore processing are becoming larger, with drum diameters around 1.5 m and widths exceeding 4 m. Other than the trend of increasing size, the use of cryogen-free superconducting coil system is also becoming prevalent, which is the most recent technology for superconducting magnetic separators using a niobium-titanium conductor to generate the magnetic field without the need for helium to cool the coil (Norrgran et al. 2009).

4.6.2 Electrostatic Separation

Electrostatic separation (ESS) is mainly used for mineral sands application, though earlier patents suggest applications in iron ore and phosphate separations. The relative electric conductivity property of the mineral surfaces is exploited to make the separation. For example, conductive minerals such as ilmenite, monazite, and rutile are separated from nonconductive silica and zircon present in mineral sands. ESS works best on clean surfaces, and in many cases the feed needs to be washed and classified for efficient separation. ESS was also successfully implemented at several precious metals smelters recovering metal prills from slag, typically after the Merrill Crowe process. In addition, this technology has been successfully applied in gold mines to recover gold and silver from slag (Hearn 2014). Dry grinding is typically used to liberate metal prills, de-dusting and then ESS to recover these prills for direct smelting, rather than the conventional wet milling and gravity tabling followed by drying before sending these materials for smelting (Maki and Taylor 1987).

4.6.3 Magnetic Pulse Treatment

The process of magnetic impulse treatment applies for refractory gold ores in ferriferous quartzite and can provide higher recovery at low-energy consumption. The process is conducted by transporting ore or pulp through a section of dielectric pipe where a series of electromagnetic coils is mounted and generates regularly impulses of an electromagnetic field with a frequency of 50 Hz. The ideal application is ahead of grinding and enhancing iron ore grinding and gold recovery.

4.6.4 Nanosecond High-Power Electromagnetic Pulses (HPEMP)

Nanosecond HPEMP treatment has been investigated in Russia since 1997. It has shown two important effects: first, the loosening of the mineral structure due to electrical breakdown mechanism and, second, the development of thermomechanical stresses at the boundary between the dielectric and conductive mineral components.

The nanosecond HPEMP treatment showed promising results in the treatment of gold refractory ores due to high efficiency and selectivity in the disintegration of mineral complexes.

The application of high-power electromagnetic pulses (HPEMP) irradiation in dressing of resistant gold-containing ores appears attractive as this technique provides for a significant increase in precious metal recovery (30–80% for gold and 20–50% for silver), therewith helping reduce both energy consumption and the cost of products (Chanturiya et al. 2001).

References

Adams, Mike D. 2016. In *Gold ore processing: project development and operations*, ed. M.D. Adams, 2nd ed. Amsterdam: Elsevier.

Aksoy, B.S., and B. Yarar. 1989. Natural hydrophobicity of native gold flakes and their flotation under different conditions. In *Processing of complex ores*, ed. G.S. Dobby and S.R. Rao. New York: Pergamon Press.

Albuquerque, L. et al. 2008. Application of high frequency screens in closing grinding circuits. In *Proceedings of the Vth international mineral processing seminar, Procemin 2008*, Santiago, Chile.

Allen, Jon. 2013. *Stirred milling machine development and application extension*. QUEBEC : Metso

AMIRA. 1993. An investigation of the Hilton Tower Mill Circuit. Method and benefits of fine grinding ores.

Arvidson, B.R., and D. Norrgran. 2014. In *Magnetic separation in mineral processing and extractive metallurgy—100 years of innovation*, ed. C.G. Anderson, R.C. Dunne, and J.L. Uhrie. SME.

Arvidson, B.R., and H. Wotruba. 2014. In *Ore sorting in mineral processing and extractive metallurgy—100 years of innovation*, ed. C.G. Anderson, R.C. Dunne, and J.L. Uhrie. SME.

Bamber, A.S. 2008. *Integrated mining, pre-concentration and waste disposal systems for the increased sustainability of hard rock metal mining*. PhD thesis, University of British Columbia.

Bartram, K., and M. Kowalczyk. 2009. New developments in sensor based ore sorting. In *Proceedings of 48th conference of metallurgists*, Sudbury, 421–432.

Beer, G. 1994. Processing options for the treatment of preg-robbing ores at the Macraes Gold Mine. In *Proceedings 6th AusIMM extractive metallurgy conference*, AusIMM, 203–208.

Blin, P., and A. Dion Ortega. 2013. A 2013 high and dry. CIM.

Brent, G.F., et al. 2013. Ultra-high-intensity blasting—a new paradigm in mining. In *Proceedings of world gold conference, Brisbane*, 395–400.

Bulatovic, S.M. 1997. Flotation behaviour of gold during processing of porphyry copper-gold ores and refractory gold bearing sulfides. *Minerals Engineering* 10: 895–908.

Burns, R., and A. Grimes 1986. The application of pre-concentration by screening at Bougainville Copper Limited. In *Proceedings of the AUSIMM mineral development symposium*, Madang, Papua New Guinea.

Castro, E., J. Lopez, and D. Switzer 2009. In *Maximum classification and efficiency through classification using the recyclone, recent advances in mineral processing plant design*, ed. D. Malhotra, P.R. Taylor, E. Spiller, and M. LeVier. SME.

Chanturiya, Valentin, et al. 2001. Use of high-power electromagnetic pulses in processes of disintegration and opening of rebellious gold-containing raw material. *Journal of Mining Science* 37: 427–437.

Chryssoulis, S.L. 2001. Using mineralogy to optimize gold recovery by flotation. *JOM* 53: 48–50.

Clark, B.H. 2007. The Derrick stack sizer: Revolutionary advancements in wet screening. In *Proceedings of the 39th Canadian mineral processors conference*, 413–428.

Douglas, W., and L. Semenyna. 2013. Magnetic recovery of gold bearing iron oxides at Barrick Goldstrike's roaster. In *Proceedings world gold 2013,* Brisbane, 79–85.

Dunne, R. 2005. In *Flotation of gold and gold-bearing ores in advances in gold ore processing*, ed. M.D. Adams. Amsterdam: Elsevier Publication.

Dunne, R., et al. 2007. Boddington gold mine—an example of sustaining gold production for 30 years. In *World Gold 2007 conference, AusIMM,* 213–230.

Ellis, S., and M. Gao. 2002. The development of ultra fine grinding at KCGM. In *SME annual conference*.

Erickson, M.T. 2014. In *Innovations in comminution equipment: crushers, high pressure grinding rolls, semi-autogenous grinding, ball mills and regrind mills in mineral processing and extractive metallurgy: 100 years of innovation*, ed. C.J. Anderson, R.C. Dunne, and J.L. Uhrie. SME.

Foggiatto, B., et al. 2014. The economics of the large scale sorting. In *Proceedings of the XXVII international mineral processing congress*, Santiago, Chile.

Franco, J.J.C., et al. 2015. Coarse gold recovery using flotation in a fluidized bed. In *Proceedings of the 13th Canadian mineral processors conference*, Ottawa.

Fuerstenau, D.W. 2007. In *A century of developments in the chemistry of flotation processing in froth flotation: a century of innovation*, ed. M.C. Fuerstenau, G.J. Jameson, and R.H. Yoon. SME.

Gillot, P. 2006. Pit-to-plant optimization at Morila gold mines. In *SAG 2006*, Vancouver.

Gleencore. 2015. IsaMill™ breaking the boundaries.

Gorain, B.K. 2012. Separation of copper minerals from pyrite using air-metabisulfite treatment application global.

Gorain, B.K. 2013. Developing solutions to complex flotation problems. In *Proceedings of the 34th Canadian Mineral Processors Conference,* Ottawa, 293–312.

Hart, S., et al. 2011. In *Commissioning and ramp up of the HPGR circuit at Newmont Boddington Gold SAG 2011*.

Hearn, S. 2014. I*nnovations in electrostatic separation in mineral processing and extractive metallurgy: 100 years of innovation*, ed. C.J. Anderson, R.C. Dunne, and J.L. Uhrie. SME.

Holloway, A.R., L. Jones, and S.R. Lawrence. 2009. Pre-concentration of platinum group and base metal ores by dense medium separation. In *Proceedings of 48th conference of metallurgists*, Sudbury, 373–384.

Hukki, R.T., and H. Allenius. 1968. *A quantitative investigation of closed grinding circuit*. SME AIME

Jameson, G. 2014. Experiments on the flotation of coarse composite particles. In *Proceedings of the XXVII international mineral processing congress*, Santiago.

Jankovic, Alex, et al. 2015. Evaluation of dry grinding using HPGR in closed circuit with an air classifier. *Minerals Engineering* 71: 133–138.

Kanchibotla, S. 2014. In *Mine to mill value chain optimization—role of blasting in mineral processing and extractive metallurgy: 100 years of innovation*, ed. C.J. Anderson, R.C. Dunne, and J.L. Uhrie. SME.

Kappes, R., D. Brosnahan, and J. Gathje. 2010. The effect of mineral liberation on the floatabilities of pyrite, arsenopyrite and arsenian pyrite for Carlin Trend ores. In *Proceedings XXV international mineral processing congress*.

Kappes, R., C. Fortin, and R. Dunne. 2011. The current status of the chemistry of flotation in industry. In *Proceedings world gold 2011, COM 2011*, Montreal, 385–395.

Kelsey, C., and J. Kelly. 2014. Super-fine crushing to ultra-fine size the "IMP" super-fine crusher. In *Proceedings of the XXVII international mineral processing congress*, Santiago, Chile, 239–251.

Klein, B., W.S. Dunbar, and M. Scoble. 2002. Integrating mining and mineral processing for advanced mining systems. *CIM Bulletin* 95 (1057): 63–68.

Koski, S., J.L. Vanderbeek, and J. Enriquez. 2011. Cerro Verde concentrator-four years operating HPGR's. In *SAG 2011*, Vancouver, BC.

Lakshmanan, V.I., et al. 2016. In *Innovative process development in metallurgical industry*, ed. V.I. Lakshmanan, Roy Raja, and V. Ramachandran. Cham: Springer International.

Lessard, J., J. deBakker, and K. McHugh. 2014. Development of ore sorting and its impact on mineral processing economics. *Minerals Engineering* 65: 88–97.

Letho, H. et al. 2013. Outotec HIGmills; a fine grinding technology. In *23rd International mining congress & exhibition of Turkey*, Antalya.

Lloyd, P.J.D. 1979. An integrated mining and extraction system for use on Witwatersrand gold mines. *Journal of the SAIMM* 79 (1): 135–148.

Logan, A., and N. Krishnan. 2012. Newcrest technology—step change. In *Proceedings of the xxvi international mineral processing congress*, New Delhi, India, Paper # 1085, 3025–3037.

Major, K. 2002. Types and characteristics of crushing equipment and circuit flowsheets. In *Mineral processing plant design, practice and control*, 566–583. SME.

Maki, T.D., and J.B. Taylor. 1987. Precious metal slag treatment using an electrostatic separator. In *Small mines development in precious metals,* 245–252.

Mazzinghy, D.B., et al. 2014. Vertical agitated media mill scale-up and simulation. *Minerals Engineering* 73: 69–76.

McCullough, W.E., R.B. Bhappu, and J.D. Hightower. 1999. Copper ore preconcentration by heavy media separation for reduced capital and operating costs, Cobre 99.

Metso Minerals Industries Inc. 2016. VERTIMILL® Grinding Mills & Stirred Media Detritor.

Mohanty, J.K., et al. 2000. Mineralogy and pre-concentration of the chromite overburden of the Sukinda Ultramafic Belt. *CIM Bulletin* 93 (1038): 37–43.

Morrell, S., U.J. Sterns, and K.R. Weller. 1993. Application of population balance models to very fine grinding in Tower mills. In *Proceedings of the XVIII international mineral processing congress*, Sydney, 61–66.

Mosher, J.B. 2005. Comminution circuits for gold ore processing. In *Advances in gold ore processing*, ed. M. Adams, 253–276. Guildford, Australia: Elsevier.

Munro, P.D., et al. 1982. The design, construction and commissioning of a heavy media plant for silver-lead-zinc ore treatment–Mount Isa Limited. In *Proceedings of the XIV international mineral processing congress*, Toronto, 1–20.

Nagaraj, D.R., and R.S. Farinato. 2014. Major innovations in the evolution of flotation reagents. In *Mineral processing and extractive metallurgy: 100 years of innovation*, ed. C.J. Anderson, R.C. Dunne and J.L. Uhrie. SME Publication.

Napier-Munn, T.J., J. Bosman, and P. Holtham. 2014. In *Innovations in dense media separation technology in mineral processing and extractive metallurgy: 100 years of innovation*, ed. C.J. Anderson, R.C. Dunne, and J.L. Uhrie. SME.

Norrgran, D., Ashton, D., and Fears, P. 2009. Installation and operation of the Eriez cryogen-free superconducting magnet at Sibelco. In *Mineral processing plant design—an update conference*. Arizona: Tucson.

Palaniandy, S., et al. 2014. VertiMill performance updates in secondary and regrind duties at Cannington Mine, performance updates in secondary and regrind duties at Cannington Mine Aussim 2014, Brisbane.

Pease, J., et al. 2015. Minerals processing: a step change in mining productivity. AusIMM Bulletin.

Pyke, P., et al. 2006. Application of HPGR technology in processing gold ore in Australia. In *International autogenous and semiautogenous grinding technology*, Vancouver.

Ralston, J., Fornasiero, D., and S. Grano. 2007. Pulp and solution chemistry, in froth flotation: a century of innovation, ed. M.C. Fuerstenau, G.J. Jameson, and R.H. Yoon. SME.

Richart, T.A. 1912. *The flotation process*. San-Francisco: Mining and Scientific Press.

Rutter Jon. 2017. *CRC ORE–grade engineering and GE view*. CRC Ore Ltd.

Schena, G.D., R.J. Gochin, and G. Ferrara. 1990. Pre-concentration by dense media separation an economic evaluation. *Transactions of the Institute of Mining and Metallurgy* 99: C21–C31.

Sepro Mineral Systems. 2019. https://seprosystems.com/.

Shi, F., W. Zuo, and E. Manlapig. 2003. Characterization of pre-weakening effect of ores by high voltage electric pulses using single particle tests. *Minerals Engineering* 50–51: 69–76.

Shi, F., et al. 2014. A potential application of high voltage pulse technology in a gold-copper ore grinding circuit. In *Proceedings of the XXVII international mineral processing congress*, Santiago, 68–77.

Simmons, G.L. 1997. Flotation of auriferous pyrite using Sante Fe Pacific's gold N2TEC flotation process. In *SME annual meeting*, Preprint No. 97-27.

Sivamohan, R., and E. Forssberg. 1991. Electronic sorting and other pre-concentration methods. *Minerals Engineering* 4 (7–11): 797–814.

Smart, R., W. Skinner, and A. Gerson. 2007. *Surface characterization and new tools for research; froth flotation*. SME.

Sonic Samp Drill. 2014. SonicSampDrill. http://www.sonicsampdrill.com/news/sorting-ore-with-the-speed-of-light-1000-tons-per-hour-.htm.

Swash, P.M. 1925. A mineralogical investigation of refractory gold ores and their beneficiation with special reference to arsenical ores. *Journal South African Institute of Mining and Metallurgy* 88 (5): 25–28.

Taggart, A.F. 1925. Flotation of gold and silver. In *Handbook of mineral dressing*. New York: Wiley.

Van der Meer, F.P., and S. Strasser. 2012. Case study of dry HPGR grinding and classification in ore processing. In *9th international mineral processing conference (Procemin 2012)*, Santiago.

Vanderbeek, J.L., et al. 2006. In *HPGR implementation at Cerro Verde, SAG 2006,* Vancouver, BC, 45–61.

Chapter 5
Gold and Silver Extraction

V. I. Lakshmanan, R. Roy, and B. Gorain

Gold is considered a store of value and has been a sought-after metal since the dawn of human civilization. It was known to Egyptians as early as 3100 BC. The value of gold was reported to be two-and-one-half times the value of silver in the code of Menes, the founder of the first Egyptian dynasty. A gold mine in Nubia in Nile valley is depicted in the Turin Papyrus Map of 1320 BC. Methods for large-scale extraction of gold were developed by Romans, who made long aqueducts to enable them to sluice large alluvial deposits for gold recovery. The feed material for gold recovery is now hard rock ores for most of the commercial gold recovery operations.

Gold is called a noble metal as it does not oxidize under ordinary conditions. Gold is very ductile, which makes it easy to shape as desired. The density of pure gold is 19.3, which made it easy to separate free gold from waste rock material with a density of about 2.5 by gravity separation. Many simple devices were used for this purpose such as the gold pan, rocker, and sluice box. To recover gold from a gold-bearing mineral, mercury was used to combine with gold. Mercury forms an amalgam with gold, and gold was recovered by separating it from mercury by retorting. This process is no longer in wider use and has been replaced by cyanide process. The basis of cyanide process is the solubility of gold in sodium or potassium cyanide and the ability to recover gold from low-grade ore economically. Depending on the complexity of gold extraction from the ores, gold ores are divided in non-refractory gold ores and refractory gold ores. Recovery of gold from refractory gold ores is complex compared to easier recovery of gold from non-refractory gold ores.

V. I. Lakshmanan (✉) · R. Roy
Process Research Ortech Inc., Mississauga, ON, Canada
e-mail: llakshmanan@processortech.com

B. Gorain
Hindustan Zinc Limited, Udaipur, Rajasthan, India

V. I. Lakshmanan, B. Gorain (eds.), *Innovations and Breakthroughs in the Gold and Silver Industries*, https://doi.org/10.1007/978-3-030-32549-7_5

5.1 Leaching of Non-refractory Ores

Examples of non-refractory gold ores include placer, free milling and oxidized ores. Gold particles in the non-refractory gold ores are not locked, and nearly 95% of the gold from non-refractory gold ores can be recovered by gravity concentration and/or direct cyanidation. The treatment options for non-refractory gold ores are shown in Fig. 5.1. Depending on the ore, gold can be recovered by (1) gravity separation, (2) crushing and cyanide/non-cyanide leaching, and (3) crushing/grinding and gravity concentration/flotation followed by cyanide/non-cyanide leaching (LaBrooy et al. 1994).

5.1.1 Cyanide Processing

Before the development of cyanide process, amalgamation with mercury was the primary process for gold production. Though the use of cyanide to leach gold was known since 1783, the process was patented by MacArthur and the Forest brothers during 1887–1888. The cyanidation process was first commercialized at the Crown Mine (New Zealand) in 1889 (Rose and Merloc 2008; Marsden and Sass 2014), and the process soon spread all over the world.

5.1.1.1 Cyanide Leaching

Gold leaching by cyanidation takes place by the following reactions:

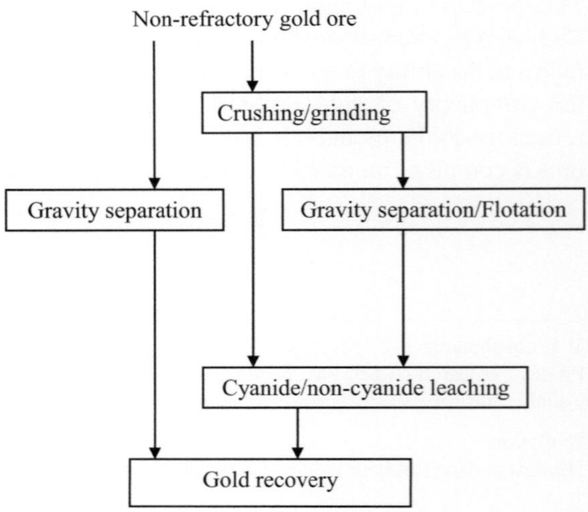

Fig. 5.1 Processing of non-refractory gold ores (adapted from LaBrooy et al. 1994)

$$2Au + 4CN^- + O_2 + 2H_2O = 2Au(CN)^-_2 + H_2O_2 + 2(OH^-)$$

$$4Au + 8CN^- + O_2 + 2H_2O = 4Au(CN)^-_2 + 4(OH^-)$$

The first reaction is the dominant reaction and a smaller amount of gold is dissolved by the second reaction. According to both reactions, the presence of oxygen is necessary for gold dissolution.

The cyanidation process resulted in resurrection of a declining gold industry. It facilitated the development of the gold mine at Witwatersrand in Transvaal, South Africa. Within 10 years from 1888 to 1898, the gold production in South Africa increased nearly ten times from less than 300,000 ounces to over three million ounces (Stanley 1987). Many new developments in gold processing took place in South Africa that significantly impacted the gold industry around the globe. As a consequence, new process flow sheets involving cyanide leaching and zinc precipitation were developed, which replaced the existing amalgamation and gravity separation processes. A continuous process was developed that replaced batch processing in vats of separate sand and slimes circuits. This was a result of many innovative process developments such as:

1. Introduction of tube milling to replace stamp milling in 1904. This allowed for higher throughput along with a finer and uniform particle size distribution, which resulted in improved liberation and recovery (Rickard 1907).
2. Mechanical advancements in solid-liquid separation during 1904 to 1908. This included the development of Dorr's thickener, CCD circuit, Butter's vacuum filter, and Oliver's drum filter (Bain 1910).
3. Development of leach tanks such as Brown's air-agitated leach tanks in New Zealand and "Pachuca" tanks in Mexico.

Further process developments included the recovery of gold and silver from cyanide solutions by cementation. This is done by the use of zinc dust instead of zinc shavings in zinc boxes. Also, before cementation the pregnant solutions are deaerated along with the addition of a small amount of soluble lead salt (Clennell 1915). This process is referred to as "Merrill-Crowe" process, named after the developers of this technology.

The use of activated carbon was the next major development in gold and silver processing. The use of wood charcoal to recover gold from chlorination leach solution was patented in the late 1800s and was applied commercially at Mount Morgan and other operations in Australia. The process was also investigated by the US Bureau of Mines (USBM) along with many other researchers during the 1920s and 1930s. The use of granular activated carbon was implemented at the San Andreas de Copan plant in Honduras in 1949. The loaded carbon was sold to smelters for gold and silver recovery. Activated carbon was also used commercially in 1952 at the Carlton Mine in Cripple Creek, Colorado. This mine also had a reactivation kiln for carbon regeneration, which allowed the carbon adsorption methods to be more economical than the Merrill-Crowe process. This important development led to the acceptance of the carbon-in-pulp (CIP) process worldwide (Marsden and Sass 2014). The CIP process was applied at a large scale at the Homestake Mine in South Dakota in 1973. With further development in the carbon elution process at Anglo American Research Laboratory (AARL) and Mintek, many plants using CIP process were commissioned during the early 1980s. This included plants at Modderfontein, President Brand, Randfontein Estates, Western Areas, Rand Mines, Beisa and 11 other installation in South Africa (Boydell 1984), plants at Kambalda, Norseman and Haveluck in Australia, and the Jerritt Canyon Mine in the USA. Along with the effective carbon elution techniques, the development of the interstage screen and regeneration of carbon using acid wash and thermal reactivation techniques enabled the rapid rise of the CIP circuits.

As the understanding of the CIP process increased, it was combined with the leaching step, which led to the development of carbon-in-leach (CIL) circuits. By the early 1980s, the CIP/CIL flow sheets became the new standard for gold and silver leaching. Though silver loading on carbon was low, it was still applied to treat primary silver deposits as well. Initially the carbon consumption was approximately 200 to 300 g/t. With improvements in pipe and agitator design, preconditioning of carbon, rejection of fines, and other modifications, the carbon consumption was reduced to 30 to 50 g/t. A typical CIL plant is shown in Fig. 5.2 (Gorain et al. 2015).

A parallel development was the resin-in-pulp (RIP) technology, which was implemented in the former USSR during the late 1960s and 1970s. The RIP process

Fig. 5.2 Carbon-in-leach plant (Gorain et al. 2015)

is not popular in the gold and silver industry as the process for resin elution and regeneration is complex. Golden Jubilee in South Africa and Penjom in Malaysia are the only commercial applications outside of the former USSR.

5.1.1.2 Cyanide Detoxification

There was an accidental release of 50 to 100 tons of cyanide in cyanide-bearing tailing solution from the Aural mining operation in January 2000 near Baia Mare in Romania. This initiated an immediate mission to prevent such events from occurring again and resulted in the formation of a steering committee to establish a framework for the use of cyanide in the mining industry. As a result, the International Cyanide Management Code was developed in May 2002. Later in the same year, the International Cyanide Management Institute (ICMI) was formed. The ICMI has the responsibility of administrating the code through a multi-stake team including representatives of mining companies. In February 2004, the Industry Advisory Group (IAG) was reconstituted with an aim to coordinate and facilitate code implementation by a group of gold producers, cyanide manufacturers, and transportation companies.

In order to meet the cyanide code guidelines (50 mg/L, weak acid dissociable cyanide, WAD) in tailings storage facilities, different methods are used to comply with the code. The natural decay of cyanide through the leach circuit allowing an acceptable discharge to the tailings facilities is the preferred method. The recovery and recycling of cyanide-bearing solutions are facilitated by the use of tailings thickener and filtration systems. This method requires extreme care in monitoring cyanide concentrations through the leach circuit in the absence of any other form of cyanide destruction. The detoxification efforts can be complemented by the use of dry stacked tailings such as at La Coipa in Chile, which allows the recycling of cyanide.

Another possibility is the cyanide destruction using biological oxidation. This has been successfully applied at the Homestake Mines in South Dakota since 1984. It was later implemented at other Homestake operations (Whitlock and Mudder 1986). The most cost-effective method of cyanide destruction is the use of sulfur dioxide and air with copper as catalyst if natural degradation is not possible. This process was developed by Inco in the early 1980s.

If rapid detoxification is required such as in circuits with limited residence time availability, other method can be employed. Such methods include the use of hydrogen peroxide and Caro's acid and have been used successfully at different operations. In these methods, both free and WAD cyanides are converted to cyanate very rapidly due to strong oxidizing conditions. Another cyanide destruction method developed by Maelgwyn Mineral Services (MMS) is called MMS CN-D Process™, which uses MMS's proprietary Aachen Reactors. This process uses oxygen, carbon, selected reagents, and catalysts to bring cyanide levels down to well below 50 ppm WAD levels (Adams and Glen 2011).

LIXKill process developed by Barrick Gold uses Caro's acid to destroy free and WAD cyanide in tailings reclaimed solution before it is fed to the mill and autoclaves (Pekrul 2007). Due to high consumption and costs along with the inclination to avoid any chloride into the cyanide leach system, the use of alkaline chloride and hypochlorite oxidation of cyanide has been discontinued.

5.1.1.3 Cyanide Recovery

A process to recover cyanide was used commercially at Flin Flon in Canada to treat cyanide-bearing process solution from about 1930 to 1995 (Marsden and Sass 2014). This process called the AVR (acidification, volatization, recovery) process involves acidifying cyanide-bearing solutions to generate hydrogen cyanide, volatilization with air to remove hydrogen cyanide (HCN) from solution followed by adsorption, and recovery of HCN gas into a caustic scrubber system. The modified AVR was used at Golden Cross in New Zealand, Delamar silver mines in Idaho during the 1990s and more recently at Rio Paracatu in Brazil and Cerro Vanguardia in Argentina. The process modification included high efficiency packed bed adsorption towers for improved stripping of HCN either from solution or directly from slurry, thus avoiding solid-liquid separation (Goldstone and Mudder 1988).

The AVR process has not been used extensively in the industry. There are two reasons for this, the main reason being the already existing provisions in the gold plants. Most gold plants have provisions for cyanide recycling such as tailings thickening to reclaim cyanide-containing solution along with effective management and control of cyanide concentration down the leach circuits. The second reason is the perceived safety issues associated with the generation of HCN gas, though the AVR process has been used safely for decades.

5.1.1.4 Cyanidation Treatment of Copper-Gold Ores

Today a significant portion of gold production comes from the processing of ores containing other valuable metals such as copper. Cyanide-soluble copper-containing ores such as chalcocite, covellite, bornite, cuprite, malachite, and azurite poses significant metallurgical challenges such as low gold recoveries during cyanidation, high cyanide consumption, and detoxification costs. Moreover, weak acid dissociable (WAD) cyanide tends to stabilize in the presence of copper in tailings in a form that is toxic to wildlife and less amenable to natural degradation processes.

The development of the SART (sulfidization, acidification, recyle, and thickening) process was a major development in the treatment of copper-gold ores with significant amount of cyanide-soluble copper. The SART process was originally developed by SGS Lakefield Research and Teck for the Lobo-Marte project (MacPhail et al. 1998) and involves copper sulfide precipitation with sodium hydrosulfide (NaHS) and sulfuric acid. This results in the release of cyanide as HCN. The solids are separated from the solution using thickening and filtration to produce a high-grade copper sulfide solid. The thickener overflow solution is re-neutralized with lime, which converts hydrogen cyanide to calcium cyanide.

The first commercial SART plant was started operating in 2004 at Newcrest's Telfer Mine in Western Australia (Barter et al. 2001). Newmont has been operating a SART plant at their Yanacocha facility in Peru since December 2008 (Guzman and Acosta 2009). Other major SART operations to treat pregnant gold-bearing solutions produced by heap leaching are located at Maricunga in Chile, Lluvia de Oro in Mexico, and Gedabek in Azerbaijan. The commercial supplier of the SART technology is BioteQ, who have designed two commercial SART plants (Lawrence and Lopez 2011; Littlejohn et al. 2013).

5.1.2 Non-cyanide Processing

Due to the toxicity of cyanides, finding a substitute of cyanide is an active area of research. A promising alternative to cyanide leaching is thiosulfate leaching. Newmont has conducted the testing and demonstration of the concept of thiosulfate leaching of a bio-oxidized double refractory ore using heap leaching (Wan et al. 1993; Wan et al. 1994; Wan and Brierley 1997). Barrick has also investigated the thiosulfate process to treat Goldstrike preg-robbing ores at a feasibility level after conducting many initial bench and pilot-scale level testwork. Barrick has used the ammonium thiosulfate-copper system for leaching and anionic exchange resin for extraction of the gold complex (Fleming et al. 2003). Barrick obtained reasonably high gold recoveries with ammonium thiosulfate in the treatment of preg-robbing ores. However, the key challenge was the need for a cost-effective elution process to make it possible to strip gold from the resin. Due to the environmental concerns with groundwater contamination with the use of ammonia along with the need to overcome gold recovery challenges with recycled water-containing residual ammonia, a different source of thiosulfate was sought to enable development of an industrial-scale thiosulfate leaching process (Choi et al. 2013). Barrick initiated extensive research and development efforts during 2008, which focused on further development of the thiosulfate leaching technology using calcium thiosulfate (CaTS) as the lixiviant. The use of CaTS resulted in not only high dissolution rates but also mitigated environmental concerns from the use of ammonium salt. Work carried out on various ores showed that gold recoveries for preg-robbing ores were not compromised using CaTS compared to that with ammonium salts. Instead of activated carbon used in CIL, resins were shown to be the most effective means of gold extraction using the CaTS process.

Barrick operated a demonstration plant using CaTS process successfully at the Goldstrike's autoclave process facility for around 3 years since May 2010. During this time, approximately 1500 tons of different ore types were treated in this demonstration plant to ensure robustness of the process. This has demonstrated that the thiosulfate process can be commercially viable for the treatment of ores associated with high preg-robbing behavior. Figure 5.3 shows the thiosulfate leaching facility

Fig. 5.3 The thiosulfate facility at Goldstrike in Nevada, USA (Gorain et al. 2015)

Fig. 5.4 An aerial view of Barrick's Goldstrike operation (Nevada, USA) along with the thiosulfate processing facility (Gorain et al. 2015)

at Goldstrike, and Fig. 5.4 shows an aerial view of the overall Goldstrike complex including the thiosulfate leaching facility. The thiosulfate process needs to be applied to other ore bodies to better understand the limitations and opportunities with this technology to treat various complex ores.

5.2 Oxidative Pretreatment

As the grade of non-refractory gold ores is falling across the globe and reserves are depleting, the interest in the treatment of refractory gold ores is increasing. Refractory ores contain gold locked as solid solution in the matrix of various gold-bearing iron sulfides, such as pyrite, arsenopyrite, and arsenian pyrite. A pretreatment is required prior to cyanidation to release gold in leach solution. Other minerals such as tellurides and stibnites also display varying degrees of refractory behavior. Refractory ores do not respond well to direct cyanidation as the recovery is very low making the process uneconomical. The refractory gold ore needs to be of higher grade, or gold price needs to be higher in order to be viable for producing gold. As the flow sheet to extract gold from refractory ores is complex, higher capital and operating costs are associated with refractory gold processing. Furthermore, the cost and availability of utilities (electricity and water) along with water quality (especially chloride content) is important as refractory gold treatment has a greater demand for both from an operating perspective. Operating costs such as reagent consumptions and maintenance costs given the extreme operating conditions in refractory gold treatment are important considerations for process economics.

Figure 5.5 shows the pretreatment options for processing of refractory gold ores. There are five pretreatment routes involving varying degrees of oxidation of sulfides that are present in ore or in concentrate: roasting, pressure oxidation, bio-oxidation, ultrafine grinding, and chemical oxidation.

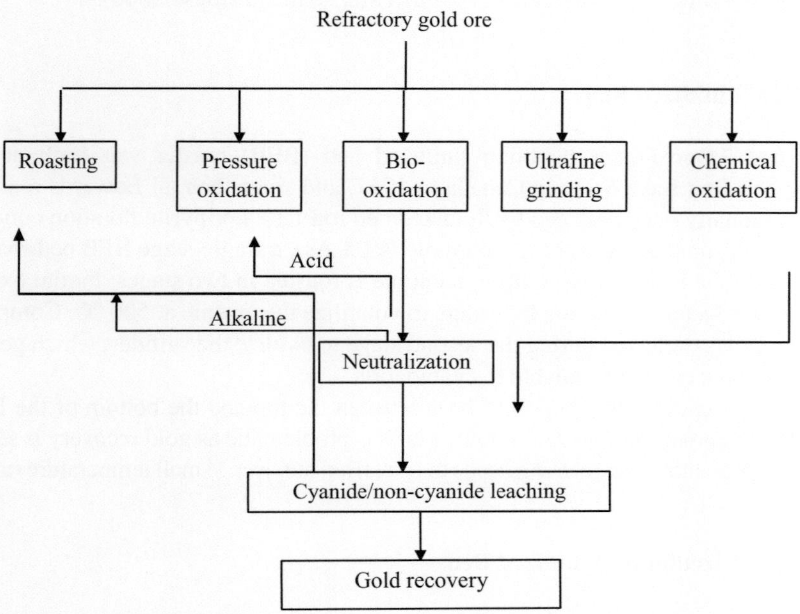

Fig. 5.5 Processing of non-refractory gold ores (adapted from LaBrooy et al. 1994)

5.2.1 *Roasting*

Until 25 years ago, roasting was the standard process option for pretreating a refractory gold flotation concentrate. Roasting as a pretreatment process for refractory gold flotation concentrates has been practiced for more than a 130 years. The application of roasting has decreased significantly as the environmental restrictions have increased over the years and alternative processes have become available. Before cyanidation and flotation processes were introduced, gold recovery and extraction included combinations of crushing and grinding of the ore, followed by roasting, fine grinding, chlorination, and precipitation of gold. The following equipment have been used for roasting over the years (Thomas and Cole 2005):

5.2.1.1 Rotary Kiln and Multiple Hearths

Rotary Kiln and multiple-hearth roasters were in use before 1940s for the treatment of refractory ores. The most common roaster in the gold industry was the single Edward roaster, which was developed in the 1890s in Australia, and became the preferred roaster around the world (Moss 1918; Kadenhe and Makende 1987; Loosley 1993). Initially, wood, gas and coal were used to heat the roasters, but the development of autogenous roasting of ores containing sulfur resulted in a process that required no additional external fuel. The minimum amount of sulfur needed for autogenous roasting depends on the moisture content in the feed. A dry feed requires at least 8% sulfur for autogenous roasting (Marsden and House 2006).

5.2.1.2 Fluidized Bed

The first TDorr-Oliver bubbling fluidized bed (BFB) roaster was built at the Campbell Red Lake Mine in Canada in 1947, and since then all Edwards roasters have gradually been replaced by fluidized bed roasters. For pyrite flotation concentrates with small amounts of arsenopyrite (<1% As), a single-stage BFB bed roaster is used, while bulk arsenopyrite concentrate is roasted in two stages. Partial oxidation is conducted during the first stage to volatilize the arsenic at 500 °C. Complete oxidation is conducted during the second stage to oxidize the sulfides, which generates a porous calcine amenable to cyanidation.

There is uneven distribution of heat between the top and the bottom of the BFB roaster (Hammerschmidt et al. 2005), which is problematic as gold recovery is sensitive to temperature and roasting needs to be carried out over a small temperature range.

5.2.1.3 Circulating Fluidized Bed

The circulating fluidized bed (CFB) roasters were first used in the alumina industry. Inco and Falconbridge commercialized the CFB roaster technology for the processing of pyrrhotite to recover nickel, copper, cobalt, and iron during the 1950s and

1960s. In these roasters, calcine is recirculated to provide longer residence time to eliminate sulfur in the coarser size fractions. This technology was used to treat refractory gold ores in the 1990s (Bunk and Bruyns 1993; Kosich 1992: Lahti 1996; Thomas and Cole 2005; Hammerschmidt et al. 2005; 2011). CFB roasters did not have the temperature differential issue of BFB roasters and also offered higher capacity. However, for treating a flotation concentrate with a high content of arsenic, the preferred roasting option is still the two-stage BFB roaster (Goode 1993; Hammerschmidt et al. 2005).

5.2.1.4 Oxygenated Fluidized Bed

A two-stage "whole" ore roasting process using oxygen as a fluidizing gas was developed by Freeport-McMoRan (Bunk and Bruyns 1993). There are many advantages to the use of oxygen in roasting over air. The use of oxygen dramatically reduces the heat losses present in the exit gases as there is no need to heat nitrogen that is present in the air. Thus, expensive to maintain "intensive" heat and recovery systems are not required in the oxygenated fluidized bed roasters. The use of pure oxygen also facilitates the oxidation of sulfur and preg-robbing carbon at lower temperatures. The first roaster using oxygen injection was built at Big Springs, USA, in 1989 (Ericsson 2012). The use of dry grinding of the feed results in minimizing the heat loss associated with water removal as steam in the oxygenated fluidized bed roasters (Thomas and Cole 2005). This allows the ores with lower sulfide content to be treated in roasters. An oxygen roasting circuit using dry grinding and Freeport-McMoran Inc. oxygenated roasting technology was commissioned Barrick Goldstrike in 2000. A photograph of the Barrick Goldstrike roaster in Nevada, USA, is shown in Fig. 5.6 (Gorain et al. 2015).

5.2.2 Pressure Oxidation

The pressure oxidation was first used in the aluminum industry for production of alumina using Bayer's process in the 1890s. Nearly 60 years later, this process was applied to treat complex base metal sulfide ores to meet the increased demand for base metals. The improvements in materials of construction of autoclaves to function in acidic conditions at high temperatures and pressures along with the availability of high-purity oxygen made it possible to use pressure oxidation for the treatment of base metal ores (Thomas 2005).

Chemical Construction Company, then a subsidiary of American Cyanamid Company, pioneered the development of hydrometallurgical processes to treat nonferrous metal ores and concentrates during 1946 and 1955. Sherritt Gordon Mines Limited, a Canadian copper and nickel mining company, was its first customer and bought the rights to pressure oxidation-based hydrometallurgical technologies from Chemical Construction Company in 1955. The first application of the pressure

Fig. 5.6 Barrick's Goldstrike roaster in Nevada, USA (Gorain et al. 2015)

oxidation process was to treat nickel sulfide concentrates at Fort Saskatchewan, Alberta in Canada. Subsequently, two pressure-leaching plants were built to treat cobalt concentrates (Berezowsky et al. 1991).

The pressure oxidation technology was then applied to treat zinc concentrates at Cominco zinc operation in Trail, British Columbia in Canada during the 1970s. Several new acid-pressure-leaching plants were built to treat nickel-copper mattes for the platinum mining industry during the 1980s (Thomas 2005). It was also applied to treat nickel laterites at Moa Bay in Cuba (1959) and at Cawse and Murrin-Murrin in Western Australia during the late 1990s.

The investigation of pressure oxidation to treat refractory gold and silver ores was conducted during 1970s and 1980s. During this period, some pilot plant tests were carried out. The first commercial application of pressure oxidation for gold recovery was at the Homestake-McLaughlin mine to treat refractory pyrite ores in 1985. This was closely followed by the application of pressure oxidation to treat gold-bearing sulfide flotation concentrates at Sao Bento in Brazil during 1986. An alkaline pressure oxidation facility was commissioned at the Barrick Mercur operation in Utah in 1988 (Thomas 2005). This led to the commissioning of a number of pressure oxidation plants including Goldstrike, Getchell, Lone Tree, and Tween Creeks during 1988 to 2000. Out of them, Barrick's Goldstrike operation was the largest by tons of ore treated.

Both whole ore and concentrate can be used as feed for pressure oxidation. If the ore is amenable to flotation, then concentrates can be used as feed. If ore has poor flotation response, then whole ore can be used as feed. Depending on the mineralogy and the carbonate content of the ores, either acidic or alkaline pressure oxidation process can be used for whole ore treatment. For autogenous operation, the minimum sulfide sulfur target for autoclave feed ranges from 5 to 7%. Figure 5.7 shows a schematic of a typical pressure oxidation circuit to treat pyrite concentrates.

Recovery of gold is typically higher for acidic pressure oxidation compared to that for alkaline pressure oxidation. The conversion of ferrous sulfate to ferric sulfate is an important factor as ferrous sulfate increases cyanide consumption during leaching. However, silver is precipitated as argentojarosite during acidic pressure oxidation, which typically requires a lime-boil pretreatment stage for recovering silver. Optimum temperature for acid autoclaving is in the range from 175 to 230 °C

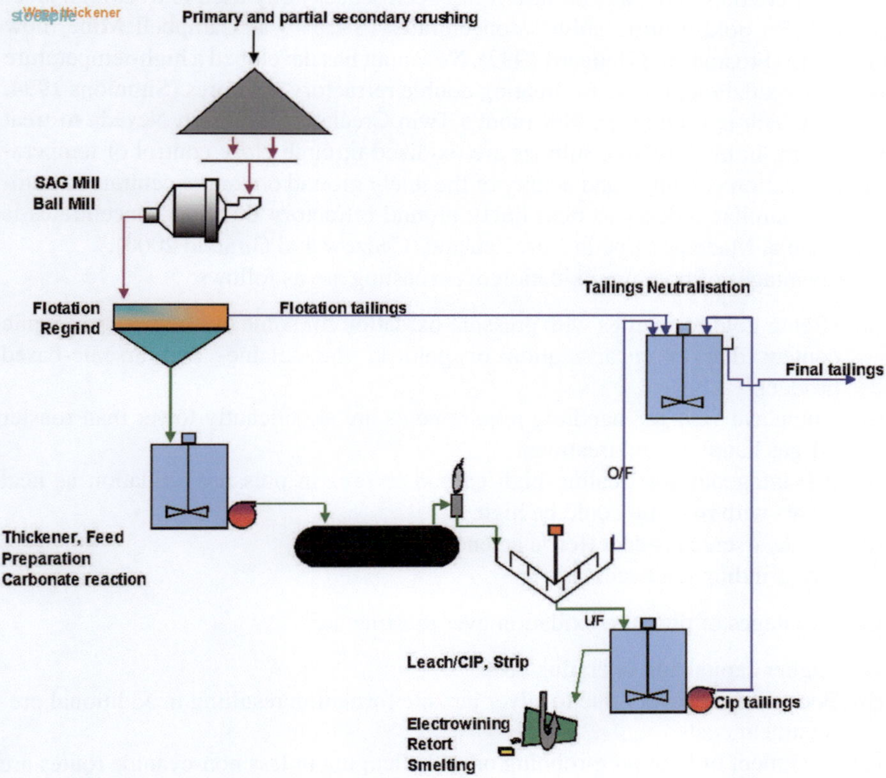

Fig. 5.7 Schematic of a typical pressure oxidation circuit treating pyrite concentrates (Gorain et al. 2015)

as this prevents production of elemental sulfur, which is known to prevent complete oxidation and reduce gold recovery.

Alkaline pressure oxidation has significant advantages for treating high carbonate ores. However, the alkaline chemistry results in production of oxidation products (iron oxides) and encapsulation of unoxidized sulfides, which lowers gold recovery (Thomas 2005). Alkaline pressure oxidation is normally preferred for ores with high carbonate content from a material construction point of view due to lower capital and operating costs.

The acidic pressure oxidation for concentrate treatment leverages high sulfide content to allow for an autogenous reaction. Thus, there is no need for splash/flash tower on autoclave discharge for heat recovery. Autoclave discharge passes to a countercurrent decantation (CCD) wash circuit to allow recycling of thickener overflow to acidify the autoclave feed and to allow a partial recycle of thickener underflow to autoclave feed. This prevents the formation of elemental sulfur as an intermediate oxidation product (Berezowsky et al. 1991).

Pressure oxidation has a major advantage over roasting in treatment of arsenic-bearing gold ores and concentrates. It has been successfully used to treat high arsenic (~10%) gold-bearing sulfide concentrates (S~18%) at Campbell Mine, now Goldcorp (Frostiak and Haugard 1992). Newmont has developed a high-temperature pressure oxidation process for treating double refractory gold ores (Simmons 1994, 1996), which is used at the Newmont's Twin Creeks operation in Nevada to treat whole ore. In this process, sulfides are oxidized through close control of temperature, oxidation potential, and acidity of the finely ground ore or concentrate in autoclave. A similar process to treat finely ground refractory flotation concentrates is also used at Macraes Mine in New Zealand (Cadzow and Giraudo 2000).

Advantages of pressure oxidation over roasting are as follows:

(a) Higher gold recoveries with pressure oxidation for some ores with high arsenic content due to encapsulation of gold in the calcine- and arsenic-based products.
(b) Autoclave vent gas handling requirements are significantly lower than roaster off gas handling and treatment.
(c) Advantageous for treating high carbonate ores in pressure oxidation as heat losses with roasting could be high.
(d) Stable arsenic product (ferric arsenate).
(e) Dry grinding not necessary.

Disadvantages of pressure oxidation over roasting are:

(a) Higher capital and operating costs
(b) Poor silver recovery due to silver jarosite formation resulting in additional pre-treatment costs
(c) Treatment of high preg-robbing ores challenging unless non-cyanide routes are used such as the use of thiosulfate process at Goldstrike

5.2.3 Bio-oxidation

Interest in bio-oxidation treatment of refractory gold ores increased during the 1970s and 1980s. The BIOX® process for the pretreatment of refractory sulfide ores was initially developed at the GENCOR Process Research in the 1970s, and successful pilot plant investigations were conducted at Fairview, South Africa, and Equity Silver Mines in British Columbia. A 10 tpd plant was commissioned at Fairview to process flotation concentrates in 1986, which is still successfully operating. A bio-oxidation circuit to partially oxidized sulfide concentrates ahead of the pressure oxidation circuit to increase throughput was used at Sao Bento in Brazil during 1991. A720 tpd agitated bio-oxidation process to treat sulfide flotation concentrates was commissioned by Ashanti-Obuasi in Ghana in 1994. This facility was later expanded to 960 tpd and is being operated successfully.

The first licensed BIOX® plant was commissioned in 1992 by Harbour Lights in Western Australia. The plant had a 40 tons per day capacity and was successfully operated till 1994. The plant was closed after the ore and stockpile reserves were depleted. Other BIOX® plants include the Suzdal BIOX® plant in Kazakhstan, the Fosterville BIOX® plant in Australia (early 2005), the Bogoso BIOX® plant in Ghana, the Jinfeng BIOX® plant in China (early 2007), BIOX® reactors at Bogoso, and the Kokpatas BIOX® plant in Uzbekistan, which is currently the largest BIOX® plant in the world.

Another bio-oxidation technology is the BACOX process developed by Mintek. Its first commercial operation was the Youanmi mine in Western Australia in 1994 (Miller and Brown 2005; Budden and Bunyard 1994). Its next commercial application was at the Beaconsfield Gold Mine in Tasmania in 1998 for sulfide treatment (Holder 2007). The bacterial oxidation circuit installed at Beaconsfield uses a combination of mesophilic iron and sulfur oxidizing cultures. The Laizhou Gold Metallurgy Plant (renamed to "Tarzan Biogold") of Shandong Province, China, also uses the BACOX technology (Miller et al. 2004) and was commissioned in 2001 (Xie et al. 2008; Miller et al. 2004).

Another bio-oxidation technology, BIONORD® technology, was developed by the Olympiada Mining Combine in the Russian Federation (Sovmen et al. 2009). BIONORD® technology was developed to accommodate the harsh, extremely cold climatic conditions found at the Olympiada mine in the Krasnoyarsk Region of Russia.

The process technology support for many bio-oxidation plants in China (Xie et al. 2008) is provided by the Changchun Gold Research Institute (CCGRI). It uses microorganism cultures isolated from hot water springs (moderate thermophiles) and mines in South China. Bio-oxidation followed by cyanide leaching and zinc cementation is the preferred technology for the treatment of concentrates in China.

Iron, sulfur, and arsenic in high arsenic-bearing flotation concentrates are converted to ferric sulfate, sulfuric acid, and arsenic acid by bio-oxidation. The acidic solution exiting the bacterial reactors flows to a CCD circuit. The overflow solution is sent to neutralization. In general, the acidic arsenic liquors arising from bio-oxidation have high ferric iron to arsenic ratios, which is ideal for the generation of ferric arsenate. A large quantity of air is needed in the biological reactor to ensure satisfactory oxidation rates. The flotation concentrate may be ground to enhance the leach kinetics. The heat released during the oxidation process could be significant and needs to be dispersed efficiently since the bacteria function effectively within their operating temperature range. To maintain optimum temperature, the heat is collected by internal cooling coils inserted in the bio-oxidation reactors, which is then dispersed to the atmosphere through cooling towers (Ritchie and Barter 1997).

The cyanide consumption is typically high in bio-oxidation process due to the formation of polysulfides and sulfur followed by conversion to thiocyanate in the cyanide leach circuit (Lunt and Weeks 2005; Ritchie and Barter 1997; Miller and Brown 2005). The bio-oxidation process is extremely sensitive to cyanide and thiocyanate concentration and becomes extremely toxic to the bacteria at levels above 5 ppm. Therefore, the process streams need to be separated to ensure that thiocyanate does not flow into the bio-oxidation feed stream. In addition, the bio-oxidation process generates organic compounds that impact the activity of activated carbon and loading capacity of the carbon in the gold adsorption circuit (Ritchie and Barter 1997). Foaming in the cyanide leach section of a bio-oxidation plant is a problem in most bio-oxidation plants to some degree (Ritchie and Barter 1997). Anti-foaming and froth breaking agents are commonly added to deal with this problem. The CIL circuits also incorporate draft tube agitators to avoid foaming.

5.2.4 *Ultrafine Grinding*

Although ultrafine grinding is not considered as a chemical oxidation process, there is evidence to suggest that for certain ore types, grinding ores to lower than 10 μm size enables sufficient liberation of gold from gold-bearing iron sulfides. The extent of oxidation of sulfides after ultrafine grinding is not accurately known and needs further investigation. Available ultrafine grinding technologies include the Glencore technology's IsaMill and Metso's SMD Detritor. Fine grinding and ultrafine grinding have been found to be an option for treating telluride flotation concentrates before cyanide leaching [Liddell and Dunne 1988; Ellis and Gao 2002]. The ultrafine grinding has also been used in the Albion process for treating refractory ores (Hourne and Turner 2012).

Ultrafine grinding at Kalgoorlie Consolidated Gold Mines (KCGM) was the first commercial application of this technology for oxidative pretreatment before leaching. This led to a gold recovery of over 90% (Ellis 2003), which paved the way for further applications of this technology at Cowal in Australia, Pogo in Alaska, and Kumtor in Kyrgyzstan for treating refractory and partly refractory gold ores.

Ultrafine grinding before pressure oxidation and bio-oxidation enhances leaching kinetics but is detrimental before roasting as it reduces the retention time in the roaster leading to inadequate oxidation of the sulfides and poor gold recovery.

5.2.5 *Chemical Oxidation*

Caro's acid can be used for pretreatment of certain gold-bearing ores that have substantial quantities of arsenic as a sulfide (Lakshmanan and Biskupski 1985). The presence of gold-containing refractory arsenopyrite inhibits solubilization of the gold, which leads to low gold recoveries from arsenic-containing ores. Roasting of arsenic-containing gold ores prior to cyanidation leads to increased yield, but such roasting often leads to undesirable atmospheric pollution. Arsenic- and antimony sulfide-containing precious metal ores can be treated with Caro's acid, i.e., peroxymonosulfuric acid (H_2SO_5), to oxidize part of the arsenic or antimony to a water-soluble arsenate form. Caro's acid is produced by reaction between 92 and 99% sulfuric acid and 50–75% aqueous hydrogen peroxide in a mole ratio of 1.5:1 to 3.5:1. The acid may be diluted with water to a concentration in the range of about 15–30%. The ore material is ground and formed into a slurry. Caro's acid is added to the slurry and agitated to oxidize the charge. The Caro's acid-treated material may be processed by cyanidation or other alternatives to recover the precious metal. The process is applicable to the recovery of gold and other precious metals not only from arsenic- and antimony-containing sulfurous ores but also from other sulfide containing arsenic- and antimony-containing materials, such as wastes and tailings.

5.3 Treatment of Double Refractory Gold Ores

A double refractory ore body contains both sulfides and carbonaceous matter (Nyavor and Egiebor 1992). The carbonaceous matter may be present as elemental carbon, graphitic or amorphous, or organic carbon, which may be in the form of hydrocarbons, humic acids, and other organic substances. Graphitic carbon is the main source of carbonaceous matter in double refractory gold ores and presents serious challenges in the recovery of gold during leaching (Afidenyo 2008).

Similar to the pretreatment of refractory gold ores, double refractory sulfide ores can be oxidized to decompose the sulfide matrix, thereby liberating gold and making the disseminated gold amenable to leaching. The oxidizing pretreatment processes may include the conventional roasting, pressure oxidation, chlorination, bacterial oxidation, and chemical oxidation as described in the previous section. However, due to the preg-robbing behavior, the carbonaceous matter needs to be either removed or passivated before cyanidation.

The preg-robbing behavior of the carbonaceous matter can be addressed by flotation (Gorain and Kondos 2012), addition of blanking agents, roasting, and competitive loading onto commercial activated carbon. If the amount of gold associated with carbonaceous matter is low, then carbonaceous matter may be floated and discarded without significant loss of value. One of the processes discussed above for refractory gold ores can then be used for further processing. Blanking agents adsorb selectively to carbonaceous ores and prevent the adsorption of gold to the carbonaceous matter. Examples of blanking agents are kerosene, fuel oil, and RV-2 (para nitro benzol azo salicylic acid). Roasting is another method to eliminate carbonaceous matter, which also oxidizes sulfides, thereby liberating gold particles. However, roasting is only successful within a narrow range of temperature as below the optimal temperature, carbon does not oxidize, and above the optimal temperature, the gold becomes difficult to extract in the subsequent cyanidation step. In addition, if the ore contains low levels of sulfide and high levels of carbonates, roasting may not be suitable or economical because it is not autogenous.

Competitive loading works by adsorbing gold onto commercial activated carbon, which can be easily separated from the leach or pulp after adsorption of gold has taken place. Carbon-in-leach (CIL) and carbon-in-pulp (CIP) processes employ activated carbon granules to recover gold from gold cyanide solutions. The treatment of double refractory gold ores is very challenging, and there is a need for the development of economical and practical methods to treat these ores.

5.4 Heap Leaching

Heap leaching of gold and silver began in 1969 at Cortez (Kappes 2005), which is still in operation with major expansions over years. The US Bureau of Mines (USBM) developed heap leaching of gold and silver in the late 1960s and early 1970s as a low-cost option for low-grade hydrothermal oxide ores. The first large-scale cyanide heap leach for gold and silver was commissioned at Newmont's Carlin Mine in 1971 to process low-grade ores below the mill cutoff grade. Since then, the use of heap leach has spread across the Western USA in Nevada, New Mexico, California, Utah, Montana, Idaho, and South Dakota (Marsden and Sass 2014).

Heap leaching allows the generation of revenue early in the life of a mining project. This helps in delaying large capital investments required for milling and additional mining operations. Gold recovery from heap leaches is typically ~70%, whereas silver recovery is typically ~55%. The economic feasibility of heap leaching is highly dependent on the type of ore and mineralogy. Typical ore types amenable to heap-leaching are sedimentary oxidized ores (e.g., Carlin, Goldstrike, Twin Creeks), low-sulfide acid volcanics or intrusives (Round Mountain in Nevada, Yanacocha in Peru), oxidized massive sulfides (Filon Sur in Spain, Hassai mine in Sudan), saprolites/laterites (West Africa and Central America), clay-rich deposits (Buckhorn Mine in Nevada, Barney's Canyon Mines in Utah), and silver-rich deposits (Coeur Rochester in Nevada and Comco in Bolivia).

Effective agglomeration, liner systems, stacking, solution distribution, and irrigation systems are important considerations for successful heap leach operations. Carbon-in-column (CIC) systems offer advantages in handling large volumes of solution and in avoiding the need for costly filtration or clarification systems. Significant cost savings have been realized by centralized carbon handling facilities for larger mining companies such as Barrick and Newmont with numerous operations in close proximity.

Heap leaching is now considered a proven technology. It can be applied beyond the conventional arid climatic regions to include extreme cold climates such as the Fort Knox operation in Alaska with temperatures down to −50 °F and Jinshan Gold Mine in China with temperatures reaching −32 °C. The common features of the heap leach operations in cold climate are the use of in-heap pregnant solution storage and the direct delivery of barren into the heap (Keane 2007).

The ore should have the following characteristics to be amenable to heap leaching:

(a) The size of the gold particles must be extremely small.
(b) The gold values must react with cyanide either by exposure through the natural porosity of the ore.
(c) The ore should be free of "cyanicides" that inhibit gold leaching.
(d) The ore must be free of carbonaceous materials that adsorb the dissolved gold values.
(e) The ore must be free of acid-generating minerals that consume lime.
(f) The ore should not contain excessive amounts of clay that cause problems in leachate percolation.

Development of a heap leaching project starts with the acquisition of samples. Representative samples based on the geology, mineralization, size, depth, and character of the ore body must be obtained. Satellite photos and maps can guide the location of sampling points. Preliminary grab samples can be used for geological studies. Core and drill chip samples can be used for laboratory tests. Bulk samples are needed for pilot studies. Sampling procedure typically requires successive size reduction. Starting with 10 kg drill core samples, samples are first reduced to less than 1 cm size and then to approximately 1 mm size. A 1 kg sample is taken from this and the size is further reduced to 0.1 mm, and a 100 g representative sample is taken from this for analysis. Mineralogical analysis shows whether the ore is suitable for heap leaching. Mineralogical information includes precious metal-containing minerals, host matrix, and grain sizes. If mineralogy is favorable, metallurgical tests are conducted to determine the amenability of ore to heap leaching. Three tests are normally performed: (1) bottle leach tests, (2) column tests, and (3) pilot tests. Additional leach tests such as agitation leaching, flotation, gravity separation, settling tests, carbon adsorption, zinc precipitation, and cyanide destruction can offer information on the association of precious metals in the ore and to substantiate the selection of the heap leaching route.

5.4.1 Bottle Leach Tests

Bottle roll leach tests are conducted at a relatively coarse size feed. Bottle leach tests offer information about recovery, rate of dissolution, and reagent consumptions. The recovery based on bottle and agitation leaches can be considered maximum recovery obtainable through heap leaching. Attrition grinding occurring during these tests generates fines which increase the recovery, but fines are not generated during heap leaching. Recoveries obtained from bottle leach tests can be higher by up to 10% compared to static heap leaching tests. Bottle leach tests can be conducted with up to 2.5 kg charges and 1 in. ore size.

The test is normally conducted on 400 g ore charge and is pulped with 600 mL of water. The ore can be crushed to pass ¾″ mesh. Lime is added to the pulp to adjust the pH to about 11. The amount of lime added to the pulp usually varies between 2 and 5 lb/ton of ore at most. Sodium cyanide at 2.0 lb/ton of solution is added. Samples are drawn at 2, 6, 12, 24, and 48 h to establish the rate of Au dissolution. Cyanide concentration, pH, and oxygen content are determined, and makeup water equivalent to the volume removed for the sample is added. Cyanide level and pH of the solution are adjusted. After 72 h, the leach solution is recovered and the ore is washed, dried, and assayed. Au recovery is determined based on the assay of the residue. The reagent consumption is calculated based on the solution analysis. The liberation size can be determined through head and residue assays of the respective screen fractions. If screen fraction assays of the residue indicate the need for fine grinding, then heap leaching is not recommended.

5.4.2 Column-Percolation Leach Tests

Small percolation leach tests are conducted to confirm recovery and reagents consumption data obtained from bottle leach tests. The column with a diameter of about 10 times the particle dimension should be chosen to minimize the wall effect. The column height should be at least five times the diameter of the column. The test is conducted by mixing an appropriate amount of lime (based upon the bottle leach test) with the ore to provide protective alkalinity during the cyanidation treatment. The ore-lime mixture is placed in a transparent plastic column, and the cyanide leach solution is applied at the top of the ore charge. CN content of the cyanide leach solution is based on the results from the bottle leach tests.

The leach solution is added at 0.003–0.005 gpm/ft^2 of the column cross-sectional area. The leach solution is allowed to trickle downward through the ore column and analyzed for precious metals content. The pregnant solution is either pumped through a carbon column or disposed. If the pregnant solution is pumped through a carbon column, the solution is reused after reagent make up. If the pregnant solution is disposed, a fresh leaching solution is prepared and applied to the heap leach column. The pH, cyanide concentration, and dissolved oxygen are determined. Leaching is allowed to continue until extraction rate is established or can be predicted based upon the leach rate during the last 15 days. The plastic column is then flushed with water and the ore is assayed. A screen analysis determines the gold content in different fractions and confirms the liberation size of the feed. Mineralogical examinations can provide information on gangue breakdown problems and residual precious metal values.

Detailed column leach tests on 400–2500 lb ore samples can be conducted to confirm heap leach feed size, better indication of leach rate on coarse feed, recovery, and reagent consumption as a function of feed size and time.

If the feed contains a large amount of clay or fines, feed agglomeration will be required. Clay and fines can cause segregation and channeling in the heap. The need for agglomeration can be indicated from the bottle leach and initial column leach tests. Problems during filtration and high moisture content of the ore while preparing the analysis samples and migration and channeling in the plastic column indicate the need for agglomeration. Generally, the feed requires crushing to −3/4″ size. If excessive fines are generated or clay hinders the flow of solution, agglomeration of the feed is recommended. During agglomeration, the clay and fine particles adhere to the coarse particles and create a coating of fine particles around coarse particles.

Optimum agglomeration requires optimum moisture content, curing time, and reagent dosage. One to three kg of agglomerates are produced for testing at various binder conditions. After 72 h curing, the agglomerates are tested on an appropriate Tyler screen and jigged 10 times in a column of water for 30 s. The agglomerates retained on the screen are dried and weighed. The retained weight is compared with the weight of the feed naturally retained on that screen. The increase in weight retained is plotted against the binder added. The break in the curve indicates the optimum binder addition to the feed. Agglomerate strength is determined by submerging the cured agglomerate in water. If the green pellet does not degrade within 24 h, the agglomerate is presumed to have supplied green strength. Crushed ore in general can be agglomerated by mixing 5–10 lb cement per ton of feed, wetting to 8–10% moisture using water or CN solution, mechanically tumbling, and curing.

5.4.3 Pilot-Scale Leach Tests

Pilot-scale leach tests can be done in large (6′ × 20–40′) column or 5000–10,000 ton heaps. The objective of these tests is not the development of a flow sheet but to train operators and confirm the selected flow sheet.

The bulk sample selected should represent the ore body. If composite cannot be obtained, at least one major ore sample should be obtained. Information on ore size, weight of the heap, and edge effects can be obtained. Column leach will require parameters considered under laboratory tests but should also correspond to operating heap leaching.

The advantages of the pilot-scale tests are:

(a) Feed samples are more representative.
(b) Commercial operation is simulated.
(c) Accurate design data is obtained.
(d) Accurate data on recovery, reagent consumption, and solution impurity buildup is obtained.
(e) Heap stability can be observed.

Disadvantages of pilot-scale tests are:

(a) Cost
(b) Large sample requirement
(c) Site preparation
(d) Preparation and assembly of solution recovery system
(e) Labor
(f) Site clearance and regeneration of natural growth conditions

5.4.4 *Preparation for Production*

Bottle and column leach and pilot tests confirm the flow sheet to be used for the chosen feed material. The results of the metallurgical testing, topographic and climatic site conditions, geotechnical and geological site conditions, and methods and rate of ore mining dictate the method of heap leaching. The operation depends on the feed source (newly mined ore, previously mined/stockpiled low-grade ore, waste rock or waste/tailings), feed preparation (none, crushing, crushing and agglomeration, or agglomeration), ore type (metallurgy, leachability), site (topography, geotechnical conditions, and site climate), pad (reusable pad, expending heap, valley), materials for impervious pad (plastic or rubber sheeting on a smooth area and underlain and overlain with a layer of washed sand and small gravel, compacted mill tailings mixed with bentonite, asphalt placed on compacted gravel and covered with an asphalt sealer, reinforced concrete or clay), solution application (pump system with sprinklers, wigglers buried pipe networks beneath a heap cover), solution collection (pipes, trenches, perforated pipes within drainage layer above the pad), pregnant solution containment (linear system, storage of solution within heap), metal recovery (Merrill-Crowe, carbon adsorption), and barren solution containment (pond).

The choice of heap leach methods (reusable pad, expending heap, valley) depends on the site and project specification. Reusable pads require loading, leaching, washing, neutralization, and removal of spent ore. The method requires predictable short leach (<60 days), flat land, suitable waste site, durable high stress environment liner, arid balanced climate, limited active area requiring covering in wet climate, and limited impact of flood storm. The method requires double ore handling and limited time for ore to mature. The main drawback of this method is the lack of flexibility and the loss of unleached ore during inefficient leach periods. A significant advantage includes limited area requirement and reduced capital cost with pad reuse.

In the expanding Carlin method of leaching, the spent feed is left in place, subsequent releaching and washing are performed, and additional lifts to the heap are performed. The method requires a relatively large area with suitable topography to prepare the pad in a staged manner. Typically ground slope should be <5–10%. To minimize discharge, a balance between evaporation and precipitation is required. Large areas are involved, and large ponds are required. The initial capital costs are low due to lower pad and liner costs. In the expanding heap method, the effect of differential settling across the heap should be considered. This can adversely affect the heap drainage and damage the liner.

In the valley leach method, the feed is placed behind a retaining structure. Leaching is conducted with subsequent tips progressing up the slope. The majority of the ore is left in contact with the leach solution during the operation and at the completion left in place. The applicability of this method depends upon the durability of the ore, requirement of stable terrain, reduced storage capacity, high-integrity (combination of synthetic and low permeability or amended soil material) liner due to hydraulic head, stable heap, and retaining structure.

This can operate under a large range of climates and can accommodate extended leach time. The storage capacity of the pore space of the heap is used to contain pregnant solution. Typically, operating valley leaches have heights of 200′ from toe to crest.

A typical procedure for constructing a pad is described below:

(a) Select a site for the pad that is on firm ground.
(b) Fill the site to a surveyed grade, using barren pit material as fill.
(c) With a bulldozer, grade and compact the pad as much as possible. At one side of the pad, provide for a drainage of minus 1–3 ft per 100 ft of pad length.
(d) At the side of the pad where the drainage has been provided, prepare three solution storage basins. Each can be 20′ × 40′ × 60′ deep. The first basin (pregnant leach solution basin) should be located approximately 10′ from the base of the pad, on the side where drainage has been provided. The second basin (barren solution basin) should be located approximately 10′ from the first basin, and the third basin (overflow basin) should be located within 10′ from the second basin.
(e) Cover the pad area with 4–6″ of fine sand and small gravel. Also cover each of the basin with 4–6″ of fine sand and small gravel. This prevents puncturing of the required plastic sheeting.
(f) Sculpture the pad base to a central diagonal line, thereby preparing a path for solutions to drain toward the drainage side of the pond. Use a 100 lb. whack hammer to compact and double tamp the pad base.
(g) Prepare a barrier, approximately 2′ high that encircles the entire pad.
(h) Once the base of the pad has been graded and compacted properly, cover it with plastic hypalon sheets that are 30 mil thick. Install a drain pipe through the plastic and into the pregnant solution basin.
(i) Cover the base and sides of the basin with plastic sheets that are 30 mil thick.
(j) Cover the plastic sheets on the pad base with 4–6″ of sand. This prevents the puncturing with coarse and angular ore that may be piled on top of sheeting.
(k) Since cyanide is used, fence the entire leaching area to prevent access and post warning signs around the leach pads and all basins.

5.4.5 Economic Analysis

Four major areas require consideration: location and analysis of ore body, mining of the ore body, milling of the ore, and environmental rehabilitation of the site. Decisions require ore grade to tonnage, mining cost vs. production rate, processing cost vs. recovery, metal value vs. time, and economic evaluation criteria vs. risk. Economic optimization of the present worth of alternative process operators reduces capital investment at the beginning, if the tax on the minimum rate of return is high.

Location and acquisition depend on the size and value of the deposit. Near-surface abandoned mines, large dumps, and sites of massive low grade of mineralization become attractive targets for acquisition. They require a low initial investment and provide a high initial cash return. Once the initial investment is recovered, the operations continue till the cash flow becomes negative.

Another major cost is mining. Advances in high capacity blast hole drilling, improved explosives, in-pit crushing, conveyor handling of waste and ore, automatic control of mining, and improvement of in-pit techniques contribute toward improved efficiency and lower cost.

Technical advances in metallurgical treatment to handle lower-grade ores, physical ore conditioning, carbon adsorption, sodium sulfide precipitation of silver, and oxidation of carbonaceous ores contribute to improved economics.

Environmental costs are associated with not only the cost of operation but also with sampling, analysis of wastes, treatment of wastes, and stabilization of wastes to prevent leakage of pollutants from the containment area.

5.4.6 Plant Design

Design responsibilities can be based upon total responsibility by the operator, operator-controlled engineering contractors, or major engineering contractor with minimum operator control. Plant location, process engineering, flow diagrams, plot plans, schemes, engineering, and drafting are major areas of design.

Construction plan is prepared from design drawings and specifications. To minimize loss of time, close attention to planning and scheduling is paid. Permitting, equipment purchases, operator training, and payments should be completed on schedule. Field construction changes are minimized and operators are involved in inspection and approval.

Operators are given courses in general safety, equipment training, process training, and cyanide safety training. Expert advice is sought to minimize loss of time and mistakes. Prior to start, operators should plan to leave the site at an environmentally acceptable situation. This includes economic tailings disposal and solution recycle. Closure plans include destruction of cyanide, containment of heavy metals, and rehabilitation of mine area.

5.5 Novel Technologies in Gold Processing

A number of reagents are currently under development that are either environmentally friendlier compared to cyanide or have higher rates of gold dissolution with possible economic advantage. Such reagents include thiocarbamide, hypochlorites, bromine-bromides, iodine-iodides (Kozin and Melekhin 2004), and chlorides (Lakshmanan et al. 2013).

5.5.1 Thiocarbamide

Thiocarbamide, also called thiourea, has the chemical formula $SC(NH_2)_2$. The most effective oxidant for leaching of gold in thiourea is ferric sulfate (Li and Miller 2006). In the presence of ferric sulfate, gold reacts with thiourea by the following reactions:

$$Au + 2SC\left(NH_2\right)_2 + Fe^{3+} = Au\left(SC\left(NH_2\right)_2\right)_2^{\ +} + Fe^{2+}$$

$$Fe^{3+}\mathrm{e}^- = Fe^{2+}$$

The amount of thiourea consumption is relatively high compared to cyanide consumption. Though the rate of gold leaching is initially faster, the rate decreases with an increase in leaching time.

5.5.2 Hypochlorites

Sodium or potassium hypochlorites can be used for leaching of gold. When chlorine is brought in contact with water or NaOH, the following reactions take place producing hypochlorites:

$$Cl_2 + \mathrm{H_2O} = \mathrm{HClO} + HCl$$

$$Cl_2 + \mathrm{NaOH} = \mathrm{HClO} + \mathrm{NaCl}$$

$$Cl_2 + 2\mathrm{NaOH} = \mathrm{NaClO} + \mathrm{NaCl} + \mathrm{H_2O}$$

Gold dissolves in the hypochlorite according to the following reaction:

$$2Au + \mathrm{HClO} + 3\mathrm{H}^+ + 3Cl^- = 2\mathrm{AuCl_2}^- + \mathrm{H_2O} + 2\mathrm{H}^+$$

The rate of gold dissolution in hypochlorite solutions is higher than in cyanide solutions and depends on the concentration of NaClO. Dundee Sustainable Technologies (DST) has developed a variation of this process that uses sodium hypochlorite with a catalytic amount of sodium hypobromite in acidic conditions to put the gold into solution (DST 2018).

5.5.3 *Bromine-Bromides*

Dissolution mechanism of gold in bromine-bromide solution is complex and depends on the solution composition. The species of gold adsorbed on gold surface depends on solution composition. Important reactions in the system are as follows (Pesic and Sergent 1993):

$$3Br_2 + 2\mathrm{e} = 2Br_3^-$$

$$Br_{2(aq)} + 2\mathrm{e} = 2Br^-$$

$$Br^{-+}\mathrm{H_2O} = \mathrm{HBrO} + \mathrm{H}^+ + 2\mathrm{e}$$

$$Br^{-+}3\mathrm{H_2O} = \mathrm{BrO_3}^- + 6\mathrm{H}^+ + 6\mathrm{e}$$

$$Br_3^{-+}3\mathrm{H_2O} = 3\mathrm{HBrO} + 3\mathrm{H}^+ + 4\mathrm{e}$$

$$Br^{-+}\mathrm{H_2O} = \mathrm{BrO}^- + 2\mathrm{H}^+ + 2\mathrm{e}$$

$$Au + 2Br^- = \mathrm{AuBr_2}^- + \mathrm{e}$$

$$Au + 4Br^- = \mathrm{AuBr_4}^- + 3\mathrm{e}$$

$$\mathrm{AuBr_4}^- + 2\mathrm{e} = \mathrm{AuBr_2}^- + 2Br^-$$

For leaching of gold encapsulated in sulfides, significantly higher gold recovery can be achieved with bromine than with cyanide (Melashvili et al. 2014). However, the bromine consumption is higher than cyanides due to the simultaneous oxidation of sulfides. The leaching of oxidized gold ores is promising for the bromine-bromide leach as the reagent consumptions are comparable. It may even be possible to have lower bromine consumption than cyanide consumption in the case of oxidized gold ores containing copper mineralization, as bromine is less reactive than cyanide with copper minerals.

5.5.4 Iodine-Iodides

Gold forms monovalent (AuI) and trivalent (AuI_3) compounds with iodide ions. Gold also forms complex compounds ($MAuI_2$ and $MAuI_4$) with iodide ions in the presence of other metals. The overall reactions for leaching of gold by iodine-iodide are as follows (Lucheva et al. 2017):

$$2Au + I_3^- + I^- = 2auI_2^-$$

$$2Au + 3I_3^- = 2AuI_4^- + I^-$$

The solubility of molecular iodine in water is 0.3 gl^{-1}, and the equilibrium constant for disproportionation of iodine in water is small.

$$I_2 + H_2O = I^- + H^+ + HIO, \quad K = 5.32 \times 10^{-13}$$

The solubility of molecular iodine in water increases significantly in the presence of KI or NaI due to the formation of triiodide ions.

$$I_2 + I^- = I_3^-$$

Thus, the presence of iodide ions is necessary for the dissolution of gold with the use of elemental iodine.

5.5.5 PRO Chloride Technology

Process Research Ortech (PRO) has developed an innovative atmospheric mixed-chloride leaching process for the recovery of gold from gold-bearing complex ores, concentrates, and tailings (Lakshmanan et al. 2013). The process flow sheet is shown in Fig. 5.8 and consists of crushing/grinding and gravity separation/flotation to first produce a concentrate. Depending on the type of ore, concentrate can then be roasted to remove sulfur and other toxic elements including arsenic. The concentrated/roasted sample is then leached with HCl and $MgCl_2$ solution at atmospheric pressure in an oxidizing condition. After liquid/solid separation, pregnant leach liquor is subjected to solvent extraction steps for selective recovery of gold from it. The gold can be extracted with different extractants such as oxime, crown ether, phosphinic acid, ester or oxide, tertiary amines, and quaternary ammonium salts with a modifier and a diluent. Among these extractants, the oxime has high selectivity of Au over Fe under certain conditions. The organic phase loaded with gold is contacted with an aqueous phase containing dilute acid for the selective removal of Fe prior to stripping of Au with thiosulfate. Gold can be precipitated from the preg

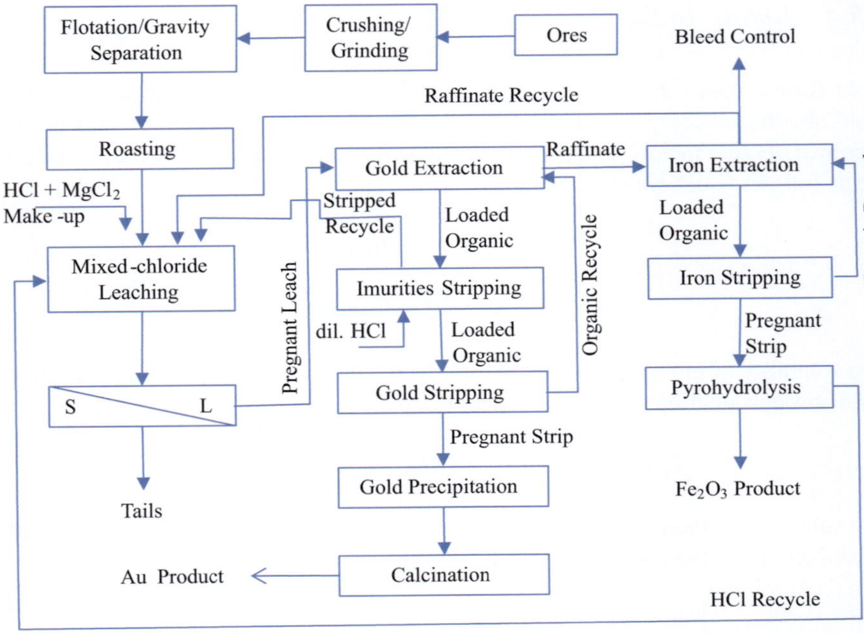

Fig. 5.8 Mixed-chloride process flow sheet for the recovery of gold from gold-bearing materials (Lakshmanan et al. 2013)

strip solution and calcined. Fe can be extracted from the raffinate of Au solvent extraction with ketone in Exxal™ 13 tridecyl alcohol and Exxsol™ D80, followed by stripping with dilute HCl solution (Lakshmanan et al. 2014). The pregnant strip solution of Fe can be used in pyrohydrolysis to produce Fe_2O_3 and HCl. Fe_2O_3 can be commercially sold and HCl can be recycled to the leaching stage.

The PRO chloride process for gold production has several advantages:

1. The extraction of gold by this process does not use cyanide, and this environmentally friendly process can be applied in jurisdictions where use of cyanide is not permitted.
2. The process offers economic advantage as process effluents and solid wastes do not need to be treated for cyanide detoxification.
3. The cost of reagent is reduced by recycling HCl from pyrohydrolysis step and iron raffinate from iron solvent extraction step to leaching stage.

5.6 Silver Processing

Silver is primarily produced from polymetallic ore deposits. It is produced as a by-product from lead-zinc mines, copper mines, and gold mines in descending order of production. In addition, silver is also produced as a primary product from some mines. Silver is contained in sulfide ores typically as argentite (Ag_2S), proustite (Ag_3AsS_3), and polybasite ($[Ag,Cu)_{16}Sb_2S_{11}]$. It is associated with other sulfide ores such as galena (PbS), sphalerite (ZnS), chalcopyrite ($CuFeS_2$), pyrite (FeS_2), and arsenopyrite (FeAsS).

After mining, silver-bearing ores are crushed, ground, and concentrated by flotation. Silver-bearing ore concentrate is subjected to specific processes depending on the major metal present in the concentrate.

For copper ores, silver along with gold and platinum-group metals accumulates in the anode slimes during electrolytic refining of copper. The anode slime is smelted in a furnace to oxidize other metals except silver, gold, and platinum-group metals, which are recovered as doré. The doré is cast to form anodes and electrolyzed in silver-copper nitrate solution to recover high-purity silver.

For lead ores, lead concentrates are roasted and smelted to produce a lead bullion. Silver is recovered from lead bullion by the Parkes process. In this process, zinc is added to molten lead bullion, which reacts with gold and silver to form insoluble compounds that are skimmed off from the top of the lead bullion. Gold and silver are recovered from the residue by cupellation. In cupellation process, the residue is heated to 800 °C to oxidize impurities such as lead. Silver is separated from gold by further processing.

For zinc ores, zinc concentrates are roasted and leached with sulfuric acid. The leach residue contains lead, silver, gold, and zinc. The residue is subjected to slag fuming, which is a process in which the residue is melted to form a slag and powdered coal or coke is blown through the slag with air to vaporize zinc and produce lead bullion-containing silver and gold. This lead bullion is processed as described above to recover silver and gold.

The processing of silver ores is similar to the processing of gold ores. For silver-containing gold ores, silver gets leached along with gold in cyanide or thiosulfate. After stripping, the enriched strip solution goes through electrolysis to recover gold and silver. The cathode-containing gold and silver is sent to the refineries to be cast as doré bars. Silver is separated from the gold-silver doré bars by chlorination of the molten doré. During chlorination, silver forms AgCl and is skimmed from the surface. Silver is leached from the silver chloride, and the leach solution goes through electrolytic refining to recover silver, which is cast into silver bars.

References

Adams, M.D., and S. Glen. 2011. P *ilot scale and continuous bench testing of the MMS CN-D™ cyanide destruction process on an Australian gold plant tailings effluent*. Presented at the world gold conference, Montreal.

Afidenyo, J.K. 2008. Microbial pre-treatment of double refractory gold ores. Master's thesis, Queen's University, Kingston, On, Canada.

Bain, H.F. 1910. *More recent cyanide practice*. San Francisco: Mining and Scientific Press.

Barter, J., G. Lane, D. Mitchell, R. Kelson, R. Dunne, C. Trang, and D. Dreisinger. 2001. Cyanide management by SART. In: Cyanide: social, industrial, and economic aspects, eds. C. Young,

L. Tidwell, and C. Anderson. Annual meeting of The Minerals, Metals and Materials Society, February, 549–562.

Berezowsky, R.M.G.S., M.J. Collins, D.G.E. Kerfoot, and N. Torres. 1991. The commercial status of pressure leaching technology. *Journal of Metals* 43: 9–15.

Boydell, D.W. 1984. Developments in plant practice for the recovery of gold in South Africa. In Perth and Kalgoorlie regional conference on gold mining, metallurgy and geology, AusIMM.

Budden, J., and M. Bunyard. 1994. Pilot plant test work and engineering design for the Bac-Tech bacterial oxidation plant at the Youanmi mine. In *Proceedings international conference and workshop: applications of biotechnology to the minerals industry*, 4.1–4.7.

Bunk, S.A., and D.J. Bruyns. 1993. Review of the state of commercial art roasting. In *Randol Gold Forum*, 175–178.

Cadzow, M.D., and T.S. Giraudo. 2000. Macraes gold project: value creation through applied technology-pressure oxidation. In *Proceedings 2000 New Zealand minerals and mining conference*, 275–281.

Choi, Y., J.Y. Baron, Q. Wang, J. Langhans, and P. Kondos. 2013. Thiosulphate processing from lab curiosity to commercial application. In *The proceedings of the world gold 2013 congress*, Brisbane, 26–29th September, 45–50.

Clennell, J.E. 1915. *The cyanide handbook*. 2nd ed. New York: McGraw Hill Book Company.

DST. 2018. Cyanide -free gold extraction process. http://dundeetechnologies.com/gold-chlorination.

Ellis, S. 2003. *Ultrafine grinding—a practical alternative to oxidative treatment of refractory gold ores*. AusIMM Publication.

Ellis, S., and M. Gao. 2002. The development of ultra-fine grinding at KCGM. In SME annual meeting, preprint 02-072, 1–6. SME.

Ericsson, M. 2012. Mining technology—trends and development. Polinares consortium, working paper # 29, March.

Fleming, C.A., J. McMullen, K.G. Thomas, and J.A. Wells. 2003. Recent advances in the development of an alternative to the cyanidation process: thiosulfate leaching and resin in pulp. In *Minerals and Metallurgical Processing*, vol. 20, 1–9.

Frostiak, J., and B. Haugard. 1992. Start-up and operation of placer Dome's Campbell mine gold pressure oxidation plant. *Mining Engineering* 44 (8): 991–993.

Goldstone, A.J., and T.I. Mudder. 1988. Cyanide regeneration process. US patent # 4994243.

Goode, J.R. 1993. Refractory gold ore: causes, processes, testing and plants. In *Proceedings annual SME conference*, 93-82, 1–21. SME.

Gorain, B., and P. Kondos. 2012. Method to improve recovery of gold from double refractory gold ores. US Patent 8,262,768 B2, Sep. 11, 2012.

Gorain, B.K., P. Kondos, and V.I. Lakshmanan. 2015. Innovations in gold and silver processing. In *Innovative process development in metallurgical industry*, eds. V.I. Lakshmanan, V. Ramachandran, R. Roy. Springer.

Guzman, G., and E. Acosta. 2009. Proceso SART y su Influencia en el Proceso CIC Minera Yanacocha SRL. In *PERUMIN, 2009 Convención Minera*, Arequipa, Peru, 14–18 de Septiembre.

Hammerschmidt, J., J. Günter, and B. Kersteins. 2005. Roasting of gold ore in the circulating fluidized-bed technology. In *Developments in mineral processing*, ed. M.D. Adams, 433–452. Amsterdam: Elsevier Publication.

Hammerschmidt, J., J. Günter, and A. Holmström. 2011. Fluidized bed technology for gold ore and gold concentrate. In Proceedings world gold 2011, CIMM, 171–182.

Holder, R., 2007. Beaconsfield gold mine-ironing out the bugs. In *Proceedings* 9th *mill operators conference*, AusIMM, 185–193.

Hourne, M., and D. Turner. 2012. Commercialisation of the Albion process. In *Proceedings of the ALTA 2012 gold conference*.

Kadenhe, R.M., and E.S. Makende. 1987. Review of the roasting plant operation and services available to small mines at Kwewe, Zimbabwe. In *Proceedings African mining congress*, 191–204. IMM.

Kappes, D.W. 2005. Heap leaching of gold and silver ores. In *Advances in gold ore processing*, ed. M.D. Adams, vol. 13, 456–478. Amsterdam: Elseveir.

Keane, J. 2007. Engineering and construction of a 20,000 TPD gold recovery process plant in North China. In *Precious metals processing symposium* 3–6. SME: Littleton.

Kosich, D.Y. 1992. Independence roasters surpass expectation, *Mining World News*, July, 2–3.

Kozin, L.F., and V.T. Melekhin. 2004. Extraction of gold from ores and concentrates by leaching with the use of cyanides and alternative reagents. *Russian Journal of Applied Chemistry* 77 (10): 1573–1592.

LaBrooy, S.R., H.G. Linge, and G.S. Walker. 1994. Review of gold extraction from ores. *Minerals Engineering* 7 (10): 1213–1241.

Lahti, P.A. 1996. Refractory gold ore processing methods at Jerritt Canyon. In *Randol Gold Forum* 96, 225–233.

Lakshmanan, V.I., and J.K. Biskupski. 1985. Recovery of precious metals. US Patent 4,537,628, Aug. 27, 1985.

Lakshmanan, V.I., R. Sridhar, and M.A. Halim 2013. Chloride process for the leaching of gold. US Patent 9,732,398 B2, Aug. 15, 2017.

Lakshmanan, V.I., R. Sridhar, J. Chen, M.A. Halim, and R. DeLaat. 2014. Separation of iron from value metals in leaching of laterite ores. US Patent 8916116 B2, 23 December 2014.

Lawrence, R.W. and Lopez, O., (2011). Improving the economics of gold-copper ore projects using SART technology, gold 2011, October 3–4., Montreal.

Li, Ji, and J.D. Miller. 2006. A review of gold leaching in acid Thiourea solutions. *Mineral Processing & Extractive Metallurgy Review* 27: 177–214.

Liddell, K.S., and R.C. Dunne. 1988. Ultra fine grinding effect on leach recovery from refractory gold ores. In *Proceedings Randol international gold conference*, 120–122.

Littlejohn, P., D. Kratochvil and A. Hall. 2013. Sulfidisation acidification recycling thickening (SART) for complex gold ores. In *Proceedings of world gold 2013*, 149–155.

Loosley, A. 1993. Gold ore treatment at Emperor Gold Mining Company, Ltd, Vatukoula, Fiji, in Australasian mining and metallurgy. *The Sir Maurice Mawby Memorial* 2: 953–954.

Lucheva, B., P. Iliev, and D. Kolev. 2017. Recovery of gold from electronic waste by iodine-iodide leaching. *Journal of Chemical Technology and Metallurgy* 52 (2): 326–332.

Lunt, D., and T. Weeks. 2005. Refractory gold ores case studies. In *Developments in mineral operating plants processing*, ed. M.D. Adams, 73–96. Amsterdam: Elsevier Publication.

MacPhail, P.K., C.A. Fleming and K.W. Sarbutt. 1998. Cyanide recovery by the SART process for the Lobo-Marte Project, Chile. In *Randol Gold Forum 1998*, Denver, CO, 26–29.

Marsden, J.O., and C.I. House. 2006. The chemistry of Gold extraction, 218, 2nd ed. Society for Mining, Metallurgy, and Exploration, Inc. (SME).

Marsden, J.O., and S.A. Sass. 2014. Innovation in gold and silver extraction and recovery, mineral processing and extractive metallurgy: 100 years of innovation, C.J. Anderson, R.C. Dunne and J.L. UhrieSME Publication, pp 361–387.

Melashvili, M., C. Fleming, I. Dymov, M. Manimaran and J. O'Day. 2014. Study of gold leaching with bromine and bromide and the influence of sulphide minerals on this reaction. In *COM 2014–conference of metallurgists proceedings*, Vancouver, September 28 to October 1, 2014, 13 p.

Miller, P., and A. Brown. 2005. Bacterial oxidation of refractory concentrates. In Developments in mineral processing (ed: MD Adams), 403–433, Elsevier Publication, pp 371–402.

Miller, P., F. Jiao, J. Wang. 2004. The bacterial oxidation (BACOX) plant at Laizhou Shandong Province China—the first three years of operation. In *Proceeding Bac-Min 2004*, AusIMM, 167–172.

Moss, F. 1918. The treatment of the auriferous sulphide ores of Kalgoorlie. In *Transaction Australian Institute of Mining and Metallurgy (Melbourne)*, vol. 99, 40–67.

Nyavor, K., and N.O. Egiebor. 1992. Application of pressure oxidation pretreatment to a double refractory gold concentrate. *CIM Bulletin* 85 (956): 84–90.

Pekrul, J. 2007. Update on LIXKillat the autoclave. In *Precious metals processing—advances in primary and secondary operations symposium*, October 3–6. Littleton: SME.

Pesic, B., and R.H. Sergent. 1993. Reaction mechanism of gold dissolution with bromine. *Metallurgical Transactions B* 24B: 419–431.

Rickard, T.A. 1907. *Recent cyanide practice*. San Francisco: Mining and Scientific Press.

Ritchie, I., and J. Barter. 1997. Process and engineering design aspects of biological oxidation plants. In *Proceedings, BIOMINE 97*, 14.41–14.48. AusIMM.

Rose, T.K., and W.T. Merloc. 2008. *The metallurgy of gold*. 6th ed. Wexford: Wexford Press.

Simmons, G.L. 1994. Development of low temperature pressure oxidation at Lone Tree. In SME annual meeting, preprint, # 93-86. SME.

Simmons, G.L. 1996. Pressure oxidation process development for treating carbonaceous ores at Twin Creeks. In *Randol Gold Forum 96*, 199–208.

Sovmen, V.K., A.V. Belyi, M.Yu Danneker, A. Gish, and A.N. Teleutov. 2009. Biooxidation of refractory gold sulfide concentrate of Olympiada deposit, 71–74. Zürich: Trans Tech Publications.

Stanley, G.G. 1987. The extractive metallurgy of gold in South Africa. Monograph M7, SAIMM.

Thomas, K.G. 2005. Pressure oxidation overview. In *Developments in mineral processing*, ed. M.D. Adams, 346–369. Elsevier Publication, Amsterdam.

Thomas, K.G., and A.P. Cole. 2005. Roasting developments-especially oxygenated roasting. In *Developments in mineral processing*, ed. M.D. Adams, 403–433. Elsevier Publication, Amsterdam.

Wan, R.Y., and J.A. Brierley. 1997. Thiosulfate leaching following bio-oxidation pretreatment for gold recovery from refractory carbonaceous-sulfidic ore. *Mining Engineering* 49 (8): 76–80.

Wan, R.Y., K.M. LeVier, and J.D. Miller. 1993. *Research and development activities for the recovery of gold from noncyanide solutions, hydrometallurgy: fundamentals, technology and innovations*, 4225–4436. SME-TMS.

Wan, R.Y., K.M. LeVier, and R.B. Clayton. 1994. Hydrometallurgical process for the recovery of precious metal values from precial metal ores with thiosulfate lixiviant. U.S. Patent 5,354,359.

Whitlock, J.L., and T.I. Mudder. 1986. The Homestake wastewater treatment process: biological removal of toxic parameters from cyanidation wastewaters and bioassay effluent evaluation. In *Fundamental and applied biohydrometallurgy*, ed. R.W. Lawrance. Elsevier.

Xie, J.Y., J.C. Gao, S.B. Jin, and X.G. Han. 2008. Bio-oxidation plants of refractory gold concentrate in China. In *Proceedings XXIV international mineral processing congress*, 2737–2740.

Chapter 6
Waste Management in the Gold and Silver Industry

V. I. Lakshmanan, R. Roy, and B. Gorain

Sustainable development is strategic for mining companies. In the late 1998, nine of the world's largest mining companies took an initiative to examine the role of the minerals sector in contributing to sustainable development and how that contribution could be enhanced. This initiative was called the Global Mining Initiative. Through the World Business Council for Sustainable Development (WBCSD), the International Institute for Environment and Development (IIED) was given the contract to undertake a scoping study in May 1999. The scope of this study was to understand the global challenge of sustainable development facing the mining sector and to propose the scope of a 2-year project to explore the role of mining sector in sustainable development. Following the scoping study, the IIED undertook a 2-year independent process of research and consultation project, which was named the Mining, Minerals and Sustainable Development Project (MMSD). The project was carried out between 2000 and 2002, and a final report was published in 2002. In its final report, a set of guiding principles for sustainable development was detailed as shown in Table 6.1 (MMSD 2002a).

In the environmental sphere, the minimization of waste and environmental damage was described as a guiding principle. In this chapter, the methods of waste management in gold mining are described starting with a description of how waste is generated in mining projects.

V. I. Lakshmanan (✉) · R. Roy
Process Research Ortech Inc., Mississauga, ON, Canada
e-mail: llakshmanan@processortech.com

B. Gorain (✉)
Hindustan Zinc Limited, Udaipur, Rajasthan, India
e-mail: Barun.Gorain@vedanta.co.in

V. I. Lakshmanan, B. Gorain (eds.), *Innovations and Breakthroughs in the Gold and Silver Industries*, https://doi.org/10.1007/978-3-030-32549-7_6

Table 6.1 Sustainable development principles (MMSD 2002a)

Economic sphere	1. Maximize human well-being 2. Ensure efficient use of all resources, natural and otherwise, by maximizing rents 3. Seek to identify and internalize environmental and social costs 4. Maintain and enhance the conditions for viable enterprise
Social sphere	1. Ensure a fair distribution of the costs and benefits of development for all those alive today 2. Respect and reinforce the fundamental rights of human beings, including civil and political liberties, cultural autonomy, social and economic freedoms and personal security 3. Seek to sustain improvements over time; ensure that depletion of natural resources will not deprive future generations through replacement with other forms of capital
Environmental sphere	1. Promote responsible stewardship of natural resources and the environment, including remediation for past damage 2. Minimize waste and environmental damage along the whole of the supply chain 3. Exercise prudence where impacts are unknown or uncertain 4. Operate within ecological limits and protect critical natural capital
Governance sphere	1. Support representative democracy, including participatory decision-making 2. Encourage free enterprise within a system of clear and fair rules and incentives 3. Avoid excessive concentration of power through appropriate checks and balances 4. Ensure transparency through providing all stakeholders with access to relevant and accurate information 5. Ensure accountability for decisions and actions, which are based on comprehensive and reliable analysis 6. Encourage cooperation in order to build trust and shared goals and values 7. Ensure that decisions are made at the appropriate level, adhering to the principle of subsidiarity where possible

6.1 Waste Generation in Mining

Waste generated during mining and processing of ores consists of waste rock and tailings. Waste rock consists of the overburden and mine development rock (UNEP 2002). Overburden refers to the soil and rock that covers an ore body. Overburden is removed during surface mining to access the ore body. The overburden ratio or stripping ratio is the ratio of the weight of overburden excavated to the weight of ore excavated. Surface mining typically generates 8–10 times more waste rock compared to underground mining (Das and Choudhury 2013). Mine development rock is part of host rock that contains the ore body and is removed during underground mining to access the ore body. Waste rock contains economically insufficient amount of metals and is typically stored in waste rock dumps. The objective of a sound mining plan is to minimize the amount of waste rock that needs to be removed from the host rock.

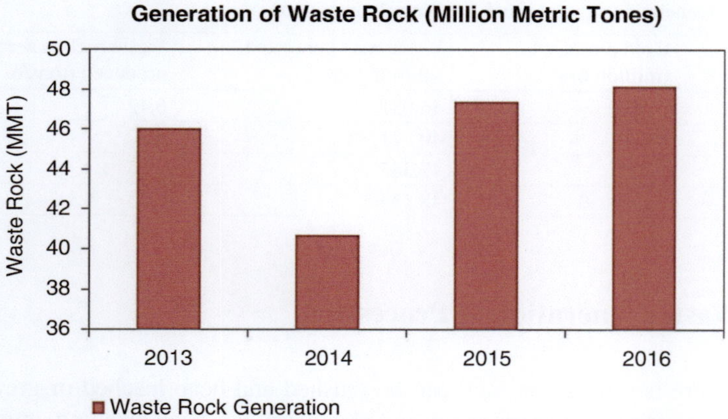

Fig. 6.1 Waste rock generated during Barrick operations

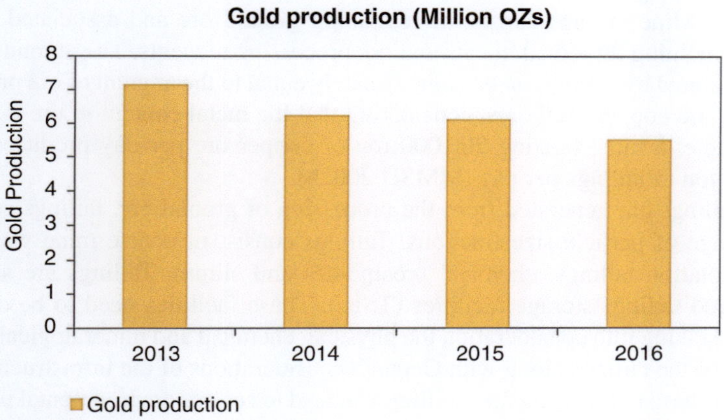

Fig. 6.2 Gold production at Barrick

Since the data on global generation of waste rock due to gold processing is not available, a rough estimate of the amount of waste rock generated for the production of a unit amount of gold can be obtained by analysing the data provided by Barrick Gold Corporation, one of the largest gold mining companies in the world, for its operations. Figures 6.1 and 6.2 show the amount of waste rock generated and amount of gold produced by Barrick between 2013 and 2016, respectively. This data on gold production and waste rock generation is listed in Table 6.2, where a column has been added to show the amount of waste rock generated per unit gold production. The amount of waste rock generated per year has been increasing since 2014, while the amount of gold produced per year has been decreasing since 2013. Waste rock generated per ounce of gold produced has been continuously increasing and was 8.73 tons per ounce of gold in 2016 for Barrick operations (Barrick 2018a, 2018b).

Table 6.2 Generation of waste rock for Barrick operations

Year	Gold produced (million ozs)	Waste rock generated (million tons)	Waste rock generated/gold produced (tons/oz)
2013	7.166	46.037	6.42
2014	6.249	40.721	6.52
2015	6.117	47.387	7.75
2016	5.517	48.183	8.73

6.2 Waste Generation in Processing

After the ore is mined, the rock can be crushed and heap leached or ground and leached depending on the ore grade and chemistry. For heap leaching, crushed ore may be placed in heap leach facilities (HLFs), where the ore is irrigated with process solutions for the recovery of precious metals. The processing of ground ores generates tailings or waste material left after the precious or value metals have been extracted. Mine tailings consist of remaining ground ore and associated process water containing dissolved metals and ore processing reagents. The amount of tailings generated by a mine can be approximately equal to the amount of ore processed for gold and copper-gold ores, considering that the metal content in ore is low. As an example, a mine treating 200,000 ton of copper ore per day produces nearly 200,000 ton of tailings per day (MMSD 2002b).

As tailings are generated from the processing of ground ore, tailings contain a wide range of particle size fractions. Tailings consist of coarse mine waste, fine clays, flotation tailings, chemical precipitates and slimes. Tailings are stored in engineered tailings storage facilities (TSFs). These facilities need to be carefully designed taking into consideration the physical, chemical and mineralogical characteristics of the tailings along with Geotech considerations of the infrastructure. The failure of these tailings storage facilities can lead to serious environmental problems in the affected areas.

According to one study, 14 billion tons of tailings were produced by the mining industry in 2010 (Jones and Boger 2012). To get a rough estimate of the amount of tailings generated for the production of a unit amount of gold, Fig. 6.3 shows the amount of tailings generated during Barrick operations between 2013 and 2016. Table 6.3 shows tailings generation, gold production and the amount of tailings generated per unit gold production for this time period. Tailings generated have been decreasing since 2014. However, as shown in Fig. 6.2, the gold production by Barrick has also been decreasing during this time period. To normalize the effect of the amount of gold production, tailings generated per ounce of gold produced has been shown in Table 6.3, and this ratio has also been decreasing since 2014 and was 27.40 tons per ounce of gold in 2016 for Barrick (2018a).

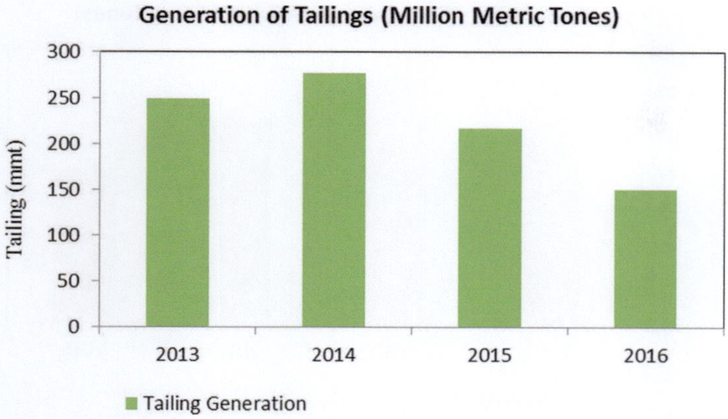

Fig. 6.3 Tailings generation for Barrick operations

Table 6.3 Generation of tailings for Barrick operations

Year	Gold produced (million ozs)	Tailings generated (million tons)	Tailings generated/gold produced (tons/oz)
2013	7.166	249.662	34.84
2014	6.249	276.675	44.28
2015	6.117	217.987	35.64
2016	5.517	151.145	27.40

Figure 6.4 combines the data on waste rock and tailings shown in Figs. 6.1 and 6.3, respectively, and shows the amount of total waste generated for Barrick operations between 2013 and 2016. Table 6.4 shows total waste generated, gold production and total waste generated per unit gold production. Total waste generated per ounce of gold produced has been decreasing since 2014 as the tailings generated per ounce of gold produced has been decreasing since 2014. The amount was 36.13 tons of total waste generated per ounce of gold produced in 2016 for Barrick (Barrick 2018a).

Even though the amount of total waste has been decreasing per unit of gold produced, it is still a huge number. The main reason for the generation of huge amount of waste generated per ounce of gold produced is the low grade of gold ores. The average grade of gold ores has been falling over the years. Between 1830 and 1900, the average grade of gold ore was around 20 g/ton and in 2000; it had dropped to around 3 g/ton (Muller and Frimmel 2010). Gold mining companies are processing lower grades of ore as the higher-grade ores are getting depleted. Table 6.5 shows the average grade of gold ores at Barrick operations. In 2016, the average grade of gold at major mines owned by Barrick was lower than 4 g/ton. There are many gold miners, including Barrick, that are processing gold ores having less than 1 g/ton of gold. One troy ounce of gold is equal to 31.10 g. Thus, for an ore containing 1 g/ton of gold, over 31 tons of tailings is generated per troy ounce of gold even at 100% recovery of gold. Thus, it has become very important to properly manage the waste generated by gold mining

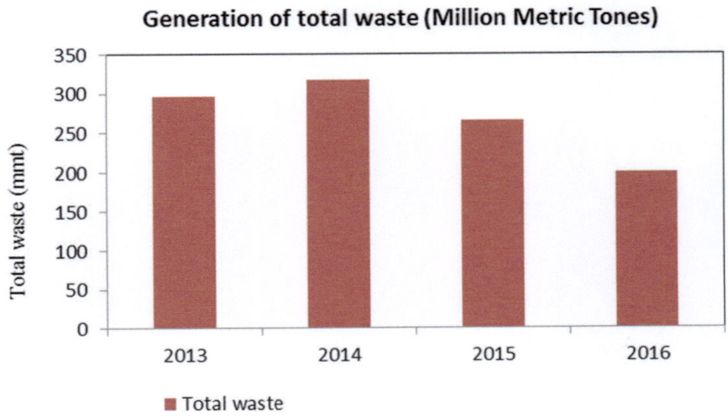

Fig. 6.4 Total waste generated during Barrick operations

Table 6.4 Total waste generated during Barrick operations

Year	Gold produced (million ozs)	Total waste (million tons)	Total waste generated/gold produced (tons/oz)
2013	7.166	295.699	41.26
2014	6.249	317.396	50.79
2015	6.117	265.374	43.38
2016	5.517	199.328	36.13

Table 6.5 Average ore grade at Barrick operations (Barrick 2018b)

Year	Ore grade (g/ton) at Goldstrike	Ore grade (g/ton) at Cortez	Ore grade (g/ton) at Pueblo Viejo	Ore grade (g/ton) at Lagunas Norte	Ore grade (g/ton) at Veladero
2013	5.01	2.59	6.14	1.06	0.94
2014	6.28	1.34	5.53	0.99	1.00
2015	6.01	1.73	4.94	1.02	0.82
2016	3.55	2.11	2.93	1.86	0.83

operations as the storage of waste is taking up a huge amount of space and creating a large environmental footprint.

6.3 Waste Disposal and Management

As discussed above, the waste from mining operations consists of waste rock and tailings. The disposal and management of waste rock and tailings is described in this section.

6.3.1 *Waste Rock Management*

Waste rock that does not present a risk to the environment due to leakage of harmful chemicals can be disposed in many ways. It can be placed in waste rock storage facilities till the facility is full and then covered with soil for growing plants and trees. Open pits or underground tunnels can also be filled with waste rock, if no environmental concerns are present. Non-reactive waste rock can also be used as construction material for road beds or tailing dams. Depending on the composition of waste rock, it may be susceptible to acid rock drainage (ARD) and/or metals leaching (ML) into the surrounding soil and water sources. Reactive waste rock can be encapsulated within non-reactive waste rock to minimize the chance of ARD and/or ML.

6.3.1.1 Acid Rock Drainage and Metals Leaching

The drainage of acid, metal or sulphates results from the oxidation of sulphide minerals due to the exposure to air or water. It is described by several terms such as acid rock drainage (ARD), acid mine drainage or acid and metalliferous drainage (AMD), mining influenced water (MIW), saline drainage (SD) and neutral mine drainage (NMD). Based on the composition of the mineral and ambient conditions, the drainage could be acidic or neutral and may contain heavy metals. Depending on the pH of the resulting solution, the drainage is classified as acid rock drainage (ARD), neutral mine drainage (NMD) or saline drainage (SD). Table 6.6 shows the difference between different types of drainages. Figure 6.5 shows the sources, pathways and receiving environment for acid rock drainage.

6.3.1.2 Acid Rock Drainage Formation

The formation of acid rock drainage (ARD), neutral mine drainage (NMD) or saline drainage (SD) results from the oxidation of sulphide minerals. When sulphide minerals are exposed to oxygen from atmosphere or oxygenated water due to mining or

Table 6.6 Classification of drainage from waste rock (INAP 2009)

Acid rock drainage	Neutral mine drainage	Saline drainage
Acidic pH	Near neutral to alkaline pH	Neutral to alkaline pH
Moderate to elevated metals	Low to moderate metals. May have elevated zinc, cadmium, manganese, antimony, arsenic or selenium	Low metals. May have moderate iron
Elevated sulphate	Low to moderate sulphate	Moderate sulphate, magnesium and calcium
Treat for acid neutralization and metal and sulphate removal	Treat for metal and sometimes sulphate removal	Treat for sulphate and sometimes metal removal

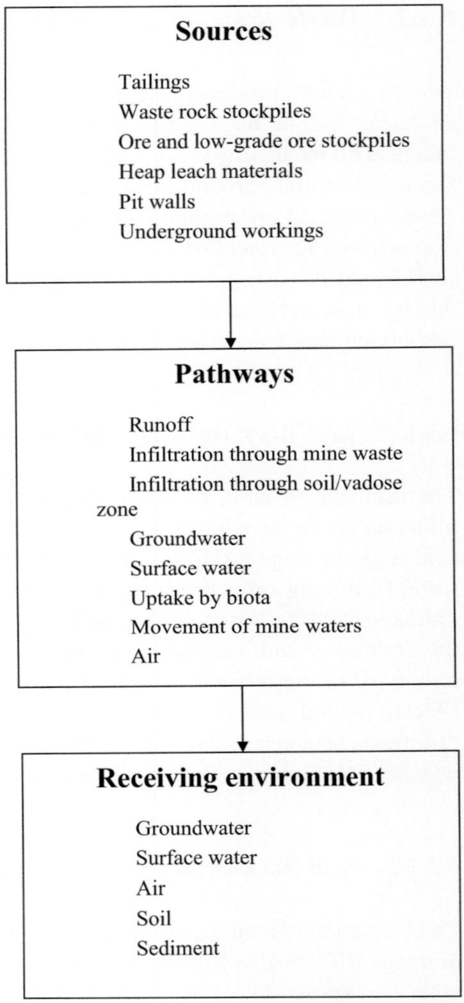

Fig. 6.5 Sources, pathways and receiving environment for acid rock drainage (adapted from INAP 2009)

mineral processing, sulphides become unstable. The reactions taking place under these conditions can be illustrated using the mineral pyrite (FeS_2), which is the most common sulphide mineral.

$$\text{Reaction}\,(1): 2FeS_2 + 7O_2 + 2H_2O = 2Fe^{2+} + 4SO_4{}^{2-} + 4H^+$$

$$\text{Reaction}\,(2): 4Fe^{2+} + O_2 + 4H^+ = 4Fe^{3+} + 2H_2O$$

$$\text{Reaction}\,(3): 4Fe^{3+} + 12H_2O = 4Fe(OH)_3 + 12H^+$$

$$\text{Reaction}\,(4): FeS_2 + 14Fe^{3+} + 8H_2O = 15Fe^{2+} + 2SO_4{}^{2-} + 16H^+$$

$$\text{Overall reaction}: 4FeS_2 + 15O_2 + 14H_2O = 4Fe(OH)_3 + 8H_2SO_4$$

Reaction (1) is the oxidation of pyrite by oxygen in the presence of water. During this reaction, sulphur is oxidized to sulphate with the release of ferrous iron. Two H^+ ions are generated for each mole of pyrite.

Reaction (2) is the oxidation of ferrous iron to ferric iron. One H^+ ion is consumed to convert one ferrous iron to ferric iron. The rate of conversion from ferrous to ferric iron can be increased by certain bacteria. The reaction proceeds slowly under stronger acidic conditions (pH 2–3) when no bacteria are present. The rate of reaction is several orders of magnitude faster under mildly acidic conditions (pH~5). This reaction is considered as a rate-limiting step.

Reaction (3) is the hydrolysis of iron, which results in the generation of three H^+ ions for each ferric ion. Ferric hydroxide precipitate is formed due to hydrolysis.

Reaction (4) is the oxidation of additional pyrite molecules by ferric iron. The reaction takes place very rapidly and continues until either ferric iron or pyrite is consumed. Overall reaction shows that two moles of sulphuric acid are produced for each mole of pyrite.

Above reactions show the formation of acid by the oxidation of pyrite. Although pyrites are dominant sulphide species in the minerals, there are other sulphide minerals present in the ore and waste rock. The oxidation of these sulphide minerals may or may not generate acid. Iron sulphides (pyrite, marcasite, pyrrhotite), sulphides with molar metal/sulphur ratios <1 and sulfosalts (e.g. enargite) generate acid by reacting with oxygen and water. Sulphides with molar metal/sulphur ratios = 1 (e.g. sphalerite, galena, chalcopyrite) do not generate acid by reacting with oxygen and water. However, all sulphides can generate acid in the presence of ferric ion. Therefore, the presence of iron sulphide is the determining factor in the potential for generating acid. The presence of carbonate minerals may result in neutralizing the acid generated by oxidation of sulphide minerals.

6.3.1.3 Acid Rock Drainage Prediction

The tests for prediction of acid rock drainage are classified in two groups: static tests and kinetic tests. The static tests are based on chemical balance and do not take rate of reactions into account. These tests assume that all the sulphur in the rocks is acid-forming pyritic sulphide and all acid-consuming components are available to neutralize acid. These tests are performed by calculating acid production potential (APP) and neutralization potential (NP). The difference between neutralization potential and acid production potential is termed Net Neutralization Potential (NPP).

The kinetic tests are conducted for the following reasons (Ritcey 2005):

(a) Confirmation of the static tests
(b) Assessment of rates of AMD potential
(c) Determination of the effect of bacterial action

(d) Assessment of the rate of depletion of neutralization capacity
(e) Estimation of metals concentration in leachate from the waste
(f) Determination of the overall biogeochemical changes
(g) Evaluation of different waste management and control strategies for the particular waste

Kinetic tests are first conducted at bench scale and are followed by large-scale testing on site. While static tests can be useful as a guide, kinetic tests are needed for better prediction of acid rock drainage potential. Kinetic tests include soxhlet, columns, humidity cell and lysimeter tests.

6.3.1.4 Acid Rock Drainage Treatment Technologies

Table 6.7 lists the various technologies that can be utilized for the treatment of acid rock drainage, which can be grouped under four categories: neutralization, metals removal, desalination and specific target pollutant treatment.

6.3.1.5 Acid Rock Drainage Management

To avoid the contamination of soil and water in the affected area, proper management plan has been developed by the International Network for Acid Prevention (INAP). The INAP is an international organization that is dedicated to understanding and

Table 6.7 Acid rock drainage treatment technologies (adapted from INAP 2009)

Neutralization	Lime/limestone process Sodium based alkali's (NaOH, Na_2CO_3) Ammonia Biological sulphate reduction Wetlands, anoxic drains Other technologies
Metals removal	Precipitation/hydroxide Precipitation/carbonates Precipitation/sulphides Wetlands, oxidation ponds Other technologies
Desalination	Biological sulphate removal Precipitation processes such as ettringite Membrane-based processes Ion-exchange processes Wetlands, passive treatment processes
Specific target pollutant treatment	Cyanide removal: chemical oxidation, biological oxidation, complexation Radioactive nuclides: precipitation, ion exchange Arsenic removal: oxidation/reduction, precipitation, adsorption Molybdenum removal: iron adsorption Other technologies

meeting the challenge of acid drainage. The INAP was founded in 1998 and is a proactive, global leader in this field. Its members include Anglo American, Antofagasta Minerals, Barrick, Freeport-McMoRan, Kinross Gold Corporation, Newcrest, Newmont, Rio Tinto, Vale and Xstrata. Based on the interactions with the global gold mining companies, the INAP has developed the Global Acid Rock Drainage (GARD) Guide (INAP 2009), which is a worldwide reference for the prevention and mitigation of acid rock drainage. The GARD Guide is periodically updated to reflect the current understanding of acid rock drainage.

Acid rock drainage management plan shown in Table 6.8 details the different activities that need to be performed at each stage of mine operation. For the management of acid rock drainage, it is important to be able to predict acid rock drainage. The approach to the prediction of acid rock drainage is shown in Fig. 6.6.

6.3.2 *Tailings Management*

The tailings management system has many components and may include many steps such as tailings treatment in the mill, slurry thickening, slurry transport, tailings impoundment, water recovery and recycle, tailings and effluent treatment and

Table 6.8 Acid rock drainage management plan (adapted from INAP 2009)

Exploration	Characterization
Assessment	Prediction
Design	Planning for avoidance
Construction	Surface water control works Groundwater control
Operation	Waste rock: special handling, segregation, encapsulation, layering, blending, re-mining, backfilling, passivation, selective mining and avoidance, hydrodynamic controls, appropriate siting of facilities, co-disposal, in-pit disposal, permafrost and freezing, bactericides, alkaline materials, organics Tailings: desulphurization, compaction, amendment, dewatering, re-mining, backfilling, passivation, selective mining and avoidance, hydrodynamic controls, appropriate siting of facilities, co-disposal, in-pit disposal, permafrost and freezing, bactericides, alkaline materials, organics Open pit: re-mining, backfilling, passivation, selective mining and avoidance, hydrodynamic controls Underground workings: re-mining, backfilling, passivation, selective mining and avoidance, hydrodynamic controls
Decommission	Dry cover for waste rock and tailings Seals for underground workings Water cover for tailings and open pit Flooding for waste rock, tailings, open pit and underground workings
Post-closure	Monitoring Maintenance Inspection Long-term collection and treatment were required

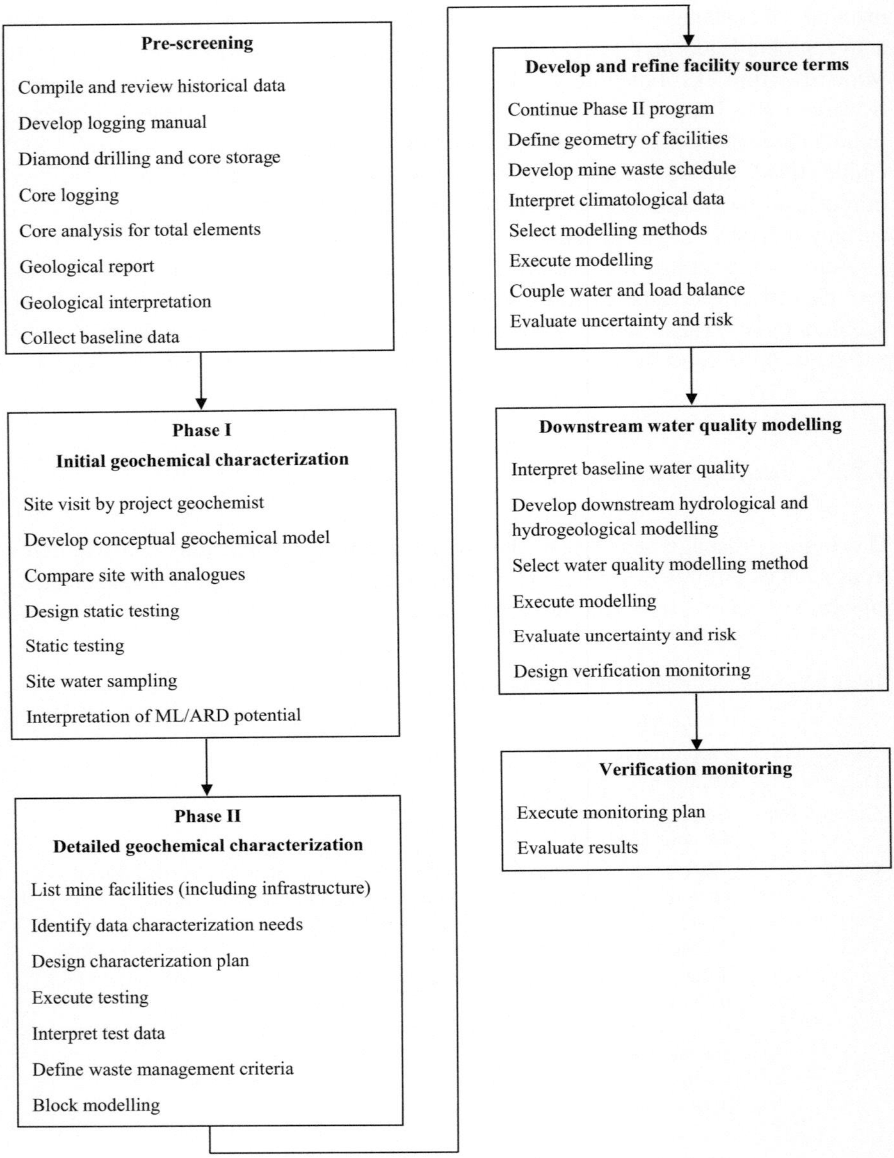

Fig. 6.6 Acid rock drainage prediction approach (adapted from INAP 2009)

evaporation and restoration of the site (Ritcey 2005). The amount of tailings produced by the mining industry in 2010 was 14 billion tons (Jones and Boger 2012). The generation of such a huge amount of tailings leaves a large environmental footprint. In terms of space, it occupies a large area for storage, which needs to

Table 6.9 Potential risks associated with mine tailings (Adiansyah et al. 2015)

Phase	Potential risks
Operation	Leaking of tailings slurry pipeline Geotechnical failure TSF overflow Seepage through containment wall Seepage infiltration to groundwater Particulate matter (PM): dust or gas emissions Interaction of wildlife or livestock with tailings Mine acid pollution into the water: groundwater and surface water
Closure	Erosion of containment wall Spillway failure Overtopping by rainfall runoff Failure of land cover system on tailings surface

be managed for environmental degradation. The management of tailings is a crucial issue in mining operations because of the potential for accidents.

6.3.2.1 Incidences of Tailings Dam Failures

Potential risks associated with mine tailings are shown in Table 6.9. A proper tailings management plan will take into account all the potential risks during the operation and closure of the tailings facility. The failure to properly manage tailings can have catastrophic consequences which can be very costly to remedy.

The failure of Santarém and Fundão tailings dams in Brazil to contain the water and sediment from iron ore extraction resulted in one of the worst environmental disasters of recent history. The two dams are owned and operated by Samarco, a privately held Brazilian mining company, which is controlled in equal parts by two shareholders: the Brazilian Vale S.A. and the Anglo-Australian BHP Billiton. The tailings dams located in the Mariana region, state of Minas Gerais, Brazil, burst on November 5, 2015, releasing about 60 million cubic meters of tailings (da Costa 2017). The tailings entered the River Doce, one of Brazil's most important rivers, and reached about 20 municipalities downstream from the mine site in the next 17 days. This had a devastating impact on water supply, fisheries activities, agriculture and tourism in the affected area. The failure of Santarém and Fundão tailings dams resulted in 19 persons losing their lives and more than 600 people losing their homes.

On January 25, 2019, another tailings dam failure took place at the Córrego do Feijão iron ore mine in Brazil operated by Vale. This dam collapse resulted in at least 58 people dead and hundreds missing. This is widely considered as a sad consequence of the lessons not learned.

Another catastrophic failure of a tailings storage facility was a leak of cyanide near Baia Mare, Romania, into the Someș River in 2000. This facility was operated by the gold mining company Aurul, which was a joint venture between the Australian company Esmeralda Exploration and the Romanian government. This spill has been called the worst environmental disaster in Europe since the Chernobyl disaster which resulted in killing large numbers of fish in Hungary and Romania as the polluted waters reached the Tisza River and then the Danube River. This toxic spill contaminated the water supplies of more than two million people (Lottermoser 2010).

The catastrophic failure of the Los Frailes tailings dam near Seville, Spain, in April 1998 released approximately 528 million gallons of pyrite sludge and another 1 billion gallons of acid water containing high concentrations of heavy metals (zinc, lead, arsenic, copper, antimony, thallium and cadmium) into the Guadiamar River affecting a 62 kilometre long section of the river, ranging from 500 to 1000 meters in width (Arenas and Méndez 2002). The area affected by the accident included 6560 acres within Doñana Nature Park and 242 acres within Doñana National Park. The cost of cleanup exceeded $225 million. It resulted in a loss of 5000 jobs in various sectors and contamination of water stream with acid, metals and metalloids. Such accidents can be avoided by proper tailings management.

Between 1910 and 2009, 218 cases of tailings accidents have taken place worldwide (Azam and Li 2010). Figure 6.7 shows the number of accidents per decade starting from 1910.

6.3.2.2 Causes of Tailings Dam Failures

Figure 6.7 shows that most accidents took place during the 1960–1980s. Out of the 218 accident, 167 accidents were analysed for causes of failure as information from the rest of the accidents were not available to make an informed conclusion. Figure 6.8 shows the number of accidents as a function of causes of tailings dam failures. These causes include unusual weather, management, foundation subsidence,

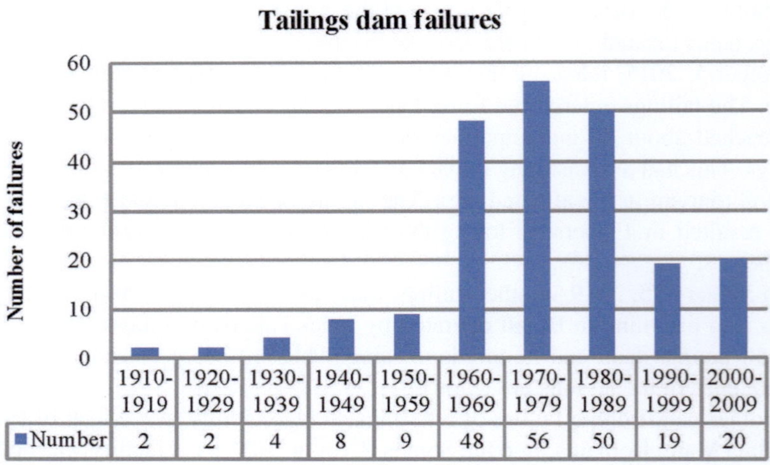

	1910-1919	1920-1929	1930-1939	1940-1949	1950-1959	1960-1969	1970-1979	1980-1989	1990-1999	2000-2009
Number	2	2	4	8	9	48	56	50	19	20

Fig. 6.7 Tailings dam failures around the world (adapted from Azam and Li 2010)

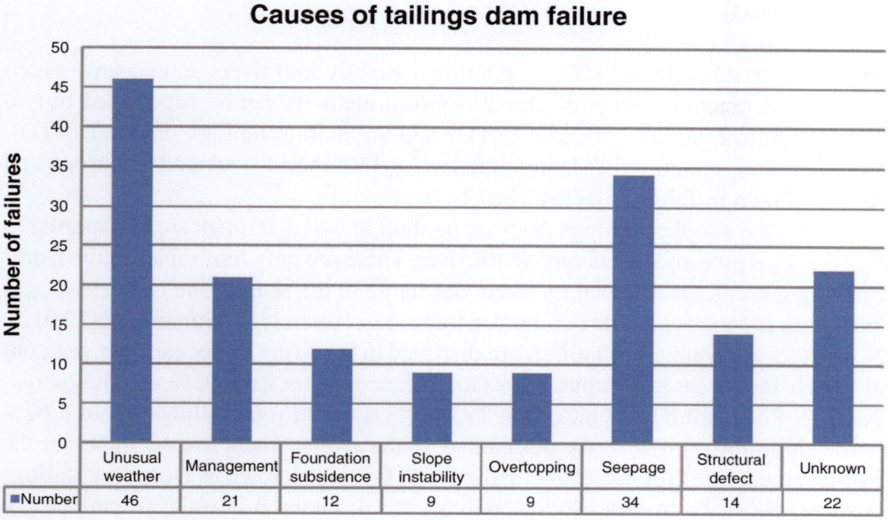

Fig. 6.8 Causes of tailings dam failures (adapted from Azam and Li 2010)

slope instability, overtopping, seepage and structural defect. For some accidents, the cause of failure could not be determined. Most accidents have taken place due to unusual weather, followed by seepage.

Tailings dams are supposed to last for a long time, but a number of factors make tailings dams more vulnerable than water retention dams. These factors are as follows (Rico et al. 2008):

1. Embankments are made from soil, coarse waste, overburden from mining operations and tailings.
2. Dams are raised as flow of solid material and effluent from mine including runoff from precipitation increases.
3. There is lack of regulations on specific design criteria.
4. There is lack of dam stability requirements regarding continuous monitoring and control during emplacement, construction and operation.
5. The cost of maintenance works for tailings dams after closure of mining activities is high.

6.3.2.3 Tailings Disposal Methods

The methods of tailings disposal can be classified in two different groups: direct disposal methods and indirect disposal methods.

Direct Disposal

Direct disposal involves discharging tailings directly into rivers and oceans, which raises environmental concerns. Direct disposal methods can be subdivided in two groups: riverine tailings disposal (RTD) and submarine tailings disposal (STD). These methods are currently being practiced at 16 mine sites located in Europe and Asia as shown in Table 6.10 (IMO 2012).

RTD is the simplest tailings disposal method in which tailings are transported to the river by a pipe and discharged in the river. There are only four mines in the world carrying out tailings disposal by direct discharge in the rivers. One of these sites is located in Indonesia and three sites are located in Papua New Guinea (IMO 2012). Nearly 295 million tons of tailings are disposed in the rivers and oceans per year, out of which Indonesia and Papua New Guinea account for over 93% of tailings disposal by RTD and STD. Out of the 295 million tpy of total tailings, around 62% (~183 million tpy) is directly discharged in the rivers. There are 12 mines in the world that use STD for disposal of tailings. STD is also known as Deep Sea Tailings Disposal (DSTD). In this method, tailings are discharged into the ocean using a pipeline and settle to the bottom of the ocean floor. The density and temperature of the tailings product need to be controlled to prevent the tailings from flowing away from the location of tailings deposition. Earlier tailings were discharged on the ocean surface, which created close interactions between tailings and environments, both biotic and abiotic. Deep sea tailings disposal was introduced to reduce the impact of mine tailings on the biotic and abiotic environment of the ocean.

Indirect Disposal

In this method, tailings are disposed in a confined area such as an impoundment, cell or dam. Depending on the solids content of the slurry, various options for indirect disposal are exercised such as conventional tailings, tailings paste, thickened tailings and tailings cake. Out of these, the most common method of indirect disposal is conventional tailings. When the solids content of the slurry is approximately 25–30% solids, it is termed conventional tailings, which can be transported in slurry form

Table 6.10 Tailings disposal by RTD and STD (IMO 2012)

	Number of riverine tailings disposal sites	Number of submarine tailings disposal sites	Mine tailings volume million tons/year
Turkey	0	1	11
England	0	1	2
Norway	0	5	7
Indonesia	1	1	127
Papua New Guinea	3	4	148
Total	4	12	295

through pipes. One of the main causes of tailings dam failures is the high water content of the slurry. The risk of tailings dam failure can be reduced by decreasing the water content of the tailings to convert conventional tailings to tailings paste, thickened tailings or tailings cake. Tailings are dewatered using vacuums and filters, which helps in water balance and also reduces the environmental impact. Tailings cake cannot be transported by pipeline. It is normally transported by conveyor or truck to tailings disposal area, where it is deposited and spread. It is then compacted to form an unsaturated tailings deposit, which is referred 'dry stack' (Davies and Rice 2001). Dry stacking has many advantages such as water conservation and no risk of tailings dam failure. However, it involves higher capital needing investment in modern filtration equipment and maintenance and operation of these equipment.

Phytostabilization can be used for long-term stabilization and containment of tailings. It works by sequestering pollutants near the roots of the plants in the soil. The growth of plants reduces erosion by wind or water. Plants can immobilize metals by adsorption around the roots, which reduces the exposure of pollutants to livestock, wildlife and humans. Trees such as *Dalbergia sissoo*, *Eucalyptus*, *Cassia siamea*, *Acacia mangium* and *Peltophorum* can be planted over overburden dumps for best results (Das and Choudhury 2013).

6.3.2.4 Life Cycle of a Tailings Facility

A proper tailings management plan begins from the project conception and planning stage and incorporates complete life cycle of the tailings facility including post-closure stage. The life cycle of a tailings facility consists of the following stages (MAC 2017):

1. Project conception and planning: This phase begins with the planning of a proposed mine. It is integrated with conception and planning for the overall site, including the mine plan and plans for ore processing.
2. Design: This phase begins once the location and best available technology for the tailings facility have been selected. During this phase, detailed engineering designs are prepared for all aspects of the tailings facility and associated infrastructure.
3. Initial construction: This phase involves the construction of structures and infrastructure that need to be in place before the deposition of tailings commences. Examples of activities during this phase include the removal of vegetation and overburden and construction of starter dams, tailings pipelines, access roads and associated water management infrastructure.
4. Operations and ongoing construction: During this phase, tailings are transported to the tailings facility. The design of tailing facility includes the provisions for raising the height of tailings dams or construction of new tailings cells as needed during the operation of the mine. The period of commercial operations of the mine may or may not coincide with the operations and ongoing construction phase of a tailings facility.

5. Standby care and maintenance: This phase begins when the mine has ceased commercial operations and the deposition of tailings into the facility is no longer taking place. The surveillance and monitoring of the tailings facility continue during this phase. As the resumption of commercial operations is expected at some point in the future, the facility and associated infrastructure are not decommissioned and the closure plan is not implemented.
6. Closure: This phase begins when the deposition of tailings into the facility has stopped permanently. Now the tailings facility and associated infrastructure are decommissioned. The key aspects of the closure plan are:

   - Transitioning for operations to permanent closure
   - Removal of key infrastructure such as pipelines
   - Changes to water management or treatment
   - Recontouring or revegetation of tailings and any containment structures or other structural elements

7. Post-closure: This phase begins when the decommissioning work has been completed, key aspects of the closure plan have been implemented and the tailings facility has transitioned to long-term maintenance and surveillance. During this phase, the responsibility for the tailings facility could transfer from the owner of the facility to jurisdictional control.

6.4 Value Generation from Waste

To make the mining industry more sustainable, there is need for value generation from the mining waste while making the mining waste environmentally inert. Yesterday's tailings is today's resource and today's waste is tomorrow's resource. Tailings need to be stored in such a way that it facilitates the recovery of value at a future date. Tailings also need to be secured properly for future processing. The development of zero discharge technology has made it easier to store tailings safely till it is processed at a future date.

6.4.1 Recovery of Gold

Now it is possible to economically recover gold from tailings having 0.3 g/ton gold by heap leaching. A number of companies are currently involved in processing tailings from previously abandoned mines as technology has advanced and gold price has risen. These companies include DRDGold, Mintails Ltd., Gold Fields, Gold One Group, Carbine resources in Australia and PanTerra Gold (Ndlov et al. 2017). Gold Fields is running a tailings treatment project (TTP) in South Africa to recover gold and uranium. Enviro Gold, a subsidiary of PanTerra Gold, is running the Las Lagunas Gold Tailings project in Dominican Republic to recover gold from Pueblo

Viejo mine tailings. The project involves ultrafine grinding, flotation, sulphide oxidation using the Albion process and extraction of gold and silver using standard carbon-in-leach cyanidation.

6.4.2 Recovery of Sulphuric Acid

The Council for Scientific and Industrial Research (CSIR) based in South Africa has developed a chemical desalination process to neutralize acid mine drainage (AMD) and recover metals and sulphate (Motaung et al. 2008; Wilsenach et al. 2008). The process consists of the following steps:

Step 1: *Pretreatment*

The first step is pretreatment using $CaCO_3$, which neutralizes the free acid and precipitates iron(III) and aluminium(III), as hydroxides.

$$H_2SO_4 + CaCO_3 = CaSO_4 + CO_2 + H_2O$$

$$2M^{3+} + 3CaCO_3 + 3H_2O = 2M(OH)_3 + 3CO_2 + 3Ca^{2+} \left(M = Fe, Al\right)$$

Step 2: *Lime treatment*

The second step is treatment with hydrated lime for removal of magnesium and partial removal of sulphate as crystals of gypsum.

$$Mg^{2+} + Ca(OH)_2 = Mg(OH)_2 + Ca^{2+}$$

$$Ca^{2+} + SO_4^{\ 2-} + 2H_2O = CaSO_4 \cdot 2H_2O$$

Step 3: *pH adjustment*

The third step is pH adjustment using carbon dioxide.

$$Ca(OH)_2 + CO_2 = CaCO_3 + H_2O$$

Step 4: *Removal of sulphate*

The fourth step is the removal of sulphate as barium sulphate.

$$Ca^{2+} + SO_4^{\ 2-} + BaCO_3 = BaSO_4(s) + CaCO_3(s)$$

Step 5: *Recovery of CaS and* CO_2

The fifth step is the processing of the $CaSO_4{\cdot}2H_2O/Mg(OH)_2$ sludge to recover CaS and CO_2.

$$CaSO_4 + 2C = CaS + CO_2$$

Step 6: *Processing of CaS*

The sixth step is the processing of CaS to produce $Ca(HS)_2$, $CaCO_3$ and H_2S.

$$2CaS + CO_2 + H_2O = Ca(HS)_2 + CaCO_3$$

$$Ca(HS)_2 + CO_2 + H_2O = CaCO_3 + 2H_2S$$

Step 7: *Recovery of BaS and CaO*

The seventh step is the processing of the $BaSO_4/CaCO_3$ sludge to recover BaS and CaO.

$$BaSO_4 + 2C = BaS + CO_2$$

$$CaCO_3 = CaO + CO_2$$

Step 8: *Processing of BaS to $BaCO_3$*

The eighth step is the processing of BaS to produce $BaCO_3$.

$$BaS + CO_2 + H_2O = BaCO_3 + H_2S$$

Step 9: *Processing of H_2S to sulphur*

The final step is the processing of H_2S to produce sulphur, which can be used to make sulphuric acid.

$$2H_2S + SO_2 = 2S + 2H_2O$$

6.4.3 Recovery of Mercury

Mercury is released to the environment during extraction of gold or silver due to its presence in many gold ores. It can be produced as a by-product, which is preferable to its release in the atmosphere or tailings. The production of mercury at a Nevada gold mine has been studied by Miller (2007). Concentration of mercury in gold ores varies from less than 0.1 mg/kg to over 100 mg/kg. Generally, 80–90% of mercury in the ore is volatilized during roasting. If one million ton of ore containing 20 mg/kg mercury is roasted, 16–18 tons of mercury becomes available for recovery. Scrubbing systems such as the calomel and hypochlorite systems can be used for the capture of mercury.

$$\text{Calomel system}: HgCl_2 + Hg = Hg_2Cl_2$$

$$\text{Hypochlorite system}: NaOCl + 2Hg + HCl = Hg_2Cl_2 + NaOH$$

Mercury remaining in the ore is recovered during further processing steps. During cyanidation, both gold and mercury form water-soluble cyanide complexes according to the following reactions:

$$4Au + 8NaCN + O_2 + 2H_2O = 4Na\left[Au\left(CN\right)_2\right] + 4NaOH$$

$$2Hg + 8NaCN + O_2 + 2H_2O = 2Na_2\left[Hg\left(CN\right)_4\right] + 4NaOH$$

These complexes are adsorbed on carbon and stripped from the carbon and electrowon to recover gold. Mercury is separated from the gold by distillation and collected as liquid mercury.

6.5 Energy and Water Management

The International Council of Mining and Metals (ICMM) has developed ten principles of sustainable development as described below (ICMM 2015):

1. Apply ethical business practices and sound systems of corporate governance and transparency to support sustainable development.
2. Integrate sustainable development in corporate strategy and decision-making processes.
3. Respect human rights and the interests, cultures, customs and values of employees and communities affected by our activities.
4. Implement effective risk management strategies and systems based on sound science and which account for stakeholder perceptions of risks.
5. Pursue continual improvement in health and safety performance with the ultimate goal of zero harm.
6. Pursue continual improvement in environmental performance issues, such as water stewardship, energy use and climate change.
7. Contribute to the conservation of biodiversity and integrated approaches to land-use planning.
8. Facilitate and support the knowledge base and systems for responsible design, use, reuse, recycling and disposal of products containing metals and minerals.
9. Pursue continual improvement in social performance and contribute to the social, economic and institutional development of host countries and communities.
10. Proactively engage key stakeholders on sustainable development challenges and opportunities in an open and transparent manner. Effectively report and independently verify progress and performance.

Continual improvement in environmental performance issues, such as energy use and water stewardship has been described as a guiding principle for sustainable development. With the increasing demand for water and energy, the feasibility of many mining projects depends on the adequate supply of water and energy throughout the life of the mine. As a good strategy for water and energy management is crucial for the success of mining projects, mining companies are taking steps to minimize energy consumption and water usage. However, both of these targets may not be satisfied simultaneously. It is often the case that water management initiatives lead to higher energy consumption (Nguyen et al. 2014). Thus, energy and water management have to be undertaken in a coupled manner.

Recently two different approaches have been developed to optimize the energy and water requirements for mining activities such as optimal mine water network design using water pinch analysis and hierarchical systems modelling.

Optimal mine water network design using water pinch analysis was developed by Gunson et al. (2010) by combining the mine water network design (MWND) approach with the water pinch analysis (WPA) approach. The WPA approach maximizes water reuse and minimizes wastewater discharge for allocation of water within a water network (Hallale 2002). The MWND approach constructs energy requirement matrices of the water system by identifying quantity and quality of potential water providers (sources) and water receivers (users) and energy demand for water processes such as pumping, treatment, cooling and heating. It then uses linear programming to minimize energy usage and select the optimal water network. This approach is suitable for greenfield water projects, but not for existing mine water system where making extensive changes to the existing mine water system may not be feasible.

The hierarchical systems model (HSM) analyses water, energy and emissions interactions in mining at different scales: subsite, site and regional levels. Water input, water output, energy input and energy output are listed for each system component at each scale, and this approach allows water use and energy use to be displayed in parallel. Based on this data, energy and water consumptions can be optimized.

6.5.1 Energy Management

Mining is an energy intensive industry considered to be one of the five largest consumers of global energy (Adiansyah et al. 2016). Energy is used not only in mining and processing activities but also in water and wastewater treatment and for residential needs. There are many opportunities in the mining industry to either minimize energy consumption or reduce pollution from energy generation by using renewable energy sources.

6.5.1.1 Tailings Water Recycling

Tailings are transported to the tailings storage facility as slurry by a pipeline, which requires energy. Water from this slurry can be recovered and pumped back to the processing plant to be used as process water. Energy is also required for the

Table 6.11 Synergy and trade-off scenario matrix (Nguyen et al. 2014)

	$\Delta V > 0$	$\Delta V = 0$	$\Delta V < 0$
$\Delta E > 0$	Trade-off	Trade-off	Not applicable
$\Delta E = 0$	Synergy	Neutral	Not applicable
$\Delta E < 0$	Synergy	Synergy	Trade-off

recycling of water from tailings slurry. Thus, the water and energy consumptions are related. The efforts to conserve energy and improve water availability can result in different scenarios as shown in Table 6.11.

1. Scenario 1: $\Delta E > 0$ and $\Delta V > 0$; there is a trade-off between higher water availability and higher energy demand.
2. Scenario 2: $\Delta E > 0$ and $\Delta V = 0$; there is a trade-off between no change in water availability and higher energy demand.
3. Scenario 3: $\Delta E > 0$ and $\Delta V < 0$; this is not an acceptable scenario as water availability decreases and energy demand increases.
4. Scenario 4: $\Delta E = 0$ and $\Delta V > 0$; there is synergy as the water availability increases and energy demand remains same.
5. Scenario 5: $\Delta E = 0$ and $\Delta V = 0$; this is a neutral scenario as there is no change in water availability and no change in energy demand.
6. Scenario 6: $\Delta E = 0$ and $\Delta V < 0$; this is not an acceptable scenario as water availability decreases and energy demand remains same.
7. Scenario 7: $\Delta E < 0$ and $\Delta V > 0$; there is synergy as the water availability increases and energy demand decreases.
8. Scenario 8: $\Delta E < 0$ and $\Delta V = 0$; there is synergy as the water availability remains same and energy demand decreases.
9. Scenario 9: $\Delta E < 0$ and $\Delta V < 0$; there is a trade-off between lower water availability and lower energy demand.

Scenarios 3 and 6 can be rejected right away as energy demand increases while water availability decreases. Scenarios 4, 7 and 8 are acceptable as either water availability increases without an increase in energy demand or energy demand decreases without a reduction in water availability. Scenarios 1, 2 and 9 require a trade-off between increased water availability and reduced energy demand. Scenario 5 is a neutral situation where neither water availability increases nor energy demand decreases.

6.5.1.2 Use of Renewable Energy

Most of remote off-grid mines use diesel for power generation, which is expensive and generates greenhouse gases. Diesel generated electricity costs over 30¢/kWh in a typical Canadian mine located in the remote regions in Canadian North (Paraszczak and Fytas 2012). Recent technological advances have reduced the cost of installing and operating power generation systems based on renewable energy sources. Generation of power from solar energy, wind energy and geothermal energy can reduce the amount of diesel used by mining companies in remote locations, thereby making mining activities more cost-effective and have lesser carbon footprint. Mining companies have started incorporating the renewable power sources as part of the energy management plans.

There are thousands of inactive or abandoned mine tailings areas around the globe that are spread over hundreds of hectares and can be converted into renewable energy generation sites to produce carbon-free clean electricity (Loftis 2010). Chevron Mining has installed a solar power generating facility in a mine tailings area that can generate 1 MW of electricity at peak output at its molybdenum mine in Questa, New Mexico, USA (Woody 2011). Barrick Gold has spent $50 million in installing a wind power generation facility in the town of La Higuera in the Coquimbo Region of Northern Chile (Barrick 2011). This facility was inaugurated in 2011 and consists of 10 wind turbines that can generate 20 megawatts of power, which is enough to supply the energy needs of 10,000 families. The wind farm can be expanded to 18 turbines, which will generate 36 megawatts of power at an additional cost of $20 million. The Lihir gold mine has installed a geothermal power plant, which is rated at 57 MW and generates 75% of the mine's power requirements (Melaku 2005). This mine, now owned by Newcrest, is located in a remote region in Papua New Guinea. Before the installation of the geothermal plant, diesel-generated power was the only source of electricity, and now there is plan to run the entire mine on geothermal power alone.

6.5.1.3 Use of Energy Storage Technology

The use of renewable energy sources and energy storage technologies can help in reducing energy costs and greenhouse gas emissions. The backup systems can also help in keeping the power supply uninterrupted in case of emergencies. Energy shortage can affect production rate and cause substantial losses. This has led to installation of energy storage technologies in combination with power generation systems. An example of this emerging trend is the installation of a lithium-ion energy storage system at Glencore Raglan Mine by Electrovaya Inc. in 2015 (Electrovaya 2015). The energy storage system was worth $0.7 million. Electrovaya designs, develops and manufactures lithium-ion batteries, battery systems and battery-related products for energy storage, clean electric transportation and other specialized applications. Electrovaya worked closely with Tuqliq Energy Corp., a specialist independent power producer, and Hatch Ltd., a global multidisciplinary management, engineering and development consultancy, for this installation. Glencore Raglan mine located at the extreme limit of Northern Quebec is one of the richest base-metal mines in the world and consumes 50 million litres of diesel annually. Delivery of diesel to remote sites is expensive, and burning of diesel as fuel generates greenhouse gases. A reduction in the amount of diesel used for energy generation can not only reduce fuel cost but also reduce the carbon footprint. The use of lithium-ion batteries for energy storage by pairing them to diesel generators can lead to reduction of diesel usage above 30% as this allows the generators to operate at maximum efficiency, which reduces diesel consumption substantially and lowers maintenance costs.

6.5.2 *Water Management*

Management of water is very important for mining activities. On the one hand, water is needed for mining and mineral processing, and, on the other hand, these activities can have adverse environmental impact on the water sources. Water is required for a range of mining and mineral processing process steps such as grinding, flotation, gravity concentration, dense medium separation and hydrometallurgical processes. Leaching of gold ores requires large quantity of water due to the lower grades of ores. Water is also required for making ore and waste slurries for transportation, cooling systems around power generation and washing equipment used in mining and mineral processing. In addition, water is required for residential usage by people working in the mines and their families. All of these activities can add up to a large amount of water needed for mining. Table 6.12 shows the amount of freshwater and saline water withdrawn by the mining industry in the USA in 2010.

As the availability of water in many mining areas may not be abundant, many countries are adopting the concept of integrated water resources management (IWRM). Such an approach is needed to ensure that mine water needs are optimized and process water is treated to impact on water resources so that clean water is available for future generations.

6.5.2.1 Impact of Mining on Water Sources

Mining activities have the potential to adversely impact groundwater and river water in the area. This can take place due to discharge of mine effluents and seepage from tailings and waste rock. Higher metal concentrations and acidity has been observed in water from many mining areas. The oxidation of sulphide minerals can increase the amount of dissolved metals in surface waters downstream from the mine. Potential impact of mining operation on water sources is shown in Table 6.13.

Table 6.12 Water withdrawal for mining in the USA in 2010 (USGS 2017)

Source	Freshwater (million gallons/day)	Saline water (million gallons/day)	Total (million gallons/day)
Surface water	1130	280	1410
Groundwater	1120	2790	3900
Total	2250	3070	5310

Table 6.13 Potential impact of mining on water contamination (adapted from (Mohapatra and Kirpalani 2017)

Mining method	Potential impact
Opencast mining: excavation not intersecting water table	1. Affecting natural surface water regime 2. Affecting groundwater recharge regime
Opencast mining: excavation intersecting water table	1. Declining of water table 2. Affecting natural springs 3. Affecting natural surface water regime 4. Affecting groundwater recharge regime
Underground mining	1. Shallow aquifers 2. Deep aquifers 3. Affecting natural surface water regime 4. Affecting groundwater recharge regime 5. Affecting groundwater flow direction 6. Drying of upper aquifers

6.5.2.2 Treatment of Arsenic

Potential sources of water contamination from mining activities are acid mine drainage and accidental release of tailings from tailings storage facilities. This can adversely affect the quality of drinking water and the life cycle of plants and animals in the affected area. The corrosion rate of metals present in bridges, railways and ships in the area can also increase due to acid mine drainage. The presence of acid can also increase the concentrations of metals and metalloids in affected water. Among the metalloids, the presence of arsenic is of greatest concern.

Arsenic is found in the form of both inorganic and organic compounds. Organic arsenic compounds are found in seafood and are considered safe as these compounds pass through the body quickly. Inorganic arsenic compounds are considered a health risk. These compounds are found in soils and groundwater. Arsenic has different valencies in these compounds, As(III) being the dominant form under reducing conditions and As(V) being the stable form under oxidizing conditions. These compounds could be naturally occurring or a result of human activities such as mining, smelting and arsenic compounds manufactured for industrial use. Inorganic arsenic compounds have been used in pesticides, paint pigments, wood preservatives and medicines in the past, but are now restricted. Arsenic from these sources can contaminate the water sources and pose a health risk to humans. Arsenic level over 10 parts per billion in water is considered toxic.

The conversion of potentially hazardous arsenic compounds into stable non-toxic phases is the most effective method to reduce the leaching of arsenic in water sources. The chemical fixation of arsenic by iron and iron compounds has been investigated by many researchers. Moore et al. investigated the use of $FeSO_4 \cdot 7H_2O$ for fixation of arsenic in the soil (Moore et al. 2000). When the molar ratio of iron to arsenic was increased from zero to two, the concentration of arsenic in the non-saturated and saturated soil solutions decreased from 554 to 15.4 μL^{-1} and 3802 to 0.64 μL^{-1}, respectively. Garcia-Sanchez et al. investigated the adsorption of arsenic by iron oxyhydroxide and aluminium hydroxide (Garcia-Sanchez et al. 2002). At pH 5, iron oxyhydroxide and aluminium hydroxide had the arsenic adsorption capacity of 76 mg/g and 122 mg/g, respectively. Bang et al. investigated the effect of dissolved oxygen on arsenic removal with elemental iron (Bang et al. 2005). The dissolved oxygen had a significant impact on arsenic removal. In the presence of oxygen, greater than 99.8% of the As(V) and 82.6% of the As(III) was removed at pH 6 after 9 h of mixing. When the dissolved oxygen was removed from the solution by purging with nitrogen gas, less than 10% of the As(III) and As(V) was removed.

The use of nanoparticles for the treatment of acid mine drainage and tailings water has been investigated by mining researchers. Kim et al. synthesized nanosized iron and nanosized magnetite coated with sodium dodecyl sulphate and used them to study the stabilization of arsenic in mine tailings. (Kim et al. 2012). Concentration of arsenic in leaching solution was determined by toxicity characteristic leaching procedure (TCLP). Nanosized magnetite was found to be more effective in arsenic immobilization. This was attributed to the enhanced mobility of magnetite due to the surface coating, which changed the surface charge of the particles and prevented the aggregation of the magnetite particles. This allowed the nano-magnetite particles to move further into soil and enhance arsenic immobilization.

Donget al. conducted experiments to investigate the mechanism of microbial-mediated arsenic mobilization from tailings sediments containing with the addition of nanoparticles (Dong et al. 2014). Additions of three different nanomaterials were investigated: SiO_2, Fe_2O_3 and Fe_3O_4. While the addition of SiO_2 increased arsenic mobilization, the addition of Fe_2O_3 and Fe_3O_4 decreased arsenic mobilization.

6.5.2.3 Cyanide Detoxification

The liquid effluents from the cyanide leach process contain cyanide compounds such as sodium cyanide, metal cyanides and thiocyanate and may also contain sulphur compounds such as sulphides, polysulphide, sulphite and thiosulphate. Relative stability of metal-cyanide compounds and complexes in water is shown in Table 6.14 in the approximate order of increasing stability. There are many different methods for the removal of cyanide from liquid effluents. The most common methods for the removal of cyanide are described below.

1. Natural degradation

   The concentration of cyanide in process effluents can decrease by natural degradation, which can be used to decrease the amount of chemicals needed for cyanide detoxification. In one study in Canada, the concentration of cyanide decreased from 68.7 to 0.08 mg/L by natural degradation during the 6-month period starting from April and ending in September (Schmidt et al. 1981). The degradation of cyanide did not take place during the cold winter months. Natural degradation of cyanide depends on many variables such as the cyanide species in solution, relative concentrations of the cyanide species, temperature, pH, aeration, amount of sunlight, presence of bacteria, pond size, depth and turbulence (Ritcey 2005).

Table 6.14 Relative stability of metal-cyanide compounds and complexes in water (Ritcey 2005)

	Types	Compounds
1.	Free cyanide	CN^-, HCN
2.	Simple cyanide compounds	
	(a) Readily soluble	NaCN, KCN, $Ca(CN)_2$, $Hg(CN)_2$
	(b) Relatively insoluble	$Zn(CN)_2$, CuCN, $Ni(CN)_2$, AgCN, $Cd(CN)_2$
3.	Weak metal-cyanide complexes	$Zn(CN)_4^{2-}$, $Cd(CN)_2$, $Cd(CN)_4^{2-}$
4.	Moderately strong metal-cyanide complexes	$Cu(CN)_2^-$, $Cu(CN)_3^{2-}$, $Ni(CN)_4^{2-}$, $Ag(CN)_2^-$
5.	Strong metal-cyanide complexes	$Fe(CN)_6^{4-}$,$Fe(CN)_6^{3-}$,$Co(CN)_6^{3-}$,$Au(CN)_2^-$,$Hg(CN)_2^-$

2. Ozonation

   Cyanide is oxidized rapidly by ozone. Depending on the stability of the metal complex, the rate of decomposition of complex cyanides varies. Cyanide complexes of nickel, zinc and copper oxidize readily, while iron cyanides do not decompose easily. The rate of decomposition of iron cyanide increases as temperature is increased. The rate of decomposition also depends on the pH of the solution and concentration of metal ions.
3. Bacterial oxidation

   Detoxification of cyanide can also be done by bacterial oxidation. At a temperature where bacteria can thrive, bacterial oxidation can reduce the concentrations of cyanides and other complexes except ferrocyanide to nearly zero. Bacterial oxidation has been used successfully by the Homestake Mining Company in South Dakota (Scott 1984). The temperature was maintained at 50–65 °F. The process uses soda ash as a source of carbon to assist nitrification and phosphorus as a trace nutrient.
4. SO_2 process

   The SO_2 process for destruction of cyanide to cyanate was developed by Inco (Devuyst et al. 1982). Oxidation of cyanide takes place according to the following reaction:

$$CN^- + O_2 + SO_2 + H_2O = CNO^- + H_2SO_4$$

   The solution containing over 50 mg/L Cu is sparged with 2–5% SO_2 in air. Copper is precipitated as the hydroxide at the end of the process. The oxidation of CN is dependent on the pH and temperature of the solution. The oxidation is very slow at pH 5–6, fast at pH 9–10 and nearly zero at pH 11. The oxidation rate increases as the temperature is increased. Instead of sulphur dioxide, soluble sulphite or metabisulphite can also be used for oxidation.
5. Acidification

   The acidification process was originally known as the Mills-Crows process, which was modified in to the AVR process (acidification-volatilization by aeration and reneutralization). The process consists of acidification of the solution containing cyanide using H_2SO_4. The HCN gas is swept by air to an absorber tower where it is contacted with lime slurry. The recovered cyanide reagent is returned to the process, which represents substantial saving in reagent costs. The following reactions take place during acidification and aeration:

$$\text{Acidification}: Na_2Ni(CN)_4 + H_2SO_{4+}xH_2O = Ni(CN)_2.xH_2O + 2HCN + Na_2SO_4$$

$$\text{Aeration}: Na_2Ni(CN)_4 + 2H_2SO_4 = NiSO_4 + 4HCN(g) + Na_2SO_4$$

6. Ion exchange

   Ion exchange resins can be used to recover cyanide from process solutions, which can then be reused. Extensive test work including pilot testing has been conducted at CANMET for cyanide recovery using ion exchange. The cyanides and metal-cyanide complexes were removed using two columns in series containing Amberlite IRA-400 anion exchange resin in the sulphate form in the primary column for the metal-cyanide complex and in the Cu(CN)2 form in the secondary column for the free cyanide.

The main advantages of the ion exchange process are as follows (Gilmore 1976):

1. Total cyanide can be reduced to 0.1 mg/L.
2. Cyanide can be recovered and reused.
3. Thiocyanate (SCN) can be reduced to less than 1 mg/ L.
4. H_2SO_4 used for desorption of cyanide is economical to use and readily available.
5. The bisulphate form of the resin is converted to the sulphate form by water for the adsorption of cyanide and heavy-metal complexes.

The disadvantages of the ion exchange process are as follows (Gilmore 1976):

1. Metals may precipitate within the resin bed and foul the resin.
2. H_2SO_4 requirement is high.
3. Capital and labour costs are high compared to chlorination or ozonation.
4. Technology is sophisticated.
5. As the hydrocyanic acid vapour is hazardous, ion exchange system should be tightly sealed. The plant area should be well ventilated and safety procedures must be followed.

References

Adiansyah, J.S., M. Rosano, S. Vink, and G. Keir. 2015. A framework for a sustainable approach to mine tailings management: disposal strategies. *Journal of Cleaner Production* 108: 1050–1062.

Adiansyah, J.S., M. Rosano, S. Vink, G. Keir, and J.R. Stokes. 2016. Synergising water and energy requirements to improve sustainability performance in mine tailings management. *Journal of Cleaner Production* 133: 5–17.

Arenas, J.M., and M. Méndez. 2002. Aznalcóllar Mine accident in Seville (Spain): lessons learnt four years later. In *Workshop on mine and quarry waste—the burden from the past*, Orta, Italy, May 27–28, 2002.

Azam, S., and Q. Li. 2010. Tailings dam failures: a review of the last one hundred years. *Geotechnical News*, 49–53.

Bang, S., G.P. Korfiatis, and X. Meng. 2005. Removal of arsenic from water by zero-valent iron. *Journal of Hazardous Materials* 121: 61–67.

Barrick. 2011. Barrick inaugurates Punta Colorada wind farm in Chile. http://barrickbeyondborders.com/environment/2011/11/barrick-inaugurates-punta-colorada-wind-farm-in-chile/.

Barrick. 2018a. *Tailings, heap leach & waste management*. Barrick Gold Corporation. http://www.barrick.com/responsibility/environment/tailings-waste/default.aspx.

Barrick. 2018b. *Annual report 2017 (and prior years)*. Barrick Gold Corporation. https://www.barrick.com/investors/annual-report/default.aspx.

da Costa, K. 2017. Corporate accountability in the Samarco chemical sludge disaster. *Disaster Prevention and Management* 26 (5): 540–552.

Das, R., and I. Choudhury. 2013. Waste management in mining industry. *Indian Journal of Scientific Research* 4 (2): 139–142.

Davies, M.P., and S. Rice. 2001. An alternative to conventional tailing management—"dry stack" filtered tailings. In *AMEC Earth & Environmental*, Vancouver, Canada, 10 p.

Devuyst, E.A., B.R. Conard, and V.A. Ettel. 1982. *Pilot plant operation of the Inco SO_2–air cyanide removal process*. Paper presented at 29th Ontario industrial waste conference, Toronto, June 1982.

Dong, G., Y. Huang, Q. Yu, Y. Wang, H. Wang, N. He, and Q. Li. 2014. Role of nanoparticles in controlling arsenic mobilization from sediments near a realgar tailing. *Environmental Science and Technology* 48: 7469–7476.

Electrovaya. 2015. Electrovaya delivers Lithium Ion 2.0 battery system for Energy Storage and Reduction of Diesel Consumption at Glencore Raglan Mine. http://www.electrovaya.com/pdf/PR/2015/PR20150129.pdf.

Garcia-Sanchez, A., E. Alvarez-Ayuso, and F. Rodriguez-Martin. 2002. Sorption of As(V) by some oxyhydroxides and clay minerals. Application to its immobilization in two polluted mining soils. *Clay Minerals* 37: 187–194.

Gilmore, A.J. 1976. The ion exchange removal of cyanide from gold mill wastes for environmental benefit. CANMET report MRP/MSL 76–26. Energy Mines and Resources, Ottawa, Canada.

Gunson, A.J., B. Klein, M. Veiga, and S. Dunbar. 2010. Reducing mine water network energy requirements. *Journal of Cleaner Production* 18: 1328–1338.

Hallale, N. 2002. A new graphical targeting method for water minimization. *Advances in Environmental Research* 6: 377–390.

ICMM. 2015. ICMM 10 principles. http://www.icmm.com/en-gb/about-us/member-commitments/icmm-10-principles/the-principles.

IMO. 2012. *International assessment of marine and riverine disposal of mine tailings*. London: International Maritime Organization.

INAP. 2009. The global acid rock drainage guide. International Network for Acid Prevention. http://www.gardguide.com.

Jones, H., and D.V. Boger. 2012. Sustainability and waste management in the resource industries. *Industrial and Engineering Chemistry Research* 51: 10057–10065.

Kim, K.R., B.T. Lee, and K.W. Kim. 2012. Arsenic stabilization in mine tailings using nano-sized magnetite and zero valent iron with the enhancement of mobility by surface coating. *Journal of Geochemical Exploration* 113: 124–129.

Loftis, B. 2010. Untapped resource: converting mine sites into renewable energy assets. *Engineering & Mining Journal* 2010: 50–56.

Lottermoser, B.G. 2010. *Mine wastes*. Berlin: Springer.

MAC. 2017. *A guide to the management of tailings facilities*. 3rd ed. The Mining Association of Canada. http://mining.ca/sites/default/files/documents/MAC-Guide-to-the-Management-of-Tailings-Facilities-2017_0.pdf.

Melaku, M. 2005. Geothermal development at Lihir—an overview. In *Proceedings of world geothermal congress 2005*, Antalya, Turkey, April 2005, 1–4.

Miller, G., 2007. By product mercury production in modern precious metals mines in Nevada. https://archive.epa.gov/mercury/archive/web/pdf/byproductmercuryproductioninmodernpreciousmetalsminesinnevada.pdf.

MMSD. 2002a. *Breaking new ground*. London: Earthscan Publications Ltd.

MMSD. 2002b. Mining for the future, Appendix A: large volume waste working paper. Mining, Minerals and Sustainable Development project, International Institute for Environment and Development, London, UK.

Mohapatra, D.P., and D.M. Kirpalani. 2017. Process effluents and mine tailings: sources, effects and management and role of nanotechnology. *Nanotechnology for Environmental Engineering* 2 (1): 1–12.

Moore, T.J., C.M. Rightmire, and R.K. Vempati. 2000. Ferrous iron treatment of soils contaminated with arsenic-containing wood-preserving solution. *Soil and Sediment Contamination* 9: 375–405.

Motaung, S., J. Maree, M. de Beer, L. Bologo, D. Theron, and J. Baloyi. 2008. Recovery of drinking water and by-products from gold mine effluents. *Water Resources Development* 24 (3): 433–450.

Müller, J., and H.E. Frimmel. 2010. Numerical analysis of historic gold production cycles and implications for future sub-cycles. *The Open Geology Journal* 4: 29–34.

Ndlovu, S., G.S. Simate, and E. Matinde. 2017. *Waste production and utilization in the metal extraction industry*, 78. Boca Raton: CRC Press.

Nguyen, M.T., S. Vink, M. Ziemski, and D.J. Barrett. 2014. Water and energy synergy and trade-off potentials in mine water management. *Journal of Cleaner Production* 84: 629–638.

Paraszczak, J., and K. Fytas. 2012. Renewable energy sources—a promising opportunity for remote mine sites? *Renewable Energies and Power Quality Journal* 1 (10): 251–256.

Rico, M., G. Benito, A.R. Salgueiro, A. Díez-Herrero, and H.G. Pereira. 2008. Reported tailings dam failures: a review of the European incidents in the worldwide context. *Journal of Hazardous Materials* 152: 846–852.

Ritcey, G.M. 2005. Tailings management in gold plants. *Hydrometallurgy* 78: 3–20.

Schmidt, J.W., L. Simovic, and E. Shannon. 1981. *Natural degradation of cyanides in gold milling effluents*. Presented at the Seminar on Cyanide and the Gold Mining Industry, Ottawa, January 22–23, Environment Canada.

Scott, J.S. 1984. *An overview of cyanide treatment methods for gold mill effluents*. Presented at symposium on cyanide and the environment, Tucson, Arizona, December 1–4.

UNEP. 2002. A threat to natural resources. United Nations Environment Program. http://www.grid.unep.ch/waste/html_file/16-17_consumption_threat.html

USGS. 2017. The USGS Water Science School. https://water.usgs.gov/edu/wumi.html.

Wilsenach, J., M. de Beer, S. Motaung, L. Bologo, V. Redebe, M. Moalusi, and J. Maree. 2008. Recovery of drinking water and by-products from gold mine effluents via alkali-barium-calcium processing. In *Science real and relevant: 2nd CSIR Biennial Conference CSIR international convention centre Pretoria, 17 and 18 November 2008, 10*.

Woody, T., 2011. Chevron mines solar energy at tailings dump. https://www.forbes.com/sites/toddwoody/2011/04/21/chevron-mines-solar-energy-at-tailings-dump/#2608e91f1b99.

Chapter 7
Financing and Development of New Mining Projects

L. Moreno

There are thousands of mineral exploration companies, and many of them are listed in stock exchanges around the world. These companies usually do not have revenues from mineral production and rely solely on external financing to advance their projects. Major gold producers usually leave the early-stage projects to junior miners and are willing to pay a premium for new projects once the exploration risk is lower and a resource has been defined.

7.1 Challenges in Financing of New Projects

New or greenfield mineral projects carry a high degree of investment risk due to the exploration, financial and development uncertainty. In addition to the inherent exploration or resource discovery risk, new projects that usually take 3–10 years or more to develop are affected by commodity prices and capital market cycles and fluctuations. Large (higher than five million ounces) and high-grade gold deposits are increasingly difficult to find, as such, some junior exploration companies have ventured to less traditional mining jurisdictions and are facing higher country risk. In fact, gold global production is projected to decline after 2021, and West Africa is expected to emerge as the second largest gold-producing region after China (S&P Global Market Intelligence 2018). In recent years, some gold majors have faced production declines; for instance, Barrick Gold Corporation, the world's largest producer, has seen its production fall by 30% between 2008 and 2017 (Fig. 7.1). A significant portion of Barrick's gold production is from mines in Nevada and South America, but in 2018 the company merged with Randgold Resources, which has gold mines in less stable countries such as the Democratic Republic of Congo (DRC).

L. Moreno (✉)
Tahuti Global Inc., Toronto, ON, Canada

V. I. Lakshmanan, B. Gorain (eds.), *Innovations and Breakthroughs in the Gold and Silver Industries*, https://doi.org/10.1007/978-3-030-32549-7_7

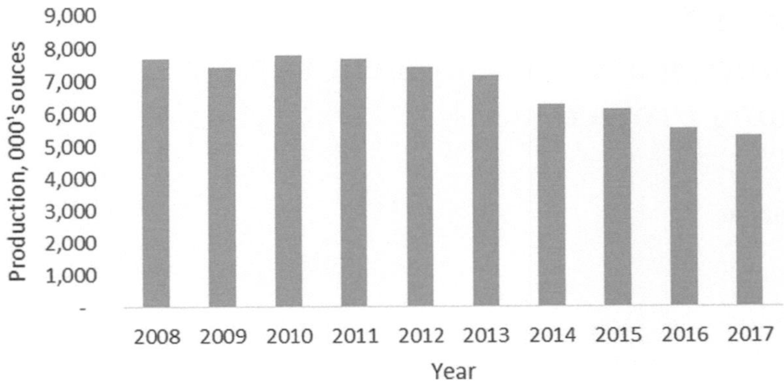

Fig. 7.1 Barrick gold production (*Source*: L. Moreno; data from Barrick)

Other challenges that mining companies are facing are the development of low-grade polymetallic gold deposits containing refractory and complex gold mineralogy that have higher processing risk. Moreover, the pursuit of deep gold deposits is also presenting higher operational and technology risk.

7.1.1 Country Risk

Country risk in the mining sector can be assessed on the basis of the level of political instability, economic uncertainty, violence or terrorism, corruption, and even armed conflicts or war. Countries afflicted by armed conflicts as expected are also negatively affected economically and politically. It should be noted, however, that international mining companies still operate in these countries despite the conflicts. One highly active mining jurisdiction that has been plagued by armed conflicts and political instability for decades is the Democratic Republic of the Congo (DRC). Since independence from Belgium in 1960, the DRC has faced numerous ethnic wars and political conflicts. However, the DRC currently supplies 60% of the world cobalt and is the largest copper producer in Africa with 4% of the global production. Some of the companies operating in the DRC include the Canadian companies Lundin Mining Corp., Anvil Mining Ltd., and Katanga Mining Ltd., the South Africa companies African Rainbow Minerals and AngloGold Ashanti, the British company Randgold Resources Ltd., the Chinese companies China Molybdenum Co., Ltd., and the Swiss company Glencore plc, among dozens of others. Afghanistan has had armed conflicts for several years too, but despite the conflicts, the United States and China have shown a great deal of interest for the country's resources. For instance, the U.S. Geological Survey (USGS) and the U.S. Task Force for Business and Stability Operations (TFBSO) entered into an agreement with the government of Afghanistan to study and assess their fuel and mineral resources; the study took place between 2009 and 2011. The report that was published in September 2011 by the Americans showed that Afghanistan has significant mineral potential (Peters 2011).

Chinese companies have bid for several concessions in Afghanistan, and in 2008 Metallurgical Group Corp. (MCC) and Jiangxi Copper secured the highly coveted Aynak copper deposit and are developing an open pit copper mine; however, the local conflicts are causing project delays. Indian companies have also bid for copper and gold projects in Afghanistan, and several Western companies have participated in the bidding for Afghan mineral projects.

Junior mining companies engaged in exploration are also found in the DRC, Afghanistan, and other conflict areas, but they are more vulnerable in these conflict areas as they usually have less political and economic influence. Corruption is also a major problem afflicting companies with new projects in developing countries. There are reports of company's exploration data being shared by geological survey officials with artisanal miners, who in turn invade exploration areas to mine near the surface gold, thus disrupting the exploration work. There are also reports of villagers colluding with corrupt police to demand money from exploration companies that want to access exploration areas during grab sampling, trenching, drilling, or surveying work. In some countries, junior mining companies sometimes are not able to renew exploration licenses, when corrupt officials illegally allocate their licenses to other interested parties for personal gain. Higher country risk also affects the ability of mining exploration companies to raise funds, as financiers usually add a premium to their lending rates, thus increasing mining companies' borrowing costs.

7.1.2 Resource Nationalism

A common theme in developing nations that are rich in natural resources is what has been coined resource nationalism and may be defined as the effort of resource nations to realize higher economic benefits (profits, jobs, etc.) from their natural resources by applying protective laws and regulations. It is not actually confined to the developing nations, but mining regulations are more volatile and subjected to sudden changes in those regions.

Proposals for changes in mineral codes are most common during elections and when prices and demand for commodities are high or projected to increase. For example, Zambia, a top producer of copper in Africa, has been in the stage of several changes in mining taxes and regulations in recent years. For example, following the close 2006 election, the ruling party in Zambia decided to increase income taxes on mining companies from 25 to 35% and royalties from 0.6 to 3% (Fjeldstad et al. 2016). In 2012, the newly elected government increased royalties again to 6%. Because of accounting manipulations and legacy tax incentives to mining companies, most of the economic benefit that Zambia has received from the mining sector has been from royalties. Therefore, in 2014, the Zambian government decided to scrap the income tax altogether and increase the royalty rate to 20%, compared with the usual royalty rates of 1–8% in the rest of the world, at that time. In 2015, following the death of then Zambian President and much lobbying and threats from major companies to suspend operations, the Zambian government decided to change the income tax rate from 0 to 20% and royalties from 20 to 9% and 6% for open pit and underground mines, respectively. More recently in September 2018, and because of increasing fiscal deficits and high external debt risk, the Zambian government announced plans to increase mining royalties yet again and replace value added tax with a nonrefundable sales tax; these regulatory changes, if enacted, would be the tenth changes in Zambia's mining code in 16 years (Mitimingi and Matthew 2018). Indonesia is currently the world's largest producer of nickel, a top-ten global producer of copper and bauxite and the fifth largest global producer of thermal coal. The country also produces zinc, gold, silver, tin, lead, and manganese. In 2012, Indonesia chocked the world by announcing major changes in the mining code, which limits the exports of unprocessed mineral ore, rocks, and coal, and requires mining companies to build domestic value-add processing facilities. Indonesia also imposed restrictions on mining companies' ownership of the country's mineral assets to 51% or less, giving the foreign companies a set divesting period between the 5th and 10th year of production (Devi and Dody 2013).

Several countries are also introducing (or increasing) free-carried interest ownership allocation for the central government or, in some cases, imposing that a free-carried interest is given to companies owned by nationals. The free-carried interest provisions make governments (and/or native-owned companies) nondilutive minority shareholders who do not need to contribute to the development of the mineral asset. For example, in 2017, Tanzania introduced a free-carried interest of 16% for the government, and in early 2018, the DRC increased the minimum government free carried interest share from 5 to 10%. It should be noted, however, that there are instances where the government of the DRC has been able to negotiate higher free carried interest through their national mining company, as much as 17.5% (The Globe and Mail, April 11 2007). South Africa, which requires resource companies to give 26% ownership to native South Africans, is considering increasing the native ownership to 30% and include a 5% free-carried interest. In some countries, the changes in regulations are targeting procurement and constructions contracts and taxes on expatriates' incomes. These are only a few examples of recently proposed changes in the mining code of developing nations. However, the announcements of aggressive mining taxes or regulations tend to have (in the near term) a negative effect on the mining and exploration companies operating in these regions, affecting their cost of borrowing, stock price, and ability to raise equity funds to advance their projects. Changes of country's mineral codes in recent years are not only limited to developing nations. For example, in Canada, the provincial government of Quebec passed a new mining bill in 2013 (Assemblee Nationale 2013), which requires mining companies to include in their final feasibility report and mining lease applications the option of value addition of minerals in Quebec. The idea is to keep more of the wealth created from Quebec minerals in Quebec. In Australia, the government introduced the Minerals Resource Rent Tax in 2012, in an attempt to increase profits from the mineral sector. Despite the recent changes in mining codes in Canada and Australia, Western nations tend to have more stable mining codes.

7.1.3 Technology and Process Metallurgy Risk

As mentioned in previous sections, large and high-grade gold deposits are increasingly difficult to find, forcing prospectors to pursue the development of difficult gold deposits such as those that are resistant to the common cyanite process, have complex mineralogy, are low grade, and yield lower recovery rates. Motivated by higher gold prices that have stayed above $1050 per ounce (between 2009 and 2018), gold prospectors have been trying to find economic ways of processing these fine grain, refractory gold ore, and low-grade deposits.

European Union nations are increasingly looking at developing their own resources, partially motivated by the growing resource nationalism around the world. However, in the case of gold, there have not been any major discoveries in Europe and many of the deposits and historical mines have low resources, are deep and costly to operate, and require complex extraction methods (Lehrberger 1995). To support the development of the gold mining industry, the European Union has sponsored several initiatives like PolymetOre, which is an industry-led initiative that includes mining companies, chemical companies, universities, and private and government laboratories. A breakthrough in extraction metallurgy of complex gold ore should increase world resources and production.

7.1.4 Economic Events and Commodity Markets

A close analysis of historical commodity prices reveals that commodities follow super cycles that run for about 25–30 years (Erdem 2016), and shorter cycles that are influenced by economic events or crisis. Since the beginning of this century, there have been two major global economic downturns: the 2000–2002 tech-bubble (also known as the dot-com crisis) and the subprime mortgage crisis that started in 2007, lasted for about 2 years, and disrupted the global banking system. Figure 7.2 shows the 18-year historical gold prices from 2000, with the recession periods of 2000 and 2007 highlighted in grey. Although gold is usually seen as a safe haven during recessions, it was also negatively affected during the downturn periods, possibly because some investors might have been forced to sell gold to cover losses on other investments.

The price of commodities is an important factor in determining the valuation and feasibility of a mining project. Therefore, during recessionary or low commodity price periods, most junior mining companies face financing challenges as the value of their projects and company's share prices fall with decreasing metals prices. Gold prices reached US$2000 per ounce in 2011 but have decreased to about US $1270

Fig. 7.2 Gold prices between 2000 and 2018 (*Source*: Macrotrends. The price on the *y*-axis is in the logarithm scale and prices are inflation adjusted)

per ounce, in 2018. Many of the projects that were feasible at the gold price per ounce of US$2000 may have to be optimized or may no longer be feasible at lower prices. As it takes years to develop mining projects, the management team of junior companies must be strategic in determining how much funds to raise and the speed of project advancement/spending, depending on the health of the financial markets.

7.2 Novel Strategies in Project Financing

Privately owned junior mining companies with no revenue have limited funding options, but those listed in stock exchanges are open to the broader public, tend to have more funding options, and often attract speculators (early-stage investors who invest in highly risk assets in anticipation for great returns in the long term). The ability of a junior mining company to raise funds depends on several factors, including:

- Stage of development of the mineral asset(s)
- Management track record and capital markets experience
- Coverage by analysts
- Stock liquidity (the liquidity of a company stock is determined by how fast it can be sold or bought in the market, without significantly affecting the current price of the stock)
- Market interest in the metal being explored (metal flavor of the season)

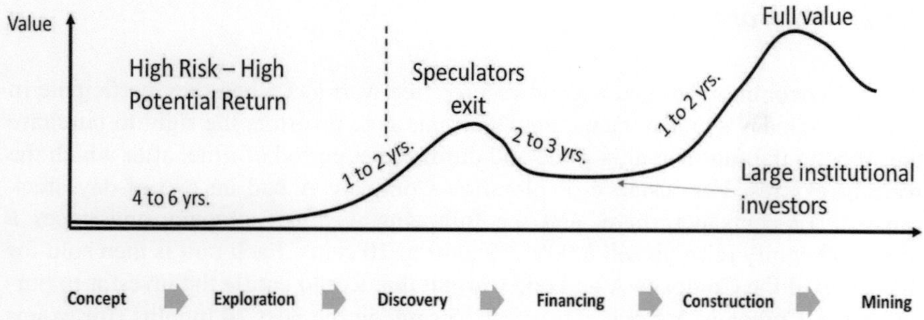

Fig. 7.3 Investment cycle of a junior mining company (*Source*: L. Moreno)

Companies with a small market capitalization (less than $15 million), with low trading volume, with inexperienced management who are not able to engage investors or investment bankers, and are exploring out of favor or uncommon metals (i.e., scandium), face huge financing challenges. On the other hand, exploration companies with advanced projects with a defined mineral resource usually have a higher stock price and higher market capitalization and are more likely to raise funds, if markets are favorable (Fig. 7.3).

Most countries have a stock exchange but the most active stock exchanges for junior mining companies are the Toronto Venture Stock Exchange (TSXV), the Australian Stock exchange (ASX), and London's Alternative Investment Market (AIM). Some companies list their company shares in more than one stock exchange.

7.2.1 Equity Financing

Junior mining companies usually raise funds by selling a percentage of their company to investors in return for cash. In the case of companies listed in the stock exchange, they sell a percentage of the company to the public by "issuing" shares (also referred to as equity or stock). When the public purchase shares of a company in the stock market, they get "a share" ownership of the company.

The funds that junior companies can raise through equity are limited by the value of the company. As such, juniors tend to raise relatively small amounts of money to preserve the share capital of founders and existing shareholders. For example, if a junior company is only worth one million dollars, founders would have to raise less than $500,000 if they want to keep control. In certain situations, companies listed in stock exchanges may require special shareholders' approval to issue and sell more than 25% of the company's shares (TSX Group Inc. 2013).

7.2.2 Warrants

It is not uncommon for investors to receive free warrants when they participate in mining company's equity financings. Warrants give investors the right to purchase the stock in the future at a fix price and during a set period of time, after which the warrant expires. For instance, exploration Company A had an end-of-day stock price of 12 cents per share, and the following day the company announces a $500,000 equity raise, to sell 5,000,000 units at 10 cents. Each unit is then sold for one share of the Company A and one warrant that would entitle the investor to purchase one Company A stock at price of 15 cents in the next 24 months (for example). So, in addition to purchasing the stock at a discount price of 10 cents, those participating in the offering also get a free warrant that gives them the right to purchase the stock at a future date. The warrants would have an intrinsic value and investors could sell the warrants or hold it until the price of the stock rises above 15 cents.

7.2.3 Project Financing

Project financing offers investors a direct ownership in the project (or mineral asset) as opposed to ownership of the company shares (equity financing). The most common sources of project financing are usually private equity firms, commodity funds, trading houses, or major mining companies. Partial ownership of a mineral asset by end users is also becoming a trend, particularly in the strategic materials space. For instance, some companies in the automotive sector have purchase interest in mining projects in the attempt to secure strategic raw materials; they sometimes request one or more positions on the board of directors of the company.

7.2.4 Debt Financing

Financing via the issuance of loans or debt is not a common option for junior mining companies with no revenues from operations. Apart from the lending schemes commonly available to small businesses like credit cards and lines of credit, most junior miners are not able to raise long-term debt to support exploration expenses. However, there are exceptions, where junior mining companies are able to secure long-term debt facilities (Critical Elements Resources 2018), sometimes by using the mineral deposit as collateral. Private equity firms, end-users, and commodity trading houses are the likely funding sources to collateralize debt of junior mining companies.

Mineral exploration companies with advanced projects and solid bankable feasibility studies usually use a combination of debt and equity to fund construction and commission of the future mine. These companies use debt-to-equity ratios like 50/50 or 70/30, but existing shareholders usually prefer funding structures with high debt-to-equity ratio to limit stock dilution (i.e., a decrease in price per share resulting from an increase in the number of total shares). Major mining companies have better access to debt financing, and when commodity markets are underperforming and there is limited upside in their stock price, debt financing may become the cheapest or the only financing option. In addition to borrowing funds from commercial banks, mining producers sometimes issue corporate bonds, which are debt instruments issued by the company and sold to investors. When the company issues bonds, the interest payments are made to investors (instead of banks), who also receive the principal amount at maturity. There are, however, hybrid debt instruments like convertible debt that are used by mining companies including those in the exploration phase.

7.2.5 Convertible Debt Instruments

A convertible debt instrument is a type of debt that allows investors or the company to convert the outstanding principal amount to a set number of shares. The conversion ratio and the time for conversion are determined at the beginning of the sale. In some cases, the issuer of the convertible debt, that is the company, has the sole power to covert the principal amount outstanding to equity (shares) of the company, but in many occasions the power to convert lies solely with the investors. The debt instrument could include provisions in which interest payments are also convertible to shares. In addition to the shares, investor could also receive equity warrants. Junior mining companies usually prefer to keep the power of conversion to limit their risk of default. Convertible debt is a more expensive financing option for mineral exploration companies and presents a higher risk of dilution (the issuance of new shares increases the total number of shares of a company causing dilution or a decrease in the price per share of existing shareholders) for shareholders.

7.2.6 Royalty Financing

Most countries require that companies exploiting oil and gas or minerals resources make royalty payments to the government once in production. As such, royalties are a well-known concept for mining companies and are usually calculated as a percentage of revenues. Government royalties on minerals usually vary from 1 to 10%, depending on the country and the metal. In some jurisdictions, particularly those with land use regulations that favor resource exploitation, the owner of the land also has royalty rights. Net smelter royalty is another common royalty in mining; it consists of a percentage of the gross revenue after certain costs have been deducted, such as insurance, transportation, mining, refining, and smelting costs, and are

Table 7.1 Examples of mining royalties (*Source*: L. Moreno)

Fixed price	20$ per ounce of gold reserves	Based on exploration results
	50$ per ounce of estimated contained gold in mill throughput	Based on estimated contained gold out of the mill, rather than recovered gold
	$100 per ounce of gold sold	Based on final gold production
Net revenues	% of gross revenues without deductions	Typical government royalties
	% of gross revenues less certain predetermined production costs	Includes, net smelter return royalty
Net profits	% of earnings before interest, taxes, and depreciation	Deductions include operating expenses, e.g., exploration expenses and general and administrative costs
	% earnings before taxes	Deductions include capital and operating expenses, but exclude taxes

usually 1–3%. Net smelter royalties have traditionally been used during the sale of a mining property by a junior miner to a mining operator. Royalty financing in the mineral sector has, however, evolved to a point where investors could lend or make investments in a company in return for royalties on future revenues or profits.

There are different types of royalties, including fixed price royalty, net revenues royalties, and net profit (Table 7.1).

Royalty calculations usually include all the products and by-products of a mine. The benefits of royalty financing include low default risk as payments are contingent on the ability to reach production, as such, if the company is not able to define economic resources to develop a mine, they would have no obligation to pay the royalty. Royalties also have lower issuing costs as intermediate brokers or banks may not be required; future royalty obligation is not recorded on the balance sheet as a liability (are off-balance sheet). Royalties are also nondilutive as no company shares are issued. Royalty deals are available to mining companies at any stage of development, but the earlier the investor enters into a royalty agreement with a junior mining company the least they would pay for the royalty. Royalty agreements remain in place even if the company sells the mineral deposit to another company. They also have an intrinsic value and a royalty deal can be sold or transferred to another investor. There are funds and investment companies that specialize in royalty financing.

7.2.7 Stream Financing

Stream financing is an arrangement that gives investors the right to purchase a portion or the total amount of a specific metal product or by-products of a mine at a set discount price in the future and for a specific term (or mine-life), in return for an upfront lump sum payment. As the name stream suggests, these deals were originally designed to purchase by-product streams, for example, precious metals by-product streams from base metal producers. This financing method is popular among commodity trading houses, investment firms, and commodity end-users who purchase or invest directly in the physical commodities (e.g., gold bars) instead of company equity. The upfront lump sum may have specific provisions specifying the use of funds (e.g., for capital expenditures only) or be left at the discretion of the company.

Currently, junior gold companies with strategic metal by-product (like cobalt) streams may secure stream deals at the early stage of development, given the high interest for cobalt for use on electric car batteries.

7.2.8 Stream Versus Royalty Financing

The main difference between stream and royalty deals is that stream investors receive the physical commodity in return for their investment, instead of cash, and may request the repayment of the upfront lump sum if production is significantly delayed or not feasible. The stream deal provides future cash flow to mining companies from the sale of the product and/or by-products, while the royalty agreement constitutes long-term outflow cash payments to the investor. Moreover, unlike royalty deals, the upfront lump sum from stream deals may be recorded on the balance sheet as debt. Royalty deals stay in place even if the mineral asset is sold to another company, they are treated as an interest on the mining license or mineral resource, but stream deals may be dissolved and the lump sum repaid (with a penalty) by the new owner of the mineral asset or mining license. Stream deals are also more common in later stages of a company's project development, but deals may be secured before a final bankable feasibility study.

7.2.9 Financing Using Tax Credits and Other Government Programs and Guarantees

Governments that are interested in encouraging investments in their countries' mineral sector provide tax and grant incentives to mineral exploration companies. Some of these incentives and deductions can be packaged and securitized by investment banks and mining companies and sold to investors.

For instance, when companies based on domestic tax rules are entitled to payable tax credits as a result of exploration and metallurgy expenses, they can securitize the expected government tax payment. For example, if Company A expects to spend \$500,000 in exploration in a year and would be entitled to \$150,000 in tax credits, they can issue the potential future credit benefit to investors in return for cash, before spending the \$500,000 in exploration and/or receiving the tax credit amount from the government. Company A could issue the tax credit benefit and at the same time offer shares of the company as well as warrants as a premium (sweetener) to investors. The $150,000 would then be recorded as tax receivable asset as well as a liability on the balance sheet of the company, until the tax credit is received from the goverment and paid to investors. This type of investment backed by government tax payments is low risk to investors who receive the full tax credit amount that they lent to the company plus free company shares and warrants that can be sold in the market if the company is publicly listed and the stock is liquid.

Canada has some of the most favorable government incentives for junior mining companies operating in the country. For instance, some provinces offer reimbursement grants to exploration companies of up to 75% for mineral exploration expenditures to a maximum cap amount of C$250,000 (in 2018) (Prospectors and Developers Association of Canada 2018). The grants and the cap amount vary depending on the stage of the development of the project and the Province. A very successful tax incentive in Canada is the Canadian Exploration Expenses (CEE) program that allows mining companies to deduct 100% of exploration expenses and 30% of preproduction expenses against taxable income. However, as most junior companies do not have revenues or income, there is a flow-through provision in the Canadian law that allows investors to claim CEE deductions earned by mining companies, against their taxable income. The CEE program allows junior mining companies to do flow-through financing and raise funds for exploration. Investors receive the promise of the company to spend the flow-through funds in exploration and receive shares of the company and warrants. In the province of Quebec in Canada, companies can claim up to 120% of exploration expenses under the CEE program.

Some countries also offer tax credits and grants against capital expenditures for the development of a mine and processing plant. Thus, if Company B has successfully raised $100 million for capital expenditures to develop a mine and plant and the law entitles them to receive, for instance, 30% tax credits, they can securitize the $30 million tax credits receivable and obtain that amount from investors before actually spending the $100 million estimated capital in full. Companies that are facing costs overruns or delays can use tax credits financing to support development. Countries that offer exploration tax incentives and grants include Canada, Australia, the United States, European Union nations, and China, among others.

7.3 Development of Green Projects: Novel Approaches

The development of mineral projects in developing countries is mostly performed by small-scale miners, and projects usually have a short time to production and carry high production risk because of limited exploration and rudimentary production techniques. The exploration and development process of green projects in the West, particularly projects owned by publicly listed companies that are regulated by the government stock market regulatory agencies, can take 5–10 years or more to develop (Fig. 7.3). Listed companies follow detailed exploration programs that generally involve sampling, air borne survey, mapping, trenching, and then drilling, which is the most expensive part of the exploration program.

Institutional investors and banks expect mineral exploration companies to complete economic studies and define economic minable reserves with a minimum of 10–15 years' mine life, and with positive net asset value (NAV). Economic reports usually include mineral resources and ore reserves, projected capital and operating costs, mine models, metallurgy results, market study, cash flow modeling, among other assets, technical and financial information.

As previously discussed, exploration companies are moving to riskier jurisdictions in search for large and high-grade deposits. However, developing countries usually do not offer the same exploration expense incentives as the Western nations, leaving exploration companies with limited financing options when capital markets are weak.

When faced with tough equity markets and limited local government support, some listed companies are using new approaches to development that includes:

1. *Acquiring or investing in small producing assets from artisanal cooperatives or small-scale miners as a source of cash flow to support the exploration costs of the main asset.*

   A small-scale miner producing 20,000 ounces of gold per year would generate $24 million in revenues at a price of $1200 per ounce; and if operating costs are 60% of revenues, they would generate earnings before interest, taxes, depreciation, and amortization (EBITDA) of $9.6 million. If projects with a similar risk profile are valued at 5× EBITDA, the project would be valued at $48 million, which is a small valuation for stock market investors. However, if the company could get, for example, $2 million in retained earnings from a small gold operation, it could invest the funds toward the exploration of another project that has the potential for a large resource.
2. *Increase metallurgy expense at early stages of the project to lower project processing risk.*

   Traditionally, mining companies are valued on the basis of the size of their resource and in situ value of the metal in the ground. Therefore, exploration companies use to spend most of the funds defining a resource and then upgrading the resource to reserves by performing metallurgical tests, market studies, etc. Some companies with complex mineralogy are opting to spend more funds toward pilot studies at the early stage of development before exhausting large funds maximizing mineral resources that may be too expensive to process.
3. *Explore the economic viability of by-products at early stage.*

   With the emergence of renewable energy technologies, mobile devices, and electric vehicles, there is an increasing demand for minor metals such as cobalt, indium, and bismuth, among other uncommon metals that are usually mined as by-products. If the extraction viability of the minor metals is proven at the early stage of development, it could attract the interest of electric metals investors and increase the valuation of the mineral asset, making it potentially more attractive to a broader variety of investors.

4. *Engage royalty and stream investors as early as possible as alternatives to equity and convertible debt financing.*

When commodity markets are weak or volatile, it affects the ability of junior mining companies to raise funds via the issuance of shares. Royalty and stream investors tend to take a long-term investment position and offer nondilutive financing options to junior companies.

References

Assemblee Nationale. 2013. An Act to amend the Mining Act (2013 Chapter 32) [Article]//First Session Fortieth Legislature—Bill No 70.

Critical Elements Resources. 2018. Management discussion and analysis. For the nine-month period ended May 31, 2018.

Devi, Bernadetta, and Prayogo Dody. 2013. Mining and development in Indonesia: an overview of the regulatory framework and policies. International Mining for Development—Action Research Report.

Erdem, Fatma Pınar. 2016. BHP Billiton Group. *Central Bank Review* 16: 137–142.

Fjeldstad, Odd-Helge, Fundanga Caleb, and Rakner Lise. 2016. The rise and fall of the mining royalty regime in Zambia, CHR. Mtchelsen Institute (CMI) Brief, Vol. 15, No. 2.

Lehrberger, G. 1995. *The gold deposits of Europe. In pre-historic gold in Europe*. Dordrecht: Kluwer Academic Publishers.

Mitimingi, Taonga Clifford, and Hill Matthew. 2018. Zambia to raise mine royalty in 10th tax change in 16 years, Bloomberg. https://www.bloomberg.com/news/articles/2018-09-28/zambia-to-raise-mine-royalties-in-bid-to-trim-budget-shortfall.

Peters, Stephen G. 2011. Summaries and data packages of important areas for mineral investment and production opportunities in Afghanistan, USGS.

Prospectors and Developers Association of Canada. 2018. Financial incentives for mineral exploration and prospecting in Canada. In *PDAC 2018*.

S&P Global Market Intelligence. 2018. Gold pipeline report.

The Globe and Mail. 2007. Lundin strikes $1.4-billion deal for Tenke Mining. Vancouver.

TSX Group Inc. 2013. Toronto Stock Exchange Company Manual.

Chapter 8
The Economics of Gold and Silver

V. I. Lakshmanan, A. Ojaghi, and B. Gorain

8.1 Supply and Demand

8.1.1 Gold Supply

The total supply of gold grew marginally by 1% in 2018, up from 4447.2 to 4490.2 ton in the world. This growth was supported by similar year-on-year increases in global mine production to a new record high and recycled gold.

Gold mining is a global business with operations on every continent, except Antarctica, and gold is extracted from mines of widely varying types and scale.

At a country level, China was the largest producer of gold in the world in 2018 and accounted for around 12% of total global production. For many years, China has been the top-producing nation. Its production fell by 6% in 2018 for the second consecutive year to 404.1 tons due to escalated efforts by the government to fight pollution and raise environmental awareness. However, the production is expected to pick up again in 2019 because of several mine upgrades at existing projects.

Australia's gold producers have capitalized on a surging price for the precious metal by delivering an all-time production record last year. With the production of 314.9 tons of gold in 2018, the gold-producing companies broke the previous record of 314.5 tons. The minerals industry produces over half of Australia's total exports and generates about 8% of GDP.

A massive 83% of European gold comes from Russia, which has been increasing its production every year since 2010. The Russia nation increased output by 25 tons

V. I. Lakshmanan · A. Ojaghi
Process Research Ortech Inc., Mississauga, ON, Canada

B. Gorain (✉)
Hindustan Zinc Limited, Udaipur, Rajasthan, India
e-mail: Barun.Gorain@vedanta.co.in

V. I. Lakshmanan, B. Gorain (eds.), *Innovations and Breakthroughs in the Gold and Silver Industries*, https://doi.org/10.1007/978-3-030-32549-7_8

in 2018 and reached 297.3 tons. The Russian government is the largest buyer of produced gold in Russia, which purchases around two-thirds of all gold, produced locally.

Gold output fell by 15 tons in the United States in 2018, and 221.7 tons of gold was produced, after three consecutive years of annual increases in production from 2015 to 2017, due to lower-grade ore mined at CC&V, Long Canyon, and Twin Creeks, as well as lower leach tons placed at Carlin, Phoenix and CC&V, partially offset by higher-grade ore milled at Phoenix, Arizona. Production was supported by project ramp-ups at the Long Canyon project in Nevada and the Haile project in South Carolina. Around 78% of American gold comes from Nevada alone.

Canada inched up two spots on the list of top gold-producing countries in 2017, with the production of 175.8 and 189 tons of gold in 2017 and 2018, respectively. Seabridge Gold stumbled upon a significant goldfield in northern British Colombia after a glacier retreated and is estimated to contain a whopping 780 metric tons. This could be a source of increased output in the coming years. Research firm Wood Mackenzie reported that Canada's gold output would increase in the next 5 years to more than 300 tons annually, an 80% growth from current levels. The rise in production would make the nation the world's second largest bullion-producing country, behind China.

Although gold output fell for the third consecutive year in Peru, by 6 tons, largely because of crackdowns on illegal mining operations in the La Pampa region. Peru remains Latin America's leading gold producer as 158.4 tons of gold was produced in 2018.

Production of gold in Indonesia fell to 136.9 tons, dropping to number 7 on the list of top global producers in 2017. The Indonesian government introduced a tax amnesty program that hoped to repatriate money from overseas, which led to production falling at new main sites as traders were reluctant to remain in the mining industry.

Ghana is Africa's second largest producer of gold and produced 130.5 tons in 2018. Bullion production rose 7 tons over the previous year and accounts for over 20% of the nation's total exports. Ghana was ranked 10th in the 2017 top-ten gold-producing countries that are shown in Fig. 8.1. According to the 2018 production reports, Ghana inched up one or two on the list.

Although production fell few tons from 2016 to 2018, Mexico remains a competitive gold source. Its gold output rose from just 50.8 tons in 2008 to 125 tons in 2018, one of the largest increases in a 9-year span. Mexico is an attractive place for mining due to a relatively low cost of regulation.

Once the top gold producer in the world by a wide margin, South Africa's gold mines have been slowing every year since 2008, with the exception of 2013 when production rose by a few tons. South Africa produced 120 tons of gold in 2018 and dropped down two spots from the world's 2017 top gold producer list (Fig. 8.2).

Gold supply is a combination of mined and recycled gold; mine production is evenly spread across continents, contributing to gold's low volatility relative to commodities. Figures 8.3 and 8.4 show 10-year average gold supply by source and region, respectively.

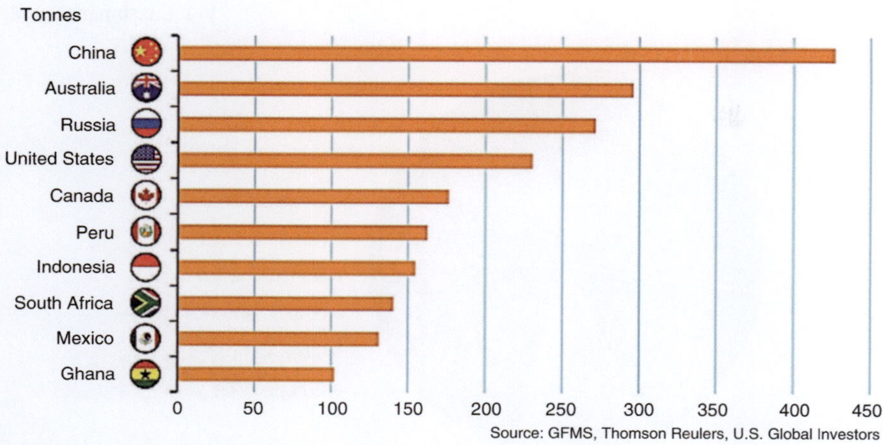

Fig. 8.1 Top-ten gold-producing countries from 2015 to 2018 (Frank 2018)

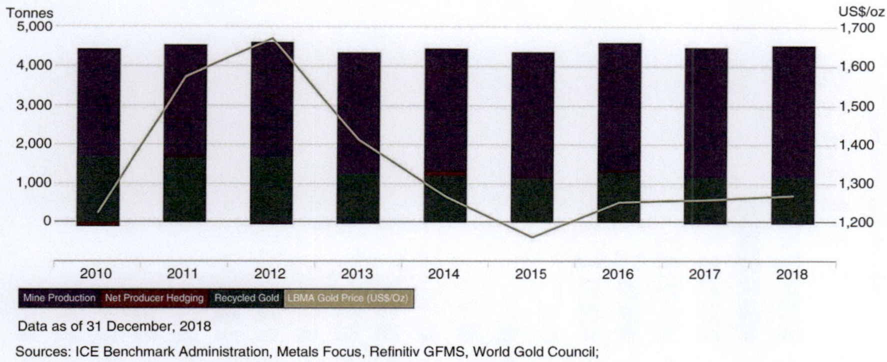

Fig. 8.2 Annual gold supply trend divided by sector from 2010 to 2018 (World Gold Council 2019)

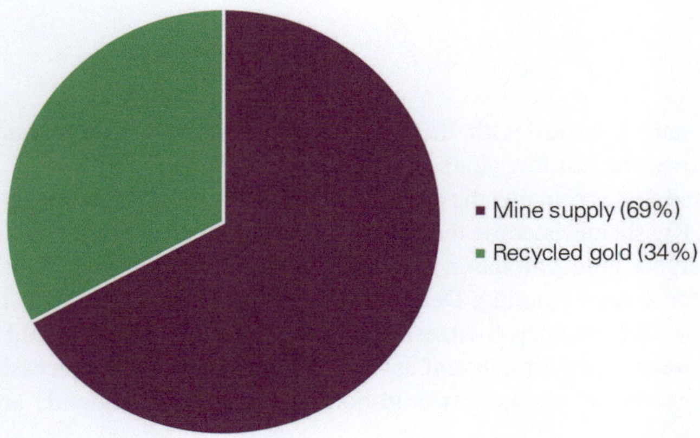

Fig. 8.3 Ten-year average gold supply by source (Gold Hub 2019)

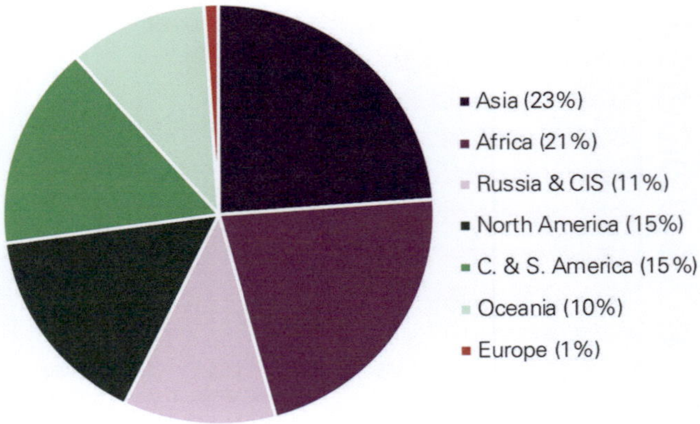

Fig. 8.4 Ten-year average gold-mine production by region (Gold Hub 2019)

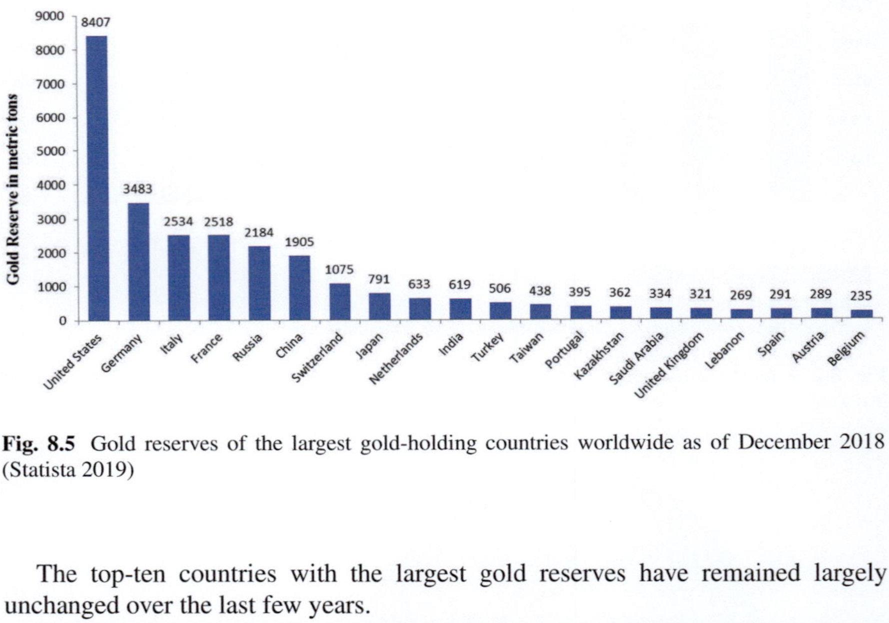

Fig. 8.5 Gold reserves of the largest gold-holding countries worldwide as of December 2018 (Statista 2019)

The top-ten countries with the largest gold reserves have remained largely unchanged over the last few years.

The United States remains the country with the largest gold reserve by a substantial margin. Russia has become the fifth largest country after six successive years of being the largest gold purchaser and, in contrast, Venezuela has been the largest gold seller for 2 years running. Despite being at the sixth place, China mines more gold than any other country. Switzerland has the largest reserves of gold per capital. Gold is a governmental investment asset and safeguards against recession or inflation. Gold reserves of the largest gold-holding countries worldwide are shown in Fig. 8.5.

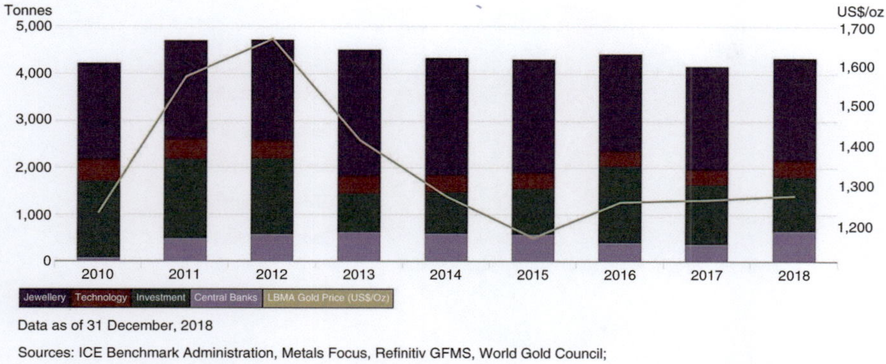

Fig. 8.6 Global gold demand from 2010 to 2018 divided by sectors (World Gold Council 2019)

8.1.2 Gold Demand

Gold's diverse uses in jewelry and technology, and by central banks and investors, mean that different sectors of the gold market rise to prominence at different points in the global economic cycle. This diversity of demand and the self-balancing nature of the gold market underpin gold's robust qualities as an investment asset. Figure 8.6 shows the demand for gold from 2010 to 2018 (World Gold Council 2019).

Central banks added 651.5 tons to official gold reserves in 2018, the second highest yearly total on record. Net purchases jumped to their highest since the end of US dollar convertibility into gold in 1971, as a greater pool of central banks turned to gold as a diversifier (Stephanie and Savage 2019).

As shown in Fig. 8.6, jewelry sector is the major driver on gold demand. Annual jewelry demand was virtually unmoved: down just 1 ton from 2017. Gains in China, the United States, and Russia broadly offset sharp losses in the Middle East. Indian demand was stable at 598 tons.

Exchange traded funds (ETFs) and similar products saw annual inflows of 68.9 tons down from 206.4 tons in 2017. Stock market volatility and signs of faltering economic growth in key markets fueled a global Q4 2018 recovery, but Europe was the only region to see net growth over the year (World Gold Council 2019).

Retail investment in gold bars and coins posted an annual growth of 4%. Coin demand surged to reach a 5-year high of 236.4 tons, the second highest on record. Demand for gold bars held steady at 781.6 tons, which shows the fifth year in succession of holding in a firm 780–800 tons range.

Marginal gains were reached in the volume of gold used in technology by increasing 2 tons in 2018 (Stephanie 2019).

Gold is bought around the world for multiple purposes whether as a luxury good, a component in high-end electronics, safe-haven investment, or a portfolio diversifier. Figures 8.7 and 8.8 show 10-year average gold demand by sources and region, respectively.

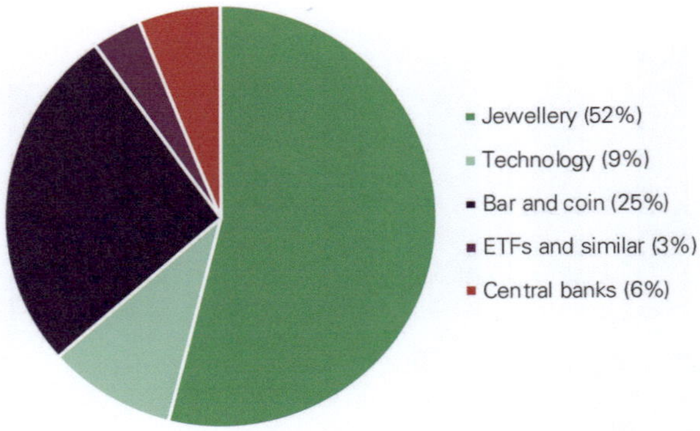

Fig. 8.7 Ten-year average gold demand by source (Gold Hub 2019)

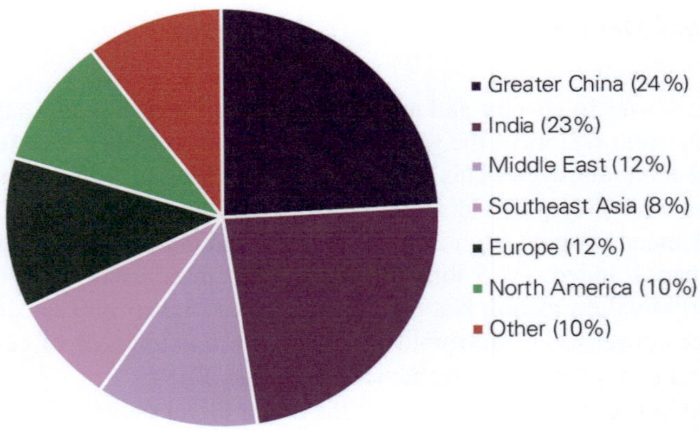

Fig. 8.8 Ten-year average gold demand by region (Gold Hub 2019)

Gold is different from almost any other asset because it appeals to both investors and consumers. Gold demand in 2018 reached 4345.1 tons, up from 4159.9 tons in 2017 and in line with the 5-year average of 4347.5 tons. A multidecade high in central bank buying drove growth, 651.5 tons in 2018, up 74% compared with 374.8 tons in 2017. Demand was bumped up in 2018, but annual inflows into these products (of 68.9 tons) were 67% lower than those in 2017. Investment in bars and coins accelerated up 4% to 1090.2 tons in 2018. Jewelry demand in 2018 was steady at 2200 tons.

Gold used in technology climbed marginally to 334.6 tons in 2018 and increased by 1% from 2017 (Stephanie 2019).

Annual gold demand has a growth of 4% by an increase of 185.2 tons in 2018, and it was driven by highest Central Bank buying in 50 years, as it is shown in Fig. 8.9.

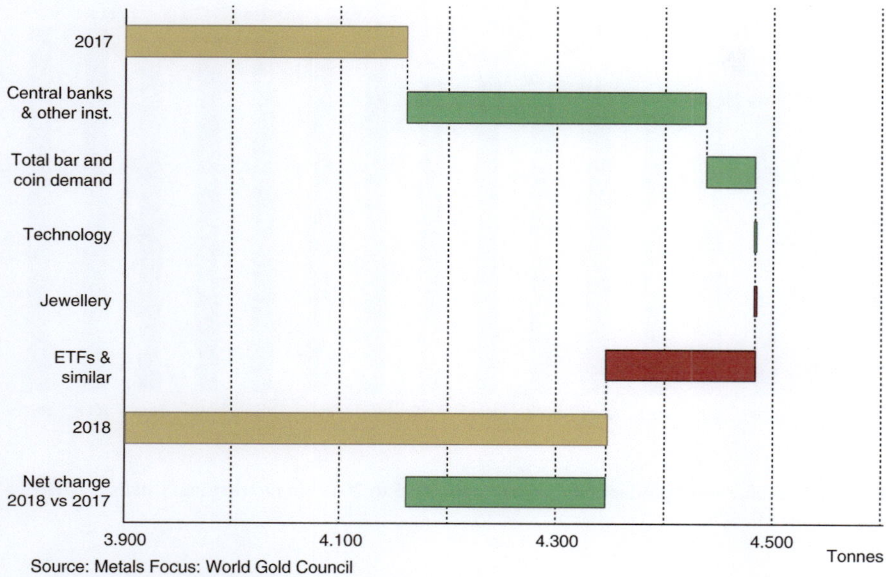

Fig. 8.9 Four percent growth in annual old demand driven by near-record Central bank buying in 2018 (World Gold Council 2019)

8.1.2.1 Indian Gold Market Primer

Unlike those in London and Shanghai, gold prices in India are not determined through a formal exchange. Rather they are determined by local jewelry associations in each state according to local market dynamics. This contributes to a fragmented market: different prices are quoted in Chennai, Mumbai, Delhi, and Kolkata, for example. The local price determination can be either national or state-specific (Market Intelligence 2018).

As India does not produce gold in significant quantities, it must import almost all the gold it consumes. As a result, gold prices in India are heavily influenced by the international gold price. The exchange rate will also affect the price at which importers will be able to source the metal.

As with any imports, taxes and duties are applied when gold enters the country, which will create a differential between the local price and the international price. Gold imports are subject to a 10% import duty and a 3% goods and service tax (GST). India's demand is connected to a broad range of local factors, including wedding seasons, religious festivals, harvests, and the monsoon. The relative strength or weakness of these events can affect demand.

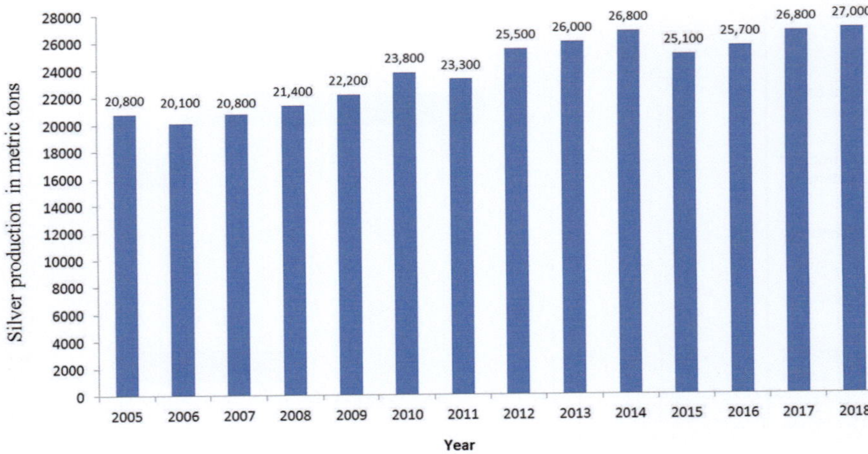

Fig. 8.10 Global mine production of silver from 2005 to 2018 (in metric tons) (Statista 2019)

8.1.3 Silver Supply

Silver mine production is forecast to decline by 2% in 2019 (The Silver Institute 2019). While a small rise from silver recovered in gold mining is expected, all other primary and by-product production is expected to fall, except for supply from lead/zinc operations, which are forecast to rise this year. Global mine silver production from 2008 to 2018 is shown in Fig. 8.10.

World's top-five silver-producing countries in 2017 are as follows:

1. **Mexico**: With the production of 6100 metric tons in 2017, Mexico is the top silver-producing country in the world. The country is home of some of the biggest silver mines in the world. The mines include Penasquito, Pitarrilla mine, and Fresnillo mine. Penasquito is a polymetallic mine containing gold, silver, lead, and zinc. Owned by Goldcorp, the open pit mine started commercial operations in 2010. The Pitarrilla mine, which is estimated to hold measured and indicated mineral resources of 497.3 million ounces of silver, was discovered by SSR Mining in 2002. Located about 60 km north-west of Zacatecas in Mexico, Fresnillo mine is regarded as the world's biggest primary silver-producing mine.
2. **Peru:** The South American nation produced 4300 metric tons of silver in 2017. Antamina mine, which is one of the world's biggest silver mines, is located 200 km from the city of Huaraz in the Andes Mountains of Peru. Silver and lead concentrates are the by-products produced by the copper–zinc mine. The Antamina mine started commercial production in 2001. It is owned and operated by Compania Minera Antamina, a joint venture among BHP Billiton (33.75%), Xstrata (33.75%), Teck (22.5%), and Mitsubishi Corporation (10%). Other major silver mines in the country include Arcata mine, Berenguela mine, Crespo mine, and Huaron mine.

3. **China** produced 3500 metric tons of silver in 2017. The country is also one of the leading consumers of silver in the world. According to a report published by the Silver Institute in September 2018, China's consumption of silver for solar applications reached an estimated 65 million ounces in 2017. In December 2018, the country announced a discovery of a huge silver deposit, which is estimated to contain a minimum of 1500 tons of silver, RT reported. The discovery was made in Henan Province and is part of the Zhonghe deposit area, which is estimated to hold 280,000 tons of lead and 320,000 tons of zinc, according to the Henan Bureau of Geo-exploration and Mineral Development.
4. **Poland:** Producing 1290 metric tons of silver in 2017, Poland stands at fourth spot in the list. The country holds some of the biggest silver mines in the world. The Polkowice-Sieroszowice mine, Lubin mine, and Rudna are the major mines in the country. The Polkowice-Sieroszowice mine is a copper–silver mine, which is owned and operated by state-owned mining enterprise KGHM. According to the KGHM website, production capacity of the mine is 12 million tons of ore annually. Mining at the site is done by using blasting technology with various room-pillar systems with roof settlement. KGHM also owns and operates the Lubin mine, which has been operational since 1968. Rudna mine is also owned by KGHM.
5. **Chile:** With the production of 1260 metric tons in 2017, Chile is one of key silver-producing countries in the world. A significant part of silver extraction in the country comes as a by-product from copper and gold mines. La Coipa gold and silver mine and Escondida copper–gold–silver mine are among the major silver mines in the country. La Coipa gold and silver mine lies within the Atacama region of Chile, while Escondida copper–gold–silver mine is located in the arid, northern Atacama Desert.

World's top-ten silver producers in 2017 are as follows:

1. **Fresnillo plc (Mexico)**: 54.2 Moz

   Fresnillo is the world's largest producer of silver. It is also the second largest producer of gold in the world. Fresnillo operates three gold and silver mines in Mexico.
2. **KGHM Polska Miedź S.A. (Poland)**: 40.0 million ounces (Moz)

   KGHM is a primary copper producer operating three mines based in Poland. The company's silver is refined and sold in the form of bullion and granulate.
3. **Glencore PLC (Switzerland):** 37.7 Moz

   Glencore is a major mining and commodity trading company based in Switzerland. It is also one of the largest companies in the world. The company produces metals, oil, coal, and minerals. It has investments all around the world, including Ecuador, Brazil, and Colombia.
4. **Goldcorp Inc. (Canada)**: 28.6 Moz

   While Goldcorp is primarily associated with the production of gold, it is also one of the largest producers and a growing silver producer. It operates four mines in Canada and one in both Argentina and Mexico.

5. **Polymetal International PLC (Russia)**: 26.8 Moz
   Russia's largest silver producer, the company owns mines in Russia, Kazakhstan, and Armenia.
6. **Compania de Minas Buenaventura S.A.A. (Peru)**: 26.4 Moz
   Buenaventura is Peru's largest holder of mining rights in the country. The company's core asset is the Uchucchacua mine.
7. **Pan American Silver Corp. (Canada)**: 25.0 Moz
   Pan American Silver operates silver mines in Latin America. The company also mines gold, zinc, lead, and copper. It has six operational mines, two in both Mexico and Peru, and one in Bolivia and Argentina each.
8. **Hochschild Mining PLC (United Kingdom)**: 19.1 Moz
   Although headquartered in the United Kingdom, Hochschild operates mines in South America, Peru, and Argentina. The company mines gold and silver.
9. **Volcan Compania Minera S.A.A. (Peru)**: 17.3 Moz
   Peru's largest silver producer, Volcan is primarily involved in processing polymetallic ores, including zinc, lead, copper, and silver. The company has 12 mines, all in Peru.
10. **Hindustan Zinc Ltd. (India)**: 16.9 Moz
    Hindustan Zinc is one of the world's largest producers of zine. It is a miner of not only zinc but also silver, lead, and cadmium.

8.1.3.1 Silver Scrap

Silver scrap supply is forecast to pick up modestly in 2019, following four consecutive years of stable scrap flows. That will be mainly a function of scrap generated not only from industrial processes but also from jewelry items, which tend to be strongly price elastic.

8.1.4 Silver Demand

8.1.4.1 Industrial Fabrication

Silver demand from industrial fabrication, responsible for approximately 60% of total demand, is forecast to rise modestly in 2019. It has expected most sectors to record reasonable growth based on the use of silver in a wide variety of applications. Silver demand from brazing alloys and solders as well as electrical and electric applications is expected to rise again. This is on the back of continued demand from the automotive sector, which uses an increasing amount of applications, such as safety features, window defogging, and infotainment systems, and for electric and hybrid vehicles. Also, growth in the use of silver in a variety of additional sectors such as water purification, chemical applications, LED lighting, flexible electronics and screens as well as antimicrobial applications in textiles was forecast.

Silver is driven significantly more by industrial demand than gold (roughly 50% industrial demand for silver versus only 10% industrial demand for gold).

8.1.4.2 Photovoltaic (PV)

Photovoltaic demand has been expanding considerably in recent years because of various countries stepping up the pace to diversify their energy-generating portfolio away from conventional fossil fuels and toward a higher share of renewable sources. Even with legislative changes in China, coupled with global overstocking and continued attempts at thrifting, PV demand will still be very supportive of silver usage, as many governments will continue to provide incentives to install more solar power. Photovoltaic solar panels remain a critical sector for silver demand as it looks like thrifting has plateaued. Solar power capacity is expected to continue to grow and that is going to have a stronger impact on demand than the effects of thrifting. Indeed, global solar capacity additions are likely to be above 100 GW per year over the 2018–2022 period. Even though growth of solar capacity additions in China slowed modestly in 2018 that slack is expected to be offset by other countries, such as India, Australia, and various European countries in the coming years.

8.1.4.3 Jewelry Demand

Jewelry demand is expected to record a solid year of growth in 2019, with Thailand set to be one of the driving forces behind the rise. In the United States, silver jewelry will remain a popular alternative to lower carat gold items, driven by many issues, but especially female self-purchases. Globally, the demand for silver jewelry is expected to continue to expand, because of its diversity of design, fine quality, and excellent retail margins.

8.1.4.4 Exchange Traded Products (ETPs)

Exchange traded products are forecast to expand by eight million ounces this year. Silver-backed ETPs are "stickier" than other precious metal products, as the majority are held by retail investors, rather than institutional investors.

8.1.4.5 Silver Physical Investment

Silver physical investment demand is likely to increase by approximately 5% in 2019. Bullion coin demand has been strong in the United States during January 2019, and sentiment in Europe were expected, which rose by 6% in 2018, and India to be supportive of global growth. Although the United States saw silver bullion demand drop to unprecedented levels in 2018, the global demand picture was a lot

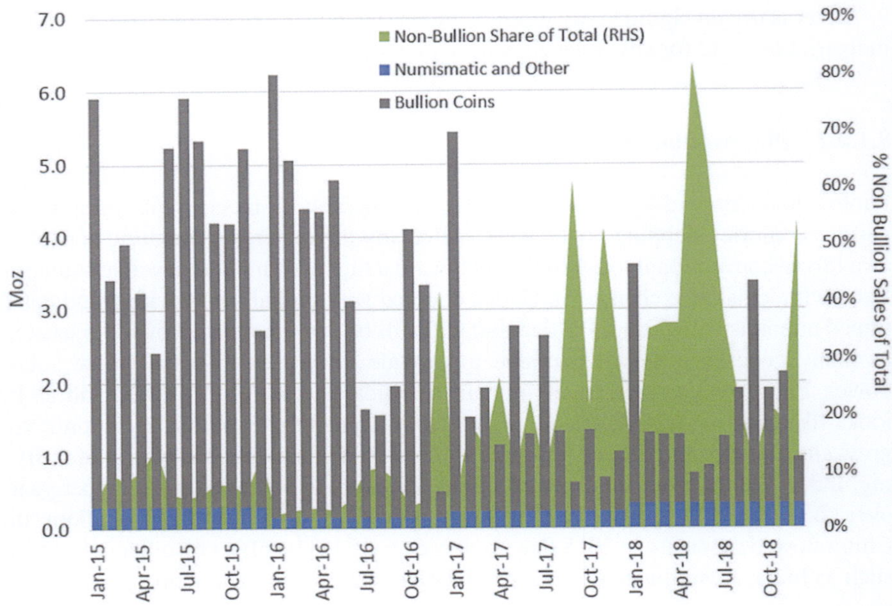

Fig. 8.11 Silver coin fabrication at the US mint: massive increase in nonbullion coin demand (*Source*: U.S. Mint) (DiRienzo 2019)

brighter. Physical silver investment was driven primarily by bar demand, which increased 53% last year. India, in particular, saw exponential growth for physical silver with demand for silver bars jumping 115%.

8.1.4.6 India Market

India is expected to continue to be one of the largest silver consumers in 2019. Silver imports reached nearly 225 million ounces (Moz) last year, which was more than 220% higher compared with 2017. Silver imports fell in India after the implementation of the demonetization plan at the end of 2016 and the introduction of the goods-and-services-tax (GST) in mid-2017, which reduced the amount of cash in circulation that was previously used for silver imports (Fig. 8.11).

8.1.5 Silver Market Balance

The silver market balance (total supply less total demand) in 2019 is projected to be the third consecutive year, within the boundaries of margin, where all the silver produced is absorbed by the various downstream sectors.

8.2 Emerging Markets for Gold and Silver

Physical demand for gold will benefit from resilient growth in emerging markets. Rising income levels and increased gold consumption in emerging economies were likely one of the drivers of the 1999–2011 boom, as China's and India's combined jewelry and gold bar demand increased from 953 metric tons per year in 1999 to 1740 metric tons per year in 2011. Today, demand from emerging markets for gold-based jewelry constitutes 45% of global gold consumption. As emerging economies recover, their currencies will likely strengthen, helping to lift physical gold demand. Acceleration of emerging economies' growth especially relative to the US growth and appreciating emerging market currencies have historically been supportive of higher gold prices.

Environmental, social, and governance issues will play an increasing role in re-shaping mining production methods.

Emerging markets are scooping up gold. Recent data from The People's Bank of China indicate that China boosted its gold reserves to 60.62 million ounces during the month of March, up from 60.26 Moz in February 2019. China has upped its gold reserves by a total of 42.9 tons.

At the same time, Russia is also beefing up its gold holdings. The World Gold Council reports that Russia bought about 274 tons of gold bullion in 2018 valued at about $11 billion. In February 2019, Russia single-handedly accounted for 1 million ounces of gold demand, roughly 6% of the world's total demand.

In addition, demand for gold in India is expected to get a boost from the wedding season each year. Gold is traditionally a popular wedding gift in India, host to roughly 20 million weddings per year with guest lists commonly in the 3000- to 6000-person range.

The gold mining industry will have to grapple with the challenge of producing similar levels of gold over the next 30 years to match the volume it has historically delivered (Magnus et al. 2018).

8.2.1 Trends That Have Reshaped Gold Demand

8.2.1.1 Emerging Markets

The economic development that emerging markets, especially China and India, have experienced for almost two decades has increased and diversified gold's consumer and investor base. As it has been shown in Fig. 8.12, China and India have doubled their gold market share in less than two decades (Fig. 8.12).

The expanding middle class in China and India, combined with broader economic growth, will have a significant impact on gold demand. Physical gold demand is strong in emerging markets, and limited supply leads to higher gold prices.

The use of gold across energy, health care, and technology is changing rapidly. Gold's position as a material of choice is expected to continue and evolve over the coming decades.

Fig. 8.12 Consumer demand and market share of India and China (Perlaky et al. 2019)

8.2.1.2 Gold-Backed ETFs

The advent of exchange-traded products reduced total cost of ownership, increased efficiencies, provided liquidity and access, and brought new interest and demand into gold as a strategic investment (Fig. 8.13).

Mobile apps for gold investment, which allow individuals to buy, sell, invest, and gift gold, will develop rapidly in India and China.

8.2.1.3 The 2008–2009 Financial Crisis

Gold has benefitted from a change in attitudes toward risk and risk management by investors. Following the Great Recession, new markets have appeared and old markets have resurfaced, lifting the demand (Fig. 8.13).

8.2.1.4 Central Banks

The expansion of foreign reserves, led by emerging markets, has resulted in net gold demand by central banks as a source of return, liquidity, and diversification. Central banks have been a steady net source of demand since 2010, led by emerging markets (Figs. 8.14 and 8.15).

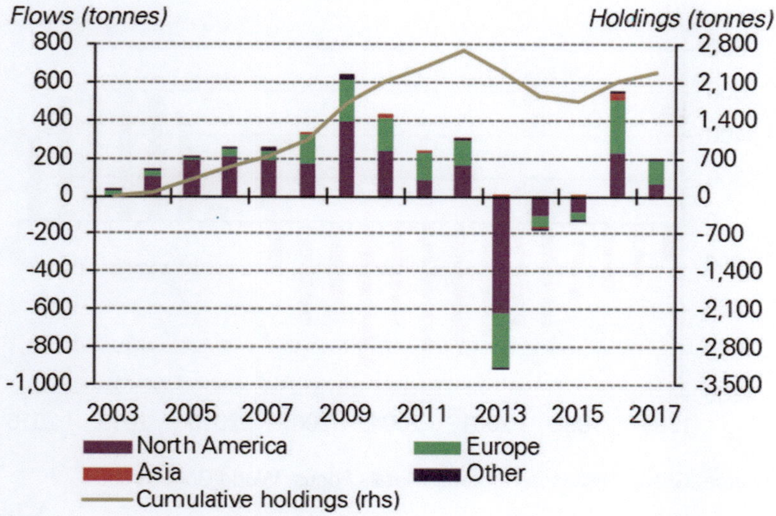

Fig. 8.13 Annual gold demand through ETFs and cumulative holding (Perlaky et al. 2019)

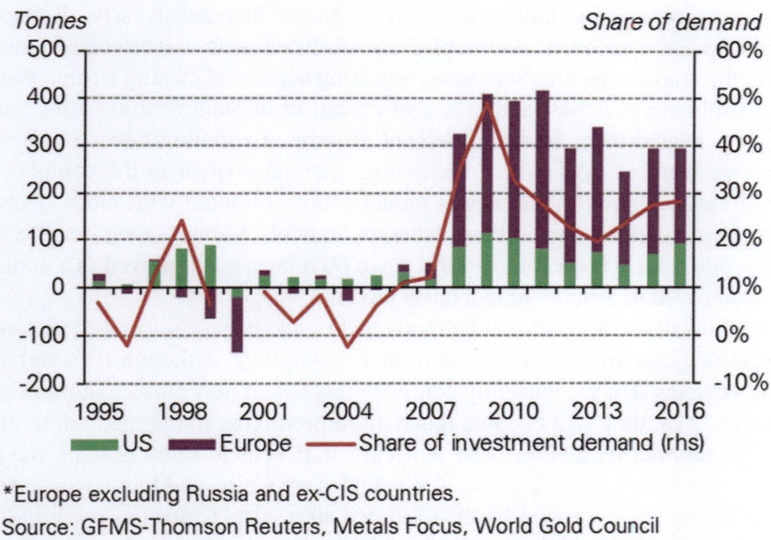

Fig. 8.14 Bar and coin demand and market in the United States and Europe (Perlaky et al. 2019)

8.2.2 *Emerging Market for Silver*

Silver is a key component in automotive manufacturing and is increasingly used in internal combustion engine (ICE) vehicles, autonomous vehicles (AV), electric (EV), and even photovoltaic (PV) technologies, that are creating new applications

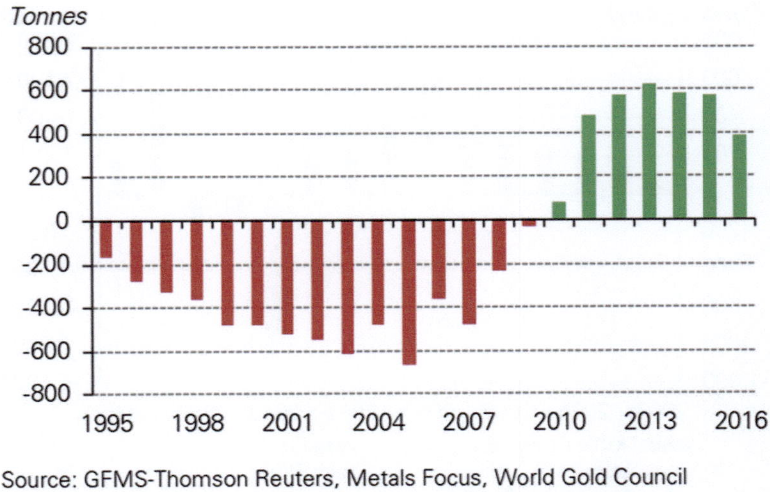

Fig. 8.15 Net global central bank gold demand (Perlaky et al. 2019)

for silver in automotive fabrication, driving future demand. Nearly all electrical connections in a modern automobile are outfitted with silver-coated contacts. Starting the engine, opening windows, adjusting seats, and closing a trunk each use silver membrane switches. Silver is also crucial to infotainment systems, window defogging, heated seats, and luminescent displays. Extending beyond established automotive features, AV and EV technologies employ silver in the complex electronic circuitry. Premium and newer models, those outfitted with more "features," tend to contain more silver than economy models because they require silver switches and relays. It is estimated that up to 60 relays are employed in a luxury car today, compared to around 30 in a basic model.

Growth in silver demand will be motivated by increasing auto sales, particularly in developing countries, and the rise of new technology. Although EVs and hybrid electric vehicles (HEVs) currently comprise just 6% of new global automotive production in 2018, they will become much more prominent in the medium term, bolstered by national regulations that will curb ICE vehicle sales. Among the many countries proposing future bans on ICE vehicles are China and India, whose emerging middle classes are major drivers of global auto sales. China has proposed a ban on sales of new ICE vehicles by 2025, while India has announced plans to become 100% electric by 2030. EVs and HEVs are projected to account for over half of global automotive silver demand by 2040. Silver is used in actuators in robotic and sensing applications, the technology behind autonomous driving. Although autonomous vehicles are often considered to belong to technology of the future rather than present, many vehicles use elements of autonomous technology today. PV technology will be another driver of silver in automotive. Although few EVs contain PV technology today, the technologies go hand in hand. Because of the scarcity of

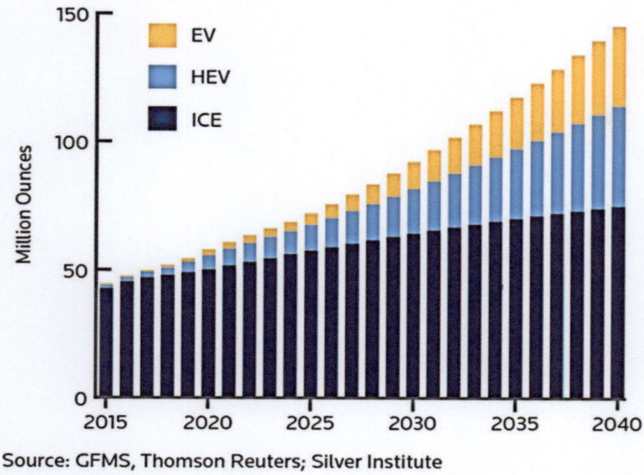

Fig. 8.16 Silver demand in an automotive industry (DiRienzo 2019)

public charging infrastructure and an overlapping client base, particularly in the United States, many EV owners are opting to charge their cars at home using solar technology (Fig. 8.16).

The silver market looks "promising" in 2019 as the supply and demand picture is expected to remain relatively stable compared with 2018.

Global silver demand hit a 3-year high in 2018, surpassing more than one billion ounces; an increase of 4% silver mine production fell for the third straight year, dropping 2% in 2018 to 855.7 million ounces (The Silver Institute 2019).

However, despite strong demand and falling mine supply, silver prices struggled, averaging the year at $15.71 an ounce, a drop of nearly 8% from 2017.

References

DiRienzo, Michae, l. 2019. *Silver market trend*. The Silver Institute.

Frank, E. Holmes. 2018. Top 10 gold producing countries in 2017. www.kitco.com.

Gold Hub. 2019. https://www.gold.org/goldhub.

Magnus, George, Lacaille Rick, and Ash Michelle. 2018. *The next 30 years for gold*. World Gold Council.

Market Intelligence. 2018. *Gold market primer; gold prices*. World Gold Council.

Perlaky, Adam, et al. 2019. *The relevance of gold as a strategic asset*. World Gold Council.

Statista. 2019. https://www.statista.com/statistics/267998/countries-with-the-largest-gold-reserves/.

Stephanie, Mackrel. 2019. World Gold Council. www.gold.org

Stephanie, Mackrell, and Kate Savage. 2019. Highest central bank buying in 50 years drives 4% growth in gold demand in 2018. www.gold.org

The Silver Institute. 2019. Silver market trends. https://www.silverinstitute.org/silver-market-trends-2019/.

World Gold Council. 2019. https://www.gold.org/goldhub/data/gold-supply-and-demand-statistics.

Chapter 9
Recycling of Gold and Silver

V. I. Lakshmanan, R. Roy, and B. Gorain

9.1 Usage of Gold and Silver

9.1.1 Usage of Gold

Gold has been a store of value for major part of human civilization. Gold was known to Egyptians as early as 3100 BCE, and the code of Menes, the founder of the first Egyptian dynasty, describes the value of gold to be two and half times the value of silver. Egyptians also made maps showing the location of gold mines. The Turin Papyrus Map of 1320 BCE shows the plan of a gold mine in Nubia in Egypt. Gold was extracted on a large scale by Romans, who made long aqueducts that enabled them to sluice large alluvial deposits for gold recovery. Now most of the gold is recovered from hard rock ores.

It is estimated that around 190,040 tons of gold has been mined throughout history (World Gold Council 2018). If all the gold existing above ground was put in a cube, the cube would be 21 m on each side. Out of all the gold mined since the dawn of human civilization, approximately two-thirds of this amount has been mined since 1950. As gold is noble and considered precious, almost all of gold mined throughout history is being kept securely or is in use. More than 90,000 tons of gold in the world has been turned into jewellery. Figure 9.1 and Table 9.1 show the different usage of gold.

V. I. Lakshmanan (✉) · R. Roy
Process Research Ortech Inc., Mississauga, ON, Canada
e-mail: llakshmanan@processortech.com

B. Gorain (✉)
Hindustan Zinc Limited, Udaipur, Rajasthan, India
e-mail: Barun.Gorain@vedanta.co.in

V. I. Lakshmanan, B. Gorain (eds.), *Innovations and Breakthroughs in the Gold and Silver Industries*, https://doi.org/10.1007/978-3-030-32549-7_9

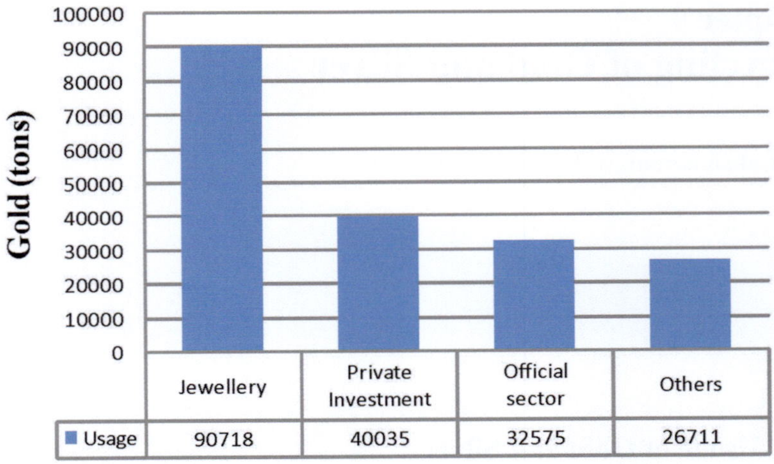

Fig. 9.1 Amount of gold in the world (World Gold Council 2018)

Table 9.1 Total amount of gold in the world at the end of 2017 (World Gold Council 2018)

Usage	Amount (tons)	Percentage (%)
Jewellery	90,718	47.7
Private investment	40,035	21.1
Official sector	32,575	17.1
Other	26,711	14.1
Total above ground stocks	190,040	
Below ground reserves	54,000	

Approximately 2500–3000 tons of gold is produced every year by global gold mining. At current estimates, there are below ground reserves of 54,000 tons of gold, which may increase as a result of future exploration activities.

Table 9.2 shows the supply and demand of gold in the world over last 5 years. While the supply of gold increased slightly from 4340 to 4415 tons between 2013 and 2017, the physical demand for gold dropped sharply from 5434 to 3988 tons between 2013 and 2017 led by drop in demand for retail investment and jewellery. Table 9.3 shows the industrial demand of gold in the world over the last 5 years. Major industrial applications of gold are in electronics and dentistry. Industrial demand for gold has dropped from 428 to 380 tons between 2013 and 2017.

Besides jewellery, investment and monetary applications, the important modern applications of gold are as follows (Forbes 2018):

1. Dentistry: Gold is used for fillings, crowns, bridges and orthodontic appliances since gold is biocompatible.
2. Conducting cables: Gold is an excellent conductor of electricity and does not corrode. High-end data transfer cables have gold plated on conducting ends for

Table 9.2 Supply and demand of gold in the world in tons (GFMS 2018)

	2013	2014	2015	2016	2017
Mine production	3076	3180	3222	3251	3247
Scrap	1303	1159	1180	1306	1210
Net hedging supply	−39	108	21	32	−41
Total supply	**4340**	**4446**	**4422**	**4590**	**4415**
Jewellery	2726	2559	2464	1953	2214
Industrial fabrication	428	411	376	366	380
Official sector	409	466	443	269	366
Retail investment	1871	1162	1160	1043	1028
Physical demand	**5434**	**4598**	**4442**	**3630**	**3988**

Table 9.3 Industrial demand of gold in the world in tons (GFMS 2018)

	2013	2014	2015	2016	2017
Electronics	306	297	267	264	277
Dental and medical	36	34	32	30	29
Other industrial	85	80	76	71	73
Total industrial demand	**428**	**411**	**376**	**366**	**380**

rapid and accurate transmission of digital data from one device to another without distortion.

3. Electronics: A small amount of gold is used in almost every electronic device such as cellular phones and computers due to its conductivity and corrosion resistance. It is typically used on edge connectors used to mount microprocessor and memory chips onto the motherboard.
4. Medicine: Gold has been used in Chinese medicine to treat furuncles, smallpox, skin ulcers and to remove mercury from skin and flesh. Nowadays, gold salts are used to relieve joint pain and stiffness.
5. Gold leaf: The sheets of Gold leaf have been used as outer cover on many monuments such as St. Michael's Cathedral in Kiev, Ukraine and many other temples in the world.
6. Aerospace: Gold-coated polyester film is used to reflect infrared radiation from space vehicles. Gold coating on the visor of astronauts' helmets filters out the sun's harmful rays.

9.1.2 Usage of Silver

Similar to gold, silver has also been a store of value for major part of human civilization. A concentrated effort to mine silver began around 3000 BCE. A 'cupellation' process to extract silver from lead–silver ores was developed by the ancient Chaldeans (modern-day Turkey) around 2500 BCE. Between 1600 and 1200 BCE, silver was mined in Laurium (near Athens). It was around 550 BCE that first silver coins were minted in the eastern Mediterranean area. Standard silver coins were used in the vast expanse of the Roman Empire from third century BCE. It is estimated that the amount of silver in the world is 1,740,000 metric tons (USGS 2018). If all the silver discovered thus far were put in a cube, the cube would be 55 m on each side.

Table 9.4 shows the supply and demand of silver in the world over last 5 years. Total supply of silver has remained close to 31,100 tons between 2013 and 2017. The physical demand for silver has dropped from 35,006 to 31,647 tons between 2013 and 2017 led by a drop in demand for silver coins and bars. Table 9.5 shows the industrial demand of silver in the world over last 5 years. Major industrial applications of silver are in electronics and photovoltaic. Industrial demand for silver has remained close to 18,660 tons between 2013 and 2017.

Silver is now an industrial metal with the use of silver in coins, silverware and jewellery accounting for 13,018 tons in 2017 compared to 18,629 tons for industrial applications. Silver has many unique properties that make it suitable for a multitude of industrial applications.

Silver has high thermal and electrical conductivities. Due to being malleable and ductile, it can be made into fine sheets and drawn out into thin, flexible wires. Silver is also resistant to corrosion and oxidation. The antimicrobial and non-toxic properties of silver have led to its use in medicine and consumer products. With the

Table 9.4 Supply and demand of silver in the world in tons (The Silver Institute 2018)

	2013	2014	2015	2016	2017
Mine production	25,605	26,989	27,838	27,635	26,500
Net government sales	246	–	–	–	–
Scrap	5940	5144	4388	4345	4295
Net hedging supply	−1082	522	243	−588	44
Total supply	**30,708**	**32,655**	**32,468**	**31,392**	**30,839**
Jewellery	6861	7041	7050	6376	6503
Coins & bars	7498	7281	9084	6463	4699
Silverware	1844	1903	1966	1630	1816
Industrial fabrication	18,803	18,545	18,138	17,938	18,629
Physical demand	**35,006**	**34,770**	**36,241**	**32,403**	**31,647**

Table 9.5 Industrial demand of silver in the world in tons (The Silver Institute 2018)

	2013	2014	2015	2016	2017
Electrical and electronics	8273	8207	7651	7274	7554
Brazing alloys and solders	1981	2074	1913	1720	1788
Photography	1571	1508	1449	1406	1368
Photovoltaic	1738	1611	1841	2466	2927
Ethylene oxide	239	156	317	317	215
Other industrial	5001	4995	4970	4755	4780
Total industrial demand	**18,803**	**18,545**	**18,138**	**17,938**	**18,629**

formulation of silver into nano-silver, silver is now being used in many home appliances. As silver is more widely available and cheaper than other precious metals, it has become economical to use silver in many industrial applications. The important modern applications of silver are as follows (Mining.com 2016):

1. Treating warts and corns: A mixture of silver nitrate and potassium nitrate is used in Caustic pencil, which is a common over-the-counter medication for warts and corns.
2. Media storage: DVDs are coated with a fine layer of silver to protect from pitting corrosion and reduce tarnishing.
3. Personal deodorant: Silver chloride is used in personal deodorants due to its antimicrobial properties. It helps to keep skin healthy and delays the onset of body odour by preventing the growth of bacteria.
4. Keeping milk safe: Earlier jugs and jars were coated with silver to keep milk fresh. Now the sealing rings contain silver-based biocide to keep milk safe from harmful bacteria.
5. Long life batteries: A silver-oxide battery uses silver (I) oxide as the cathode, zinc as the anode, and sodium hydroxide (NaOH) or potassium hydroxide (KOH) as an alkaline electrolyte. Silver-oxide batteries are being used in watches, hearing aids, digital cameras and mobile phones.
6. Engines: Ball bearings are electroplated with silver to withstand high temperatures, reduce friction and enhance the performance of many engines, including jet engines.
7. Stained glass: Silver is mixed with clay, gum or turpentine, applied to the back of painted glass and heated in a kiln during the making of stained glass.
8. Electronics and 3D printing: Silver films are being used for flexible screens. Silver nanoparticles are used to make small items by 3D printing.
9. Antimicrobial lab coats: Antimicrobial lab coats embedded with a silver-based compound are being used by doctors and other medical staff to prevent infection from microorganisms such as MRSA and *Escherichia coli*.
10. Touch screen gloves: These gloves have small amount of silver nylon woven in them and are useful for using touch screen cellular phones in winter.
11. Water purification: Ceramic filters are coated with colloidal silver to purify water.
12. Products to stop smoking: Small amounts of silver acetate can be added to products such as chewing gum, mouth sprays and lozenges to aid in smoking cessation. Silver acetate produces an extremely unpleasant taste when mixed with cigarette smoke.
13. Laundry detergent: Silver nano-particles are being used in laundry detergents. When silver ions are released in water, they kill bacteria on clothes and eliminate odours.
14. Plastics: Silver is used within catalysts used by the petrochemical industry in the production of ethylene oxide and formaldehyde, which are key ingredients in making plastics.

15. Wood preservation: Silver is being used in wood preservatives to prevent white-rot decay. Silver solutions are being used to treating wooden countertops in the kitchen to protect from bacteria.
16. Automotive industry: Silver is used in many applications in the automotive industry such as electrical contacts, distance sensors, advanced driver assistance systems for automatic parking and conductive lines on rear window.
17. Novelty explosives: A small amount of silver fulminate is used in novelty explosives known as bang snaps. Silver fulminate is also used in Christmas crackers due to its ability to detonate when friction is applied by pulling the trigger.
18. Photography: Before the advent of the digital age, silver halide was widely used in photographic plates as they are readily reduced by light. It is still being used in X-ray photography.
19. Weather modification: Silver iodide has a molecular structure similar to ice and promotes cloud densification, when released in air. Due to this reason, it is used for cloud seeding.
20. Solar panels: Silver paste is applied to solar panels as silver has the highest electrical and thermal conductivities of all metals. It significantly improves the performance-to-cost ratio of solar cells.
21. Food garnishing: Very thin silver foil is widely used to garnish sweets in South Asian cuisine. It is considered edible and harmless in very small quantities.

9.2 Life Cycle of Precious Metals

Figure 9.2 shows the life cycle of precious metals (He and Kappler 2017). After mining, precious metals are produced by either pyrometallurgical route or hydrometallurgical route. Precious metals are used in industrial, domestic and medical applications. Wastewater from mining, hydrometallurgical processes, and industrial, domestic and medical applications is treated in a wastewater treatment plant to recover precious metals. Municipal solid waste containing precious metals generated from domestic use goes through municipal solid waste incineration. Residue from incineration along with the sludge-containing precious metals is processed by bio-hydrometallurgy to recover precious metals. Rest of the municipal solid waste along with the sludge not containing precious metals goes to the landfill. Slag from pyrometallurgical process, tailings from mining and E-waste from industrial use can also be processed by bio-hydrometallurgy to recover precious metals.

9.3 Recovery of Precious Metals from E-waste

There are many sources of scraps containing gold such as Waste Electrical and Electronic Equipment (WEEE), jewellery, spent dental and orthopaedic materials and spent catalysts (Syed 2012). WEEE or E-waste products include items such as

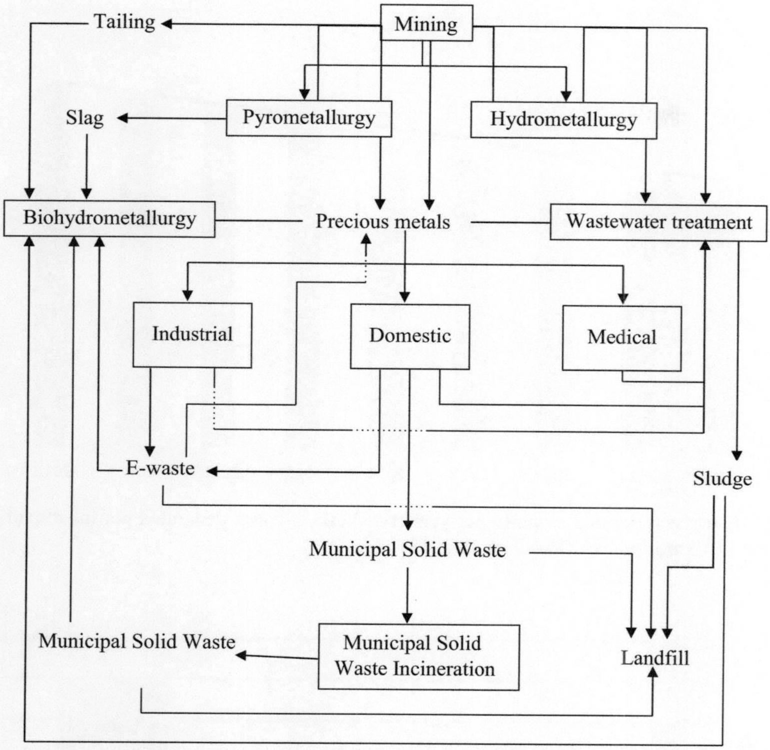

Fig. 9.2 Life cycle of precious metals (Adapted from He and Kappler 2017)

printed circuit boards (PCBs), television sets, refrigerators, personal computers, cell phones and batteries. The E-waste generation has been increasing steadily as shown in Fig. 9.3. It is projected to reach 50 million tons in 2018. The amount of E-waste per capita has also been increasing steadily despite the increase in world population as shown in Fig. 9.3 and is projected to reach 7 kg per capita in 2018. E-waste contains approximately 40% metals as shown in Fig. 9.4. E-waste contains a large number of organic materials, ceramics and metals as shown in Table 9.6.

The concentration of precious metals in E-waste is much higher than in ores as seen from Fig. 9.5. Recovery of precious metals from E-waste can be economical and also have environmental benefits. Figures 9.6 and 9.7 show the amounts of gold and silver used in electronics respectively over 10 years from 2008 to 2017.

The processing routes for recovery of precious metals from E-waste are shown in Fig. 9.8 and are described below.

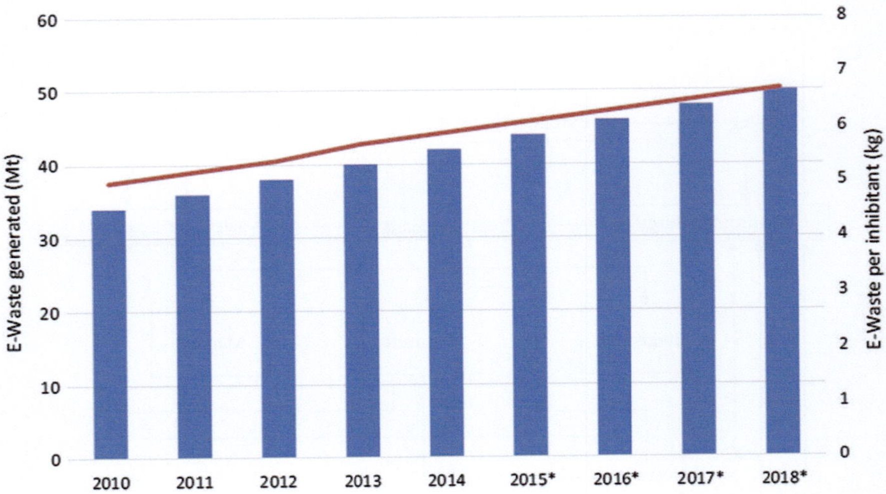

Fig. 9.3 E-waste generated globally per year (bars) and E-waste generated per inhabitant (line); data from 2015 are forecast (Birich et al. 2018)

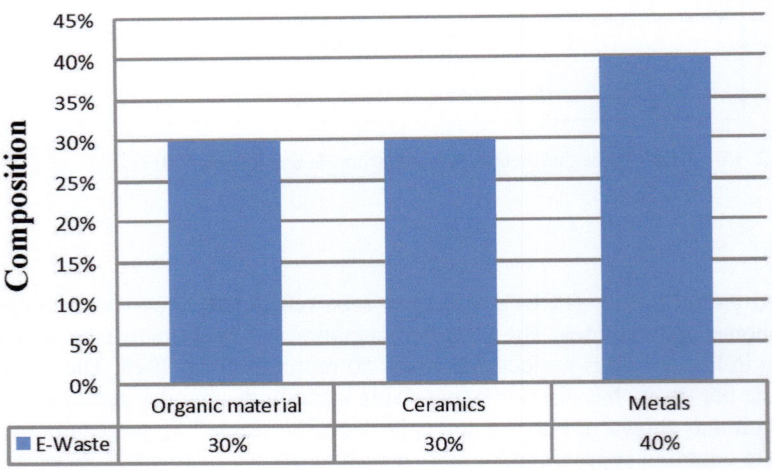

Fig. 9.4 The composition of E-waste (based on Kaya 2016)

Table 9.6 The detailed composition of E-waste (based on Kaya 2016)

	Category	Constituents
E-waste	Organic material	Polyethylene, polypropylene, polyvinyl chloride, polytetrafluoroethylene, polyphenyl ether, nylon, brominated flame retardants (polybrominated biphenyl, polybrominated diphenyl ethers, tetrabromobisphenol A), epoxy resins, fibres
	Ceramic	Silica, alumina, alkaline earth oxides, barium titanate
	Metals	Au, Ag, Pt, Pd, Ta, Ga, Cr, Hg, Cd, Be, Ni, Zn, Pb, Cu, Al

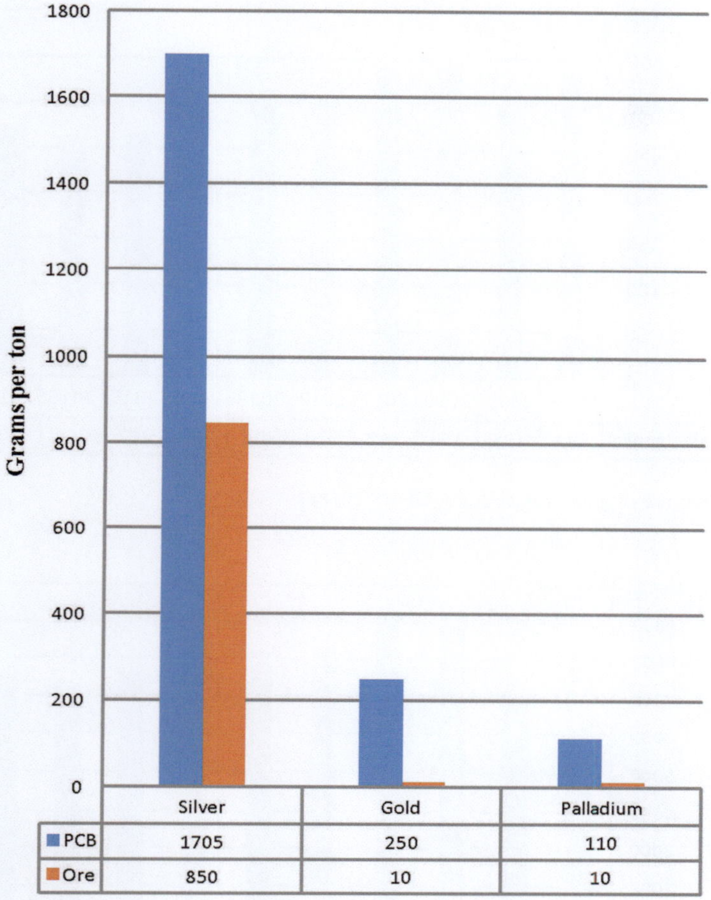

Fig. 9.5 Precious metals content in PCBs versus primary ores (adapted from Birich et al. 2018)

9.3.1 Mineral Processing

First stage in the processing of recycled E-waste is disassembly of the equipment by mechanical methods. Manual dismantling of E-waste is commonly practiced in developing countries. Low-speed high torque shear shredders (10 mm) or hammer mills are employed for primary crushing, which is followed by milling by ball mills or disc mills. Metallic and non-metallic fractions are separated by a combination of gravitational separation, magnetic separation, electrostatic separation and flotation. E-waste can be separated in three fractions by gravity separation, plastics with density less than 2.0 g/cm^3, light metals and glass with density close to 2.7 g/cm^3, and heavy metals with density greater than 7 g/cm^3. Magnetic separation is widely used

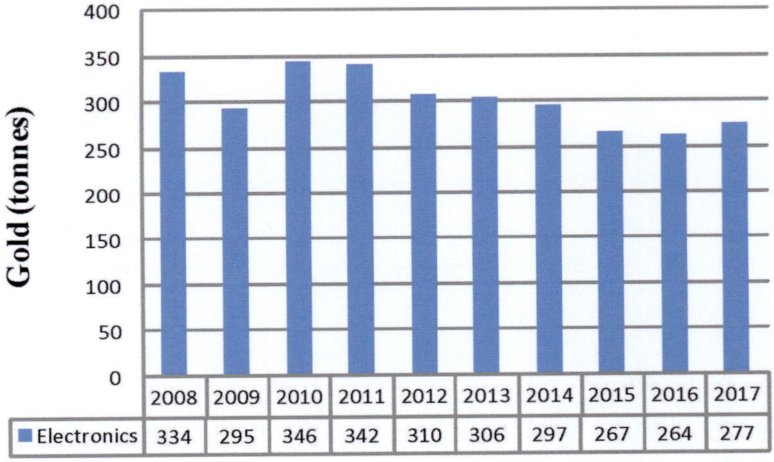

Fig. 9.6 The use of gold in electronics (GFMS 2018)

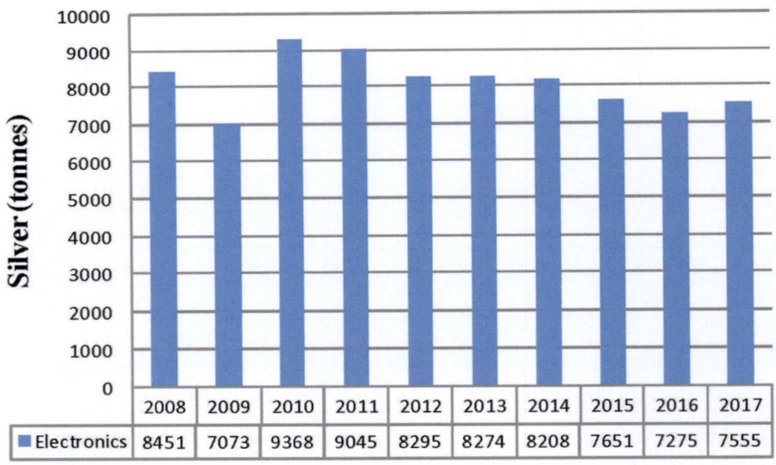

Fig. 9.7 The use of silver in electronics (The Silver Institute 2018)

to separate ferromagnetic metals from nonferrous metals and other nonmagnetic material. Electrical conductivity-based separation techniques such as corona electrostatic separation, triboelectric separation and eddy current separation are used to separate fractions based on the difference in polarity and electrical conductivity. Collectorless flotation with gasoline has been demonstrated as a promising method for separating polymers from metals (Sarvar et al. 2015).

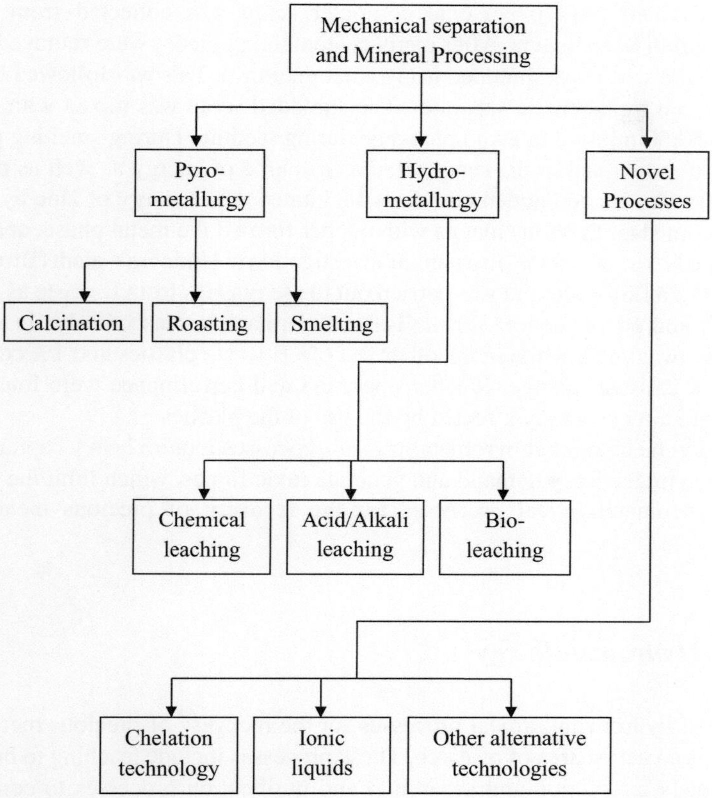

Fig. 9.8 Metallurgical approaches for metal extraction from E-waste (adapted from Chauhan et al. 2018)

9.3.2 *Pyrometallurgy*

After mineral processing, next step in recovery of precious metals can be pyrometallurgy, hydrometallurgy or some novel processes. Pyrometallurgical processes such as calcination, roasting and smelting can be employed to separate metals from rest of the material. During smelting, molten metal containing precious metals settles at the bottom of the furnace with a slag layer on top. Smelting of E-waste can generate toxic emissions, including dioxin from the plastic components, and proper care needs to be taken to deal with these emissions. Further processing such as leaching and electro-refining is carried out to separate precious metals from rest of the metals such as aluminium, copper and iron.

In a research investigation (Mark and Lehner 2000) sponsored by Boliden Minerals AB, Association of Plastics Manufacturers Europe (APME) and American

Plastics Council (APC), personal computer scrap was collected from various sources within Scandinavia. After mercury-containing pieces were removed by dismantling, the scrap was shredded using a hammer mill. This was followed by separation of iron by magnetic separator. The shredded scrap was mixed with crushed slag in a 50:50 mixture to avoid blockage during feeding. During smelting process, the plastics contained in the scrap acted as a source of energy as well as reducing agent. The plastics and added coal fines facilitated the recovery of zinc by fuming of zinc from slag. Precious metals with copper formed the metal phase, and recovery of metals was over 95%. In a similar investigation at Umicore's plant (Brusselaers et al. 2005), a full-scale trial was carried out to use plastics from E-waste as a reducing agent and energy source for the IsaSmelt furnace instead of coke. In one test, 4.5% coke was used, while in the other test 6% E-waste plastics and 1% coke were used for a 250-tons charge. Smelter operation and performance were found to be similar and not negatively affected by the use of the plastics.

It should be noted that pyrometallurgical processes require heavy capital investment, have high energy demand and generate toxic fumes, which limit the application of pyrometallurgical processes for the recovery of precious metals from E-waste.

9.3.3 Hydrometallurgy

The use of hydrometallurgical processes for the recovery of precious metals from E-waste is an active area of research. These processes include leaching to bring precious metals in solution and separation and purification processes to concentrate and separate precious metals. The application of leaching processes for recovery of precious metals from E-waste is described below.

9.3.3.1 Chemical Leaching

Chemical leaching is the leaching of gold and silver by cyanide and non-cyanide lixiviants such as thiosulphate, thiourea and halide leaching reagents. The leaching of gold and silver ores by these lixiviants has been described in Chap. 5. These lixiviants can also be used for recovery of gold and silver from E-waste.

The leaching of PCB using cyanide results in significant leaching of copper, which increases cyanide consumption. Montero et al. (2012) reported recoveries of 47.9% Au, 51.6% Ag, 48.1% Nb and 77.2% Cu in column leaching using NaCN. To improve gold and silver recovery and decrease cyanide consumption, copper can be leached from E-waste using sulphuric acid in the presence of suitable oxidant (H_2O_2 or O_2) prior to the cyanidation (Kamberovic et al. 2011).

Leaching of gold by thiosulphate is increasingly being pursued due to the toxicity of cyanide. Cupric ion (Cu^{2+}) and ammonia (NH_3) act as catalysts to improve gold recovery during thiosulphate leaching (Breuer and Jeffrey 2000). In a study to optimize the thiosulphate leaching of gold from PCB, maximum value of the initial rate of gold leaching was calculated to be 2.395 × 10^{-5} mol m^{-2} s^{-1} at 72.71 mM thiosulphate, 10.0 mM copper(II) ion and 0.266 M ammonia concentrations (Ha et al. 2014).

Leaching of gold by thiourea is another alternative to cyanide leaching. Thiourea leaching is carried out under acidic conditions as thiourea is unstable in alkaline solution. Recovery of gold by leaching in thiourea is affected by the concentration of Fe^{3+} ions in solution. Gurung et al. (2013) reported nearly four times higher extraction efficiency of gold in thiourea leaching in presence of Fe^{3+} ions compared to the absence of Fe^{3+} ions. The concentration of Fe^{3+} ions needs to be controlled carefully. At higher concentrations Fe^{3+} ions may oxidize thiourea to formamidine disulphide (FDS), which may further oxidize to sulphur and cyanamide (NH_2CN) and lower recovery of gold (Birloaga et al. 2014).

Another alternative to cyanide leaching of E-waste is halide leaching, which includes chloride, bromide and iodide leaching. Out of these, iodide leaching has attracted the attention of researchers. Sahin et al. (2015) investigated the effect of various process parameters such as halide concentration, oxidant concentration and pulp density on Au recovery from E-waste using iodine leaching method at 35 °C and pH 7. The recovery of gold decreased as percentage of solids was increased. Gold recovery was nearly 100% in a leaching solution containing 3% iodine, 1% H_2O_2 with 15% solids.

Xiu et al. (2015) investigated the recovery of Au, Ag and Pd from waste PCBs of mobile phones by supercritical water oxidation (SCWO) pre-treatment combined with iodine–iodide leaching. Pre-treatment in supercritical water followed by diluted hydrochloric acid leaching resulted in nearly complete recovery of Cu. The residue from hydrochloric acid leaching was subjected to iodine–iodide leaching, which resulted in maximum leaching efficiencies of 98.5% Au, 99% Ag and 97.2% Pd. Though high recoveries of gold and silver have been obtained, more research work is needed to reduce reagent consumption and make the iodine/iodide leaching process cost effective for recovery of gold and silver from E-waste.

9.3.3.2 Acid/Alkali Leaching

Acid/alkali leaching refers to leaching in acids such as sulphuric acid, hydrochloric acid and nitric acid or in bases such as sodium hydroxide and potassium hydroxide. As gold does not readily dissolve in acids, acid/alkali leaching has been mainly used for recovery of other metals such as copper (Yang et al. 2011; Calgaro et al. 2015), lead (Jha et al. 2012), indium (Silveira et al. 2015) and tin (Castro and Martins 2009).

Sheng and Etsell (2007) used nitric acid to dissolve base metals and then used aqua regia to dissolve gold from the residue of nitric acid leach. Gold dissolved in aqua regia was precipitated from the filtrate using ferrous sulphate. Joda and Rashchi (2012) used nitric acid for the leaching of E-waste. Silver, copper and other base metals dissolved in nitric acid, while gold and palladium reported to the residue.

The filtrate was treated with NaCl to precipitate silver chloride. The residue was leached in KOH to recover silver.

9.3.3.3 Bioleaching

Bioleaching is a process that employs microorganisms for leaching. Research in bioleaching started in the 1940s with the discovery of the role of microorganisms in the formation of acid mine drainage. Since then bioleaching has been used for extraction of uranium, copper, zinc, nickel and gold. The interest in the use of bioleaching is increasing due to the process being environmentally friendly and having lower operating cost and energy requirements.

Based on their useful range of operating temperature, the microorganisms can be classified as mesophiles, thermophiles and extreme thermophiles. The mesophiles operate well between 30 and 42 °C, examples being *Acidithiobacillus ferrooxidans*, *Acidithiobacillus caldus* and *Leptospirillum ferrooxidans*. The moderate thermophiles operate well between 45 and 55 °C, examples being *Sulfobacillus*, *Acidimicrobium* and *Ferroplasma* species, as well as *Leptospirillum ferriphilum* and *Acidithiobacillus caldus*. The extreme thermophiles operate well between 60 and 90 °C, examples being various species of *Sulfolobus*, *Acidianus* and *Metallosphaera*.

Currently, research investigations are being carried out for the application of bioleaching to recover base and precious metals from E-waste (Brandl and Faramarzi 2006; Brandl et al. 2008; Chi et al. 2010; Pradhan and Kumar 2012; Jadhav et al. 2016). In a two-step bioleaching experiments to investigate metal leaching efficiency of single and mixed culture of bacteria, three different cyanogenic bacteria strains (*C. violaceum*, *P. aeruginosa* and *P. fluorescens*) were used (Pradhan and Kumar 2012). *C. violaceum* strain exhibited the maximum bioleachability followed by *P. aeruginosa*. Recoveries of gold and silver were 69.3% and 7.1%, respectively, using *C. violaceum*. Recoveries of gold and silver increased to 73.2% and 8.4%, respectively, using a mixture of *C. violaceum* and *P. aeruginosa*.

The application of bioleaching to the processing of E-waste is limited due to very slow rate of leaching, incomplete leaching, possibility of contamination and sensitivity of microorganism for pH and temperature. These limitations need to be addressed for the application of bioleaching of E-waste at industrial scale.

9.3.4 Novel Processes

In addition to the more established processes described above, a number of novel processes are being developed for recovery of metals from E-waste. These promising new technologies are described in this section.

9.3.4.1 Chelation

Chelation is the process of formation of water-soluble metal–ligand complexes by incorporation of a metal ion into a heterocyclic ring structure. It has been used in medicines, soil remediation and metal recovery from industrial waste. A closed-loop chelation–dechelation process has recently been developed for extraction of nickel from spent catalyst (Vuyyuru et al. 2010). In this process more than 98% chelating agent was recovered which could be recycled for subsequent chelation–dechelation cycles.

The use of chelation technology for recovery of metals from E-waste is currently under development (Sharma et al. 2017). Research needs to be carried out to better understand chelation–dechelation chemistry, to develop new biodegradable chelating agents, and to recover the chelating agent for recycling in the process.

9.3.4.2 Ionic Liquids

Ionic liquids have the ability to extract hydrophobic compounds from aqueous solutions. Metal ions can be made hydrophobic by addition of extractants to form hydrophobic complexes, thereby facilitating their extraction by ionic liquids. Ionic liquids have been used in the extraction of lanthanides and actinides (Billard et al. 2011). Recently, ionic liquids have been used for the extraction of lanthanides and yttrium from aqueous solution (Shen et al. 2014), removal of platinum from aqueous solution (Stojanovic et al. 2010) and extraction of metal ions from industrial and communal wastewater (Fischer et al. 2011).

The application of ionic liquids for extraction of metals from E-waste is currently being explored (Chauhan et al. 2018). Major research challenges for this research include the development of suitable extractants and understanding the chemistry of the chosen system.

9.3.4.3 Other Alternative Technologies

Other promising technologies that are being explored for extraction of metals from E-waste are membrane filtration, photocatalysis and green adsorption (Chauhan et al. 2018). Membrane filtration is highly selective for metals, but requires periodic regeneration of membrane due to fouling. Photocatalysis is a relatively new technology for the removal or recovery of dissolved metal ions in wastewater (Chen and Ray 2001). The process involves the reduction of metals by electrons generated from photocatalysis. Green adsorption is the process for metal extraction by low-cost materials derived from agricultural sources, agricultural byproducts, agricultural residues and wastes. These green adsorbents have lower adsorption capacity compared to the commercial adsorbents such as activated carbon, but their low cost makes them an attractive choice.

9.4 Recovery of Precious Metals from Jewellery

Gold content of jewellery is designated by a Karat system or Millesimal Fineness. In Karat system, pure gold is designated 24 Karat and purity of gold alloys is measured in 1/24th weight fractions. Millesimal Fineness is a system that measures the purity of precious metals such as gold, platinum and silver by parts per thousand. Table 9.7 lists the specifications of gold purity in Karats and Millesimal Fineness. Gold is commonly alloyed with silver and copper for use in jewellery. A small amount of zinc is added to prevent the occurrence of blisters on the surface. Figures 9.9 and 9.10 show the amounts of gold and silver used in jewellery respectively over 10 years from 2008 to 2017.

After gold jewellery is recycled, first step in recovery of gold and silver is the fire assay to determine the composition. The processing steps depend on the composition of gold jewellery scrap.

Processing of gold–silver scrap with less than 15% silver content is shown in Fig. 9.11. The scrap is first melted and granulated to a fine powder. It is then dissolved in aqua regia. The preg liquor is evaporated to a syrup consistency. During evaporation, HCl is added to drive off nitrous oxide. After evaporation, water is

Table 9.7 Specification of gold purity (Wikipedia 2018)

Karats	Parts of gold	Gold purity (%)	Millesimal fineness	Remarks
8	8/24	33.3	333	Minimum standard for gold in Germany after 1884
9	9/24	37.6	376	
10	10/24	41.7	417	
12	12/24	50.0	500	
14	14/24	58.5	585	
15	15/24	62.5	625	
18	18/24	75.0	750	
20	20/24	83.4	834	
		90.0	900	One nine fine
22	22/24	91.6	916	The most widely used fineness for gold bullion coins
23	23/24	95.8	958	
		99.0	990	Two nines fine
		99.5	995	The minimum purity allowed in good delivery gold bars
24	24/24	99.9	999	Three nines fine
		99.99	999.9	Four nines fine, purity of Canadian Gold Maple Leaf and American Buffalo coins
		99.999	999.99	Five nines fine, the purest gold currently produced
		99.9999	999.999	Six nines fine, the purest gold ever produced

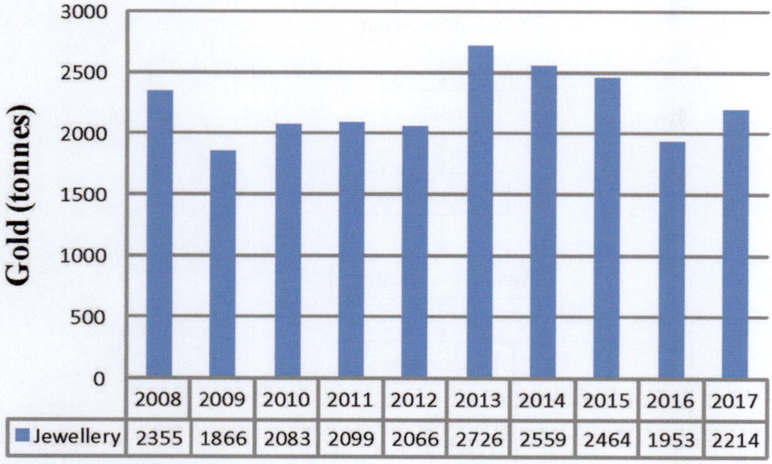

Fig. 9.9 The use of gold in jewellery (GFMS 2018)

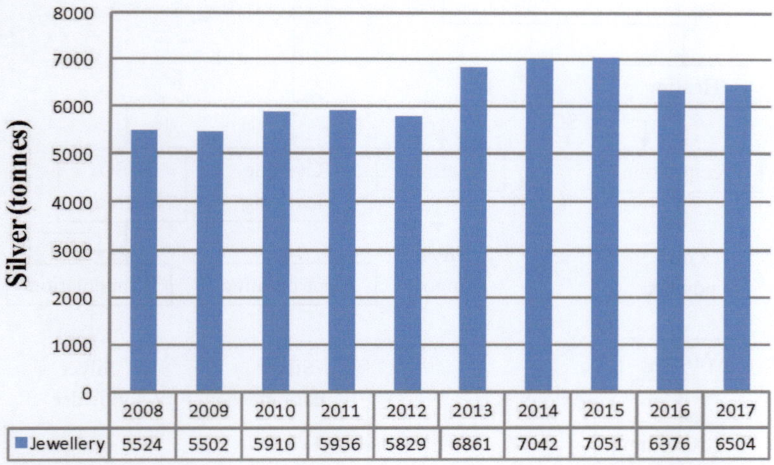

Fig. 9.10 The use of silver in jewellery (The Silver Institute 2018)

added to precipitate silver chloride. Silver chloride is filtered out and the filtrate is heated to 60 °C. The gold-containing filtrate is bubbled with SO_2 to precipitate gold. Gold precipitate is washed with dilute HCl first and then with water. Gold can then be electro-refined to increase the purity of gold. Silver chloride can be treated in three different ways to recover silver. Silver chloride can be smelted with sodium carbonate and carbon to produce silver ingot. Silver chloride can also be dissolved in cyanide and electroplated to form silver cathode. It can also be dissolved in dilute

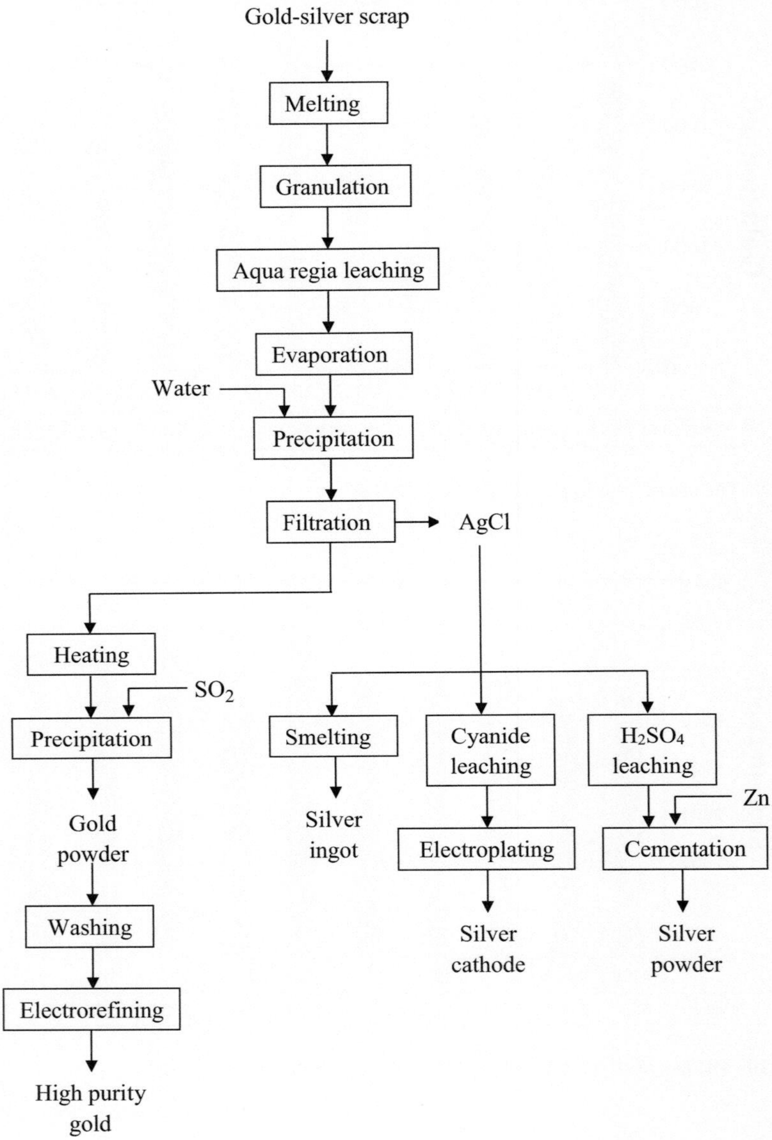

Fig. 9.11 Processing of gold-rich gold–silver scrap (adapted from Williams 1981)

sulphuric acid and silver can be precipitated by metallic zinc. High-purity silver can be obtained from the silver precipitate by electro-refining.

Processing of gold–silver scrap with greater than 80% silver content is shown in Fig. 9.12. The scrap is first melted and granulated to a fine powder. It is then leached in concentrated sulphuric acid or nitric acid to dissolve silver and separate gold as

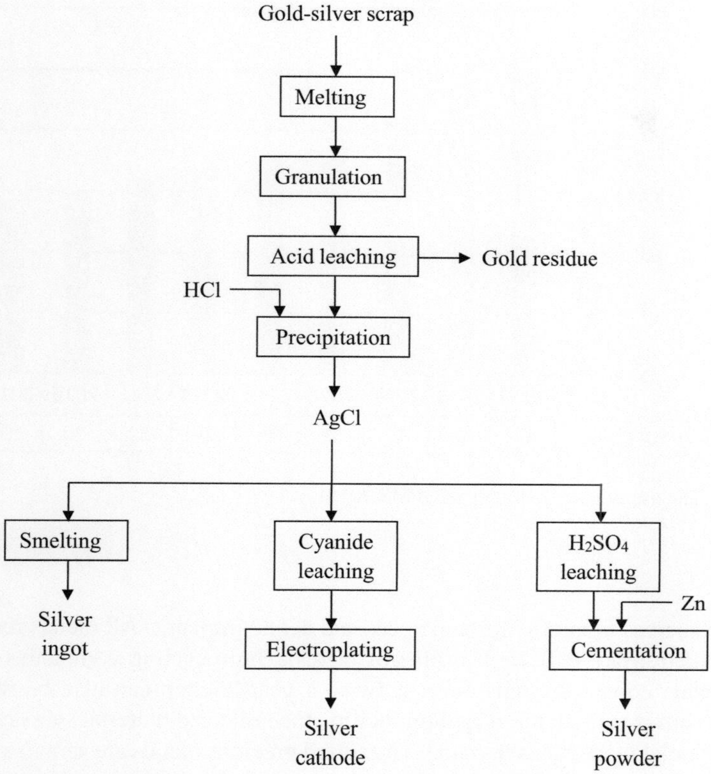

Fig. 9.12 Processing of silver-rich gold–silver scrap (adapted from Williams 1981)

residue. AgCl is precipitated from the filtrate by addition of HCl. Silver chloride is filtered out and can be treated in three different ways to recover silver as described above.

9.5 Recovery of Precious Metals from Dental Implants

Gold and gold alloys are used in conservative and restorative dentistry such as in fillings and dental implants. Gold is commonly alloyed with silver and copper and small amounts of platinum and palladium. A small amount of zinc is also added for deoxidation. Figure 9.13 shows the amount of gold used in dentistry for over 10 years from 2008 to 2017. The amount of gold used in dentistry has been decreasing steadily and is below 30 tons/year now.

The collection of gold-containing dental implants starts from crematorium. Cremation results in the incineration of biological material leaving behind medical

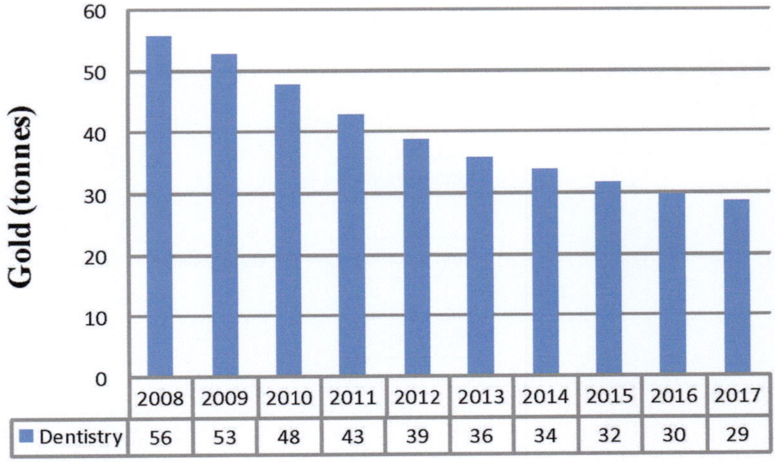

Fig. 9.13 The use of gold in dentistry (GFMS 2018)

implants such as titanium hips and knees and dental implants. All metals remaining after cremation are collected in bins or boxes. The recycling companies such as OrthoMetals collect the bins once or twice a year. Pickup can also be requested when the bin is full. At the recycling facility, the collected materials are sorted and precious metal pieces are separated. The sorted precious metals are sent to a specialized refinery for assay and recovery of precious metals.

Recovery of gold and silver from gold-silver alloys can be done as described in previous section. Processing of gold-palladium-platinum scrap is shown in Fig. 9.14. The scrap is first melted and granulated to a fine powder. It is then dissolved in aqua regia. The preg liquor is evaporated to a syrup consistency. During evaporation, HCl is added to drive off nitrous oxide. After evaporation, deionized water is added to bring gold concentration to 2–3 troy ounces per litre. The solution is heated to 60 °C and bubbled with SO_2 to precipitate gold. Gold precipitate is washed with dilute HCl first and then with water. Gold can then be electro-refined to increase the purity of gold.

The filtrate from gold precipitation containing palladium and platinum is re-evaporated to a syrup consistency. It is then diluted to approximately 70 g per litre palladium or platinum by adding deionized water. Next palladium is precipitated as palladium glyoxime by the addition of dimethylglyoxime leaving platinum in solution. Palladium glyoxime precipitate is washed with 60% alcohol first and then with 10% HCl. Palladium glyoxime precipitate is dissolved in ammonium hydroxide and then diaminopalladous chloride salt is precipitated by the addition of HCl. Diaminopalladous chloride salt is dissolved in deionized water and then palladium is precipitated by the addition of hydrazine.

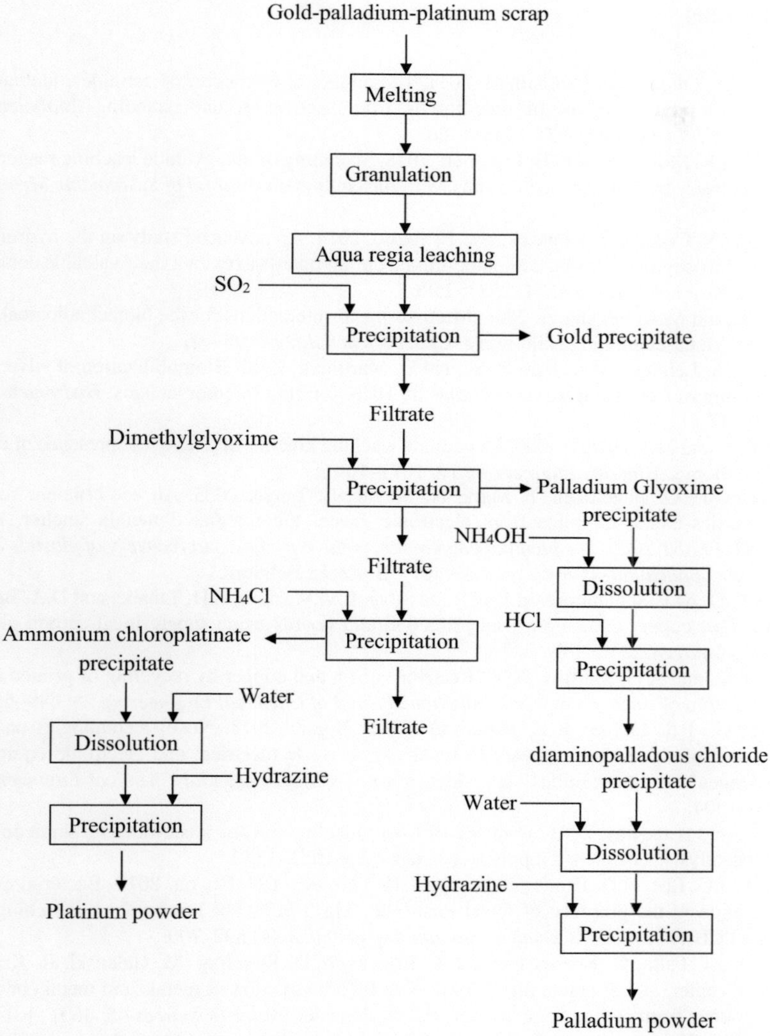

Fig. 9.14 Processing of gold–palladium–platinum scrap (adapted from Williams 1981)

The filtrate containing platinum is boiled with HCl and $NaClO_3$ and then cooled to 70 °C. Next platinum is precipitated as ammonium chloroplatinate by the addition of ammonium chloride leaving base metals in solution. Ammonium chloroplatinate precipitate is washed with a saturated solution of ammonium chloride. Ammonium chloroplatinate precipitate is dissolved in deionized water and then platinum is precipitated by the addition of hydrazine.

References

Billard, I., A. Ouadi, and C. Gaillard. 2011. Liquid–liquid extraction of actinides, lanthanides, and fission products by use of ionic liquids: from discovery to understanding. *Analytical and Bioanalytical Chemistry* 400: 1555–1566.

Birich, A., S.R. Mohamed, and B. Friedrich. 2018. Screening of non-cyanide leaching reagents for gold recovery from waste electric and electronic equipment. *Journal of Sustainable Metallurgy* 4: 265–275.

Birloaga, I., V. Coman, B. Kopacek, and F. Vegliò. 2014. An advanced study on the hydrometallurgical processing of waste computer printed circuit boards to extract their valuable content of metals. *Waste Management* 34: 2581–2586.

Brandl, H., and M.A. Faramarzi. 2006. Microbe-metal-interactions for the biotechnological treatment of metal-containing solid waste. *China Particuology* 4: 93–97.

Brandl, H., S. Lehman, M.A. Faramarzi, and D. Martinelli. 2008. Biomobilization of silver gold, and platinum from solid waste materials by HCN-forming microorganisms. *Hydrometallurgy* 94: 14–17.

Breuer, P.L., and M.I. Jeffrey. 2000. Thiosulfate leaching kinetics of gold in the presence of copper and ammonia. *Minerals Engineering* 13: 1071–1081.

Brusselaers, J., C. Hageluken, F. Mark, N. Mayne, L. Tange. 2005. An eco-efficient solution for plastics-metals-mixtures from electronic waste: the integrated metals smelter. In *5th IDENTIPLAST 2005, The biennial conference on the recycling and recovery of plastics identifying the opportunities for plastics recovery*, Brussels, Belgium.

Calgaro, C.O., D.F. Schlemmer, M.D.C.R. da Silva, E.V. Maziero, E.H. Tanabe, and D.A. Bertuol. 2015. Fast copper extraction from printed circuit boards using supercritical carbon dioxide. *Waste Management* 45: 289–297.

Castro, L.A., and A.H. Martins. 2009. Recovery of tin and copper by recycling of printed circuit boards from obsolete computers. *Brazilian Journal of Chemical Engineering* 26: 649–657.

Chauhan, G., P.R. Jadhao, K.K. Pant, and K.D.P. Nigam. 2018. Novel technologies and conventional processes for recovery of metals from waste electrical and electronic equipment: challenges and opportunities—a review. *Journal of Environmental Chemical Engineering* 6: 1288–1304.

Chen, D., and A.K. Ray. 2001. Removal of toxic metal ions from wastewater by semiconductor photocatalysis. *Chemical Engineering Science* 56: 1561–1570.

Chi, T.D., J.C. Lee, B.D. Pandey, J. Jeong, K.K. Yoo, and T.H. Huynh. 2010. Bacterial cyanide generation in the presence of metal ions (Na^+, Mg^{2+}, Fe^{2+}, Pb^{2+}) and gold bioleaching from waste PCBs. *Journal of Chemical Engineering of Japan* 44: 692–700.

Fischer, L., T. Falta, G. Koellensperger, A. Stojanovic, D. Kogelnig, M. Galanski, R. Krachler, B.K. Keppler, and S. Hann. 2011. Ionic liquids for extraction of metals and metal containing compounds from communal and industrial waste water. *Water Resources* 45: 4601–4614.

Forbes. 2018. The many uses of gold. https://www.forbes.com/sites/greatspeculations/2018/03/15/the-many-uses-of-gold/#411e8ce3279f.

GFMS. 2018. *GFMS gold survey 2018*. London: Thomson Reuters.

Gurung, M., B.B. Adhikari, H. Kawakita, K. Ohto, K. Inoue, and S. Alam. 2013. Recovery of gold and silver from spent mobile phones by means of acidothiourea leaching followed by adsorption using biosorbent prepared from persimmon tannin. *Hydrometallurgy* 133: 84–93.

Ha, V.H., J.C. Lee, T.H. Huynh, J. Jeong, and B.D. Pandey. 2014. Optimizing the thiosulfate leaching of gold from printed circuit boards of discarded mobile phone. *Hydrometallurgy* 149: 118–126.

He, J., and A. Kappler. 2017. Recovery of precious metals from waste streams. *Microbial Biotechnology* 10: 1194–1198.

Jadhav, U., C. Su, and H. Hocheng. 2016. Leaching of metals from printed circuit board powder by an Aspergillus Niger culture supernatant and hydrogen peroxide. *RSC Advances* 6: 43442–43452.

Jha, M.K., A.K. Kumari, P.K. Choubey, J.C. Lee, V. Kumar, and J. Jeong. 2012. Leaching of lead from solder material of waste printed circuit boards (PCBs). *Hydrometallurgy* 121: 28–34.

Joda, N.N., and F. Rashchi. 2012. Recovery of ultra fine grained silver and copper from PC board scraps. *Separation and Purification Technology* 92: 36–42.

Kamberovic, Z., M. Korac, and M. Ranitovic. 2011. Hydrometallurgical process for extraction of metals from electronic waste—Part II: Development of the processes for the recovery of copper from printed circuit boards (PCB). *Journal of Metallurgy* 17: 139–149.

Kaya, M. 2016. Recovery of metals and nonmetals from electronic waste by physical and chemical recycling processes. *Waste Management* 57: 64–90.

Mark, F.E., and T. Lehner. 2000. Plastics recovery from waste electrical & electronic equipment in non-ferrous metal processes. APME (Association of Plastics Manufacturers in Europe) technical report. http://depa.fquim.unam.mx/ipm/introd_ing_metymat/mat_apoyo/bal_macro_energia/plastic_recycling.pdf.

Mining.com. 2016. 21 incredible uses for silver. http://www.mining.com/web/21-incredible-uses-for-silver/.

Montero, R., A. Guevara, and E. de la Torre. 2012. Recovery of gold, silver, copper and niobium from printed circuit boards using leaching column technique. *Journal of Earth Science and Engineering* 2: 590–595.

Pradhan, J.K., and S. Kumar. 2012. Metals bioleaching from electronic waste by Chromobacterium violaceum and pseudomonads sp. *Waste Management & Research* 30: 1151–1159.

Sahin, M., A. Akcil, C. Erust, S. Altynbek, C.S. Gahan, and A. Tuncuk. 2015. A potential alternative for precious metal recovery from E-waste: iodine leaching. *Separation Science and Technology.* 50 (16): 2587–2595.

Sarvar, M., M.M. Salarirad, and M.A. Shabani. 2015. Characterization and mechanical separation of metals from computer Printed Circuit Boards (PCBs) based on mineral processing methods. *Waste Management* 45: 246–257.

Sharma, N., G. Chauhan, A. Kumar, and S.K. Sharma. 2017. Statistical optimization of heavy metals (Cu^{2+}/Co^{2+}) extraction from printed circuit board and mobile batteries using chelation technology. *Industrial and Engineering Chemistry Research* 56: 6805–6819.

Shen, Y., W. Li, J. Wu, S. Li, H. Luo, S. Dai, and W. Wu. 2014. Solvent extraction of lanthanides and yttrium from aqueous solution with methylimidazole in an ionic liquid. *Dalton Transactions* 43: 10023–10032.

Sheng, P.P., and T.H. Etsell. 2007. Recovery of gold from computer circuit board scrap using aqua regia. *Waste Management and Research* 25: 380–383.

Silveira, A.V., M.S. Fuchs, D.K. Pinheiro, E.H. Tanabe, and D.A. Bertuol. 2015. Recovery of indium from LCD screens of discarded cell phones. *Waste Management* 45: 334–342.

Stojanovic, A., D. Kogelnig, L. Fischer, S. Hann, M. Galanski, M. Groessl, R. Krachler, and B.K. Keppler. 2010. Phosphonium and ammonium ionic liquids with aromatic anions: Synthesis, properties, and platinum extraction. *Australian Journal of Chemistry* 63: 511–524.

Syed, S. 2012. Recovery of gold from secondary sources—a review. *Hydrometallurgy* 115–116: 30–51.

The Silver Institute. 2018. *World silver survey 2018*. London: Thomson Reuters.

USGS. 2018. How much silver has been found in the world? https://www.usgs.gov/faqs/how-much-silver-has-been-found-world?qt-news_science_products=0#qt-news_science_products.

Vuyyuru, K.R., K.K. Pant, V.V. Krishnan, and K.D.P. Nigam. 2010. Recovery of nickel from spent industrial catalysts using chelating agents. *Industrial and Engineering Chemistry Research* 49: 2014–2024.

Wikipedia. 2018. Fineness. https://en.wikipedia.org/wiki/Fineness.

Williams, P.E. 1981. The refining of jewellery and dental scrap. In *Symposium on recovery, reclamation and refining of precious metals*, *March 10-13*. San Diego: The International Precious Metals Institute, Inc.

World Gold Council. 2018. How much gold has been mined? https://www.gold.org/about-gold/gold-supply/gold-mining/how-much-gold-has-been-mined.

Xiu, F.R., Y. Qi, and F.S. Zhang. 2015. Leaching of Au, Ag, and Pd from waste printed circuit boards of mobile phone by iodide lixiviant after supercritical water pre-treatment. *Waste Management* 41: 134–141.

Yang, H., J. Liu, and J. Yang. 2011. Leaching copper from shredded particles of waste printed circuit boards. *Journal of Hazardous Materials* 187: 393–400.

Chapter 10
Case Study: Digital Disruption in the Mining Industry

V. I. Lakshmanan, R. Roy, and B. Gorain

10.1 Digitization in the Gold and Silver Industry

The last few decades have seen the advancement in information technology as no other period in history. The same period has also seen dramatic increase in computational power and a correspondingly significant decrease in the cost of electronics. In addition, internet and wireless telecommunication have exploded in the first decade of the twenty-first century. Advancements in these areas have resulted in several synergies to deliver innovation in diverse applications leading to an era of digital transformation. The mining industry has been relatively slow in adopting these technologies compared to many other industries such as aviation, finance or even oil and gas.

Over the last few years, there has been a surge in the usage of digital technologies by the mining companies to improve safety, operational efficiency and to reduce operational costs. Embracing digitization also calls for a change in culture, which, for most organization, is not trivial. This change in culture is all about breaking down the barriers of working in silos and move toward collaboration, integration, and partnerships not just within an operation or a company but beyond with the potential to *connect anything from anywhere to anyone across the globe*.

Attempts in the past to integrate functions in gold and silver industries such as through Mine-to-Mill or Mine-to-Market initiatives have demonstrated significant benefits to many operations. Still, sustaining these benefits has clearly been a challenge due to manpower turn over and lack of proper systems and structures that integrate information between business functions within an operation and also

V. I. Lakshmanan (✉) · R. Roy
Process Research Ortech Inc., Mississauga, ON, Canada
e-mail: llakshmanan@processortech.com

B. Gorain (✉)
Hindustan Zinc Limited, Udaipur, Rajasthan, India
e-mail: Barun.Gorain@vedanta.co.in

V. I. Lakshmanan, B. Gorain (eds.), *Innovations and Breakthroughs in the Gold and Silver Industries*, https://doi.org/10.1007/978-3-030-32549-7_10

between operations and the corporate functions. These challenges can now be addressed readily through digitization and automation, with many examples of mining companies embracing an *integrated operations* approach connecting operations remotely (BHP 2017; World Economic Forum 2017; Barrick 2017b).

10.2 Start of the Enterprise-Wide Integration Through Digitization

The advent of digital technology has opened up opportunities to integrate interdepartmental information for a unified understanding of, and across, an entire organization. The advances in instrumentation, equipment control systems, supervisory control systems (SCADA/Historian) and industrial networks have helped in the direction of enterprise wide integration. The availability of real-time information across an enterprise has given rise to numerous possibilities to improve organization performance. Some of the opportunities pursued in mining and metal extraction are as follows (Seshan and Gorain 2016).

10.2.1 Informed and Timely Decision-Making

The current state of the art allows for production performance to be linked to planning, design, and the supply chain. Integrated software tools have allowed for an organization to share not just raw process data but intelligence that is abstracted from such data to drive well-informed decisions (called "actionable intelligence"). Therefore, changes in performance happening in any part of the production/mining value chain (ecosystem) can have an instantaneous impact on the other parts making the interactions between the various stakeholders of the ecosystem dynamic and real time. This provides a better capability to all stakeholders in the ecosystem to predict outcomes that are relevant to their respective scope. Various software tools and platforms like OSI-PI, SAP, Dassault, Predix, Mine-RP and Deswick are increasingly being used by the mining companies to integrate disparate set of data and make data-driven decisions.

10.2.2 Cost Optimization

The cost of operations can now be optimized in a dynamic manner through real-time integration of information across the entire mining value chain. An example of a useful synergy in the mining industry is the one between production and maintenance, allowing companies to innovatively maximize asset availability at minimal maintenance cost. In other words, of late, maintenance has increasingly ceased to be a cost center and has become an "opportunistic" driver to improve asset, and therefore, production performance.

10.2.3 Asset Optimization

Asset availability has been one of the key performance metrics for the mining and the metallurgical industry. Availability typically reduces due to inadequate and timely information on the condition of the critical assets to perform proactive maintenance and asset health care. Predictive intelligence of assets is an evolving method to maximize asset performance and to achieve better return on physical assets. Increasingly, equipped with predictive intelligence, proactive maintenance of assets is taking over reactive maintenance. Proactive is herein referred to as a combination of preventive and predictive maintenance. Critical assets in the mining and metallurgical industries have increasingly been subject to usage-based preventive maintenance and appropriate predictive maintenance to avoid unplanned downtime. The monitoring of usage is done by the use of advanced software tools and open standards available for enterprise application integration. The usage of critical machines is tracked by a real-time software, and when they cross a threshold for a preventive maintenance call, a maintenance work order is automatically generated by the real-time software to maintain the health of the assets in good condition at all times (Enterprise Gateway 2010). These software tools deliver the capability of optimizing the maintenance functions in such a manner to achieve maximum availability of production at a minimum cost of maintenance.

10.2.4 Energy Savings

Related to the operational performance of machines is energy efficiency. Mining and metallurgical processes such as haulage, communication, pressure oxidation, and electrowinning consume a significant amount of energy. The energy intensity of a metallurgical process is always a critical factor to an organization from multiple points of view, profitability, greenhouse gas emissions and system/process degradation. Many metallurgical operations are starting to have an advanced and intelligent energy management system in place. The purpose of energy management varies from implementing operational control of energy consumption in real time in a local facility to achieving a strategic goal of the enterprise to reduce energy consumption across the portfolio. These are normally performed in two parallel "loops" (refer to Fig. 10.1): the strategic loop that allows for an aggregated and comparative evaluation of the energy performance of various sites to arrive at useful strategies and the lower-level control loop that implements the strategies via an existing control systems infrastructure.

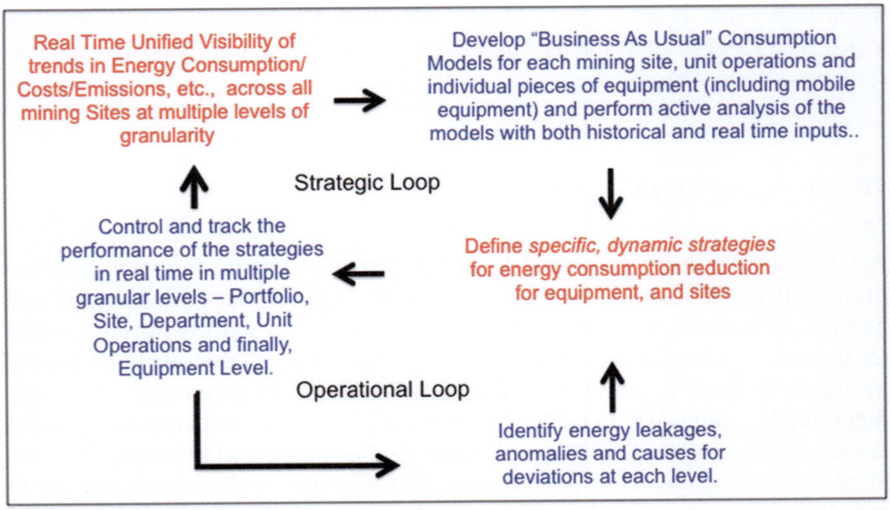

Fig. 10.1 The integration (and traceability) between strategic and operational control of energy consumption in a multisite facility

The state-of-the-art solutions for achieving improvement in energy management allow for the *dynamic* generation of energy-saving strategies in response to evolving operational conditions (as opposed to a shrink-wrapped set of pre-programmed strategies) and, furthermore, provide improved traceability (integration and visualization) of the downstream implementation of such strategies via the distributed control of multiple, individual pieces of equipment.

10.3 The Concepts of Enterprise-Wide Integration

Some mining companies are already working in this direction by integrating their operating data, scattered across multiple locations, into centralized database with dashboards and analytics capability with an aim to make a step change in improving productivity and reducing operating costs through better collaboration between different business units.

Companies like Rio Tinto and BHP-Billiton in Australia, Codelco (in partnership with Honeywell called Kairos mining) in Chile, and some other companies and operations have taken steps in the direction of a partially or fully integrated operations strategy (World Economic Forum 2017; BHP 2017; BHP-Billiton 2013; Sherring 2012; Zamora et al. 2010).

This chapter elaborates on the concepts of a digital integrated enterprise that comprises an architecture involving five logical layers. This integration includes functional elements of an organization from strategic, tactical, and operational points of view. *It allows for a simultaneous focus on both production and cost control*. This initiative provides the industry with an ability to deal with frequently changing business dynamics and is therefore strategic to many mining companies.

For instance, in the mining industry, the target for tangible improvements through this initiative could be 5–10% increase in productivity and a 15–20% reduction in operating costs as a first step. These targets are realistic and are based on improvements that have been demonstrated by some mining companies pursuing even through a simpler strategy such as *Mine-to-Mill*.

Another simple strategy is to integrate Geology, Mining, and Processing with a unified visualization of the entire enterprise focusing on a common target such as mining product quality (e.g., metal in ore, fragmentation) that delivers the best value in processing instead of just focusing on mining tonnage. This tactical goal in turn can result in the acquisition and tracking of relevant operational data in the example above, the data collected and tracked are, for instance, ore dilution, ore loss in waste, blast fragmentation, top size and hardness of rocks, mineral liberation during milling, metal recovery in process plant along with energy and material consumption for the mining and processing steps involved. The use a web-based integrated dashboard to share such data between the different stakeholders or functions along with the ability to interact and interrogate the database will allow personnel in different functions to cooperate as they all have one common goal as this goal is continuously displayed and monitored in strategic locations.

This integrated enterprise must have appropriate value drivers and KPIs. As an example, if a mining manager has indices that measure only tonnage and production cost per ton mined, then there is no incentive to change a well-established blasting practice, because such a change may increase the blasting cost per ton of ore mined. On the other hand, an integrated enterprise approach is defined holistically by a “hierarchy” of metrics at various levels of abstraction and the organization strives to optimize across the hierarchy.

Having an integrated database with estimates of key value drivers and KPIs as mentioned above allows benchmarking of an operation in terms of its unique set of KPIs. This database could receive information from a number of sources such as:

- The mine dispatch system
- The plant process data historian (e.g., OSI-PI system)
- The Laboratory Information Management System (LIMS).

An integrated enterprise strategy will require systems to measure, monitor and control the raw data that result in the estimation of the key value drivers and KPIs across the mining value chain with an emphasis on optimization of the interfaces between business units to produce the desired outcomes. The key advantages of such an approach in the mining industry are availability of more accurate, reliable, and quantified information to assist with the following applications:

(a) Mine-to-Mill reconciliation and optimization
(b) Ore stockpile management
(c) Geometallurgical modeling
(d) Dynamic mine block modeling and planning

(e) Dilution control
(f) Grade Engineering (a new concept from CRC ORE mentioned in McKee, 2013)
(g) Mill feed stabilization

10.3.1 Optimization Philosophy: Optimization of the Parts Vs. Optimization of the Whole

As discussed earlier, one of the key differentiating philosophical insights seen in the industry now as a step change from traditional approaches has been to strive for optimization across the entire organization instead of only in parts. This goal is achieved via *real-time* integration of the enterprise. This philosophy has proven to be successful in many industries, including oil and gas, manufacturing and recently in mining by bringing many fragmented solutions together.

The "optimization of the parts," no doubt, allows improving efficiency of individual units and is still an important part of the mining or any metallurgical business. But this results in potential economic benefits to be missed as the focus is on constant execution to meet the targets of the individual business units, and the improvement opportunities associated with a systemic approach come only as an afterthought (Sherring 2013). As stated earlier, focusing on parts alone is tantamount to working in various business unit silos resulting in limited conformance to life-of-mine (LOM) plan in the case of the mining industry and looking beyond the plan is often challenging.

In order to address these challenges and improve profitability, it is becoming imperative to look at a mining enterprise holistically requiring a strong interaction of professionals in various business disciplines involving Geology, Mining, Processing, Maintenance, Environment, Health and Safety, Finance, Supply Chain and Logistics, IT, HR, Community, and Public Relations. All of these professionals need to interact with a common automation platform that will allow for multiple applications to share data between them.

There are many examples that suggest that integrating different business disciplines such as Mine-to-Mill involving true integration of geology, mining, and processing functions have led to significant productivity gains and cost savings along with intangible benefits, including a more cohesive, satisfied, and performance-oriented workforce transcending the traditional boundaries (McKee 2013).

The Rio Tinto experience, as shared by Sherring (2013), suggests that an integrated operations strategy also allows a balanced focus on planning, execution, and improvements with an enabling organizational and working environment.

The Kairos Mining's Collaboration Centre (a joint venture of Codelco and Honeywell) provides real-time monitoring of plant operating data and automated control systems for multiple sites with a focus on knowledge management and long-term process performance improvements (Zamora et al. 2010).

10.3.2 Five-Layer Architecture for an Integrated Mining Enterprise

Typically, an integrated mining enterprise could have five logical layers as illustrated in Fig. 10.2. The first layer and the lowest logical layer involve raw data acquisition and monitoring of various unit processes, machines, sensors, and transducers in the production environment.

The data acquired from the lowest layer are used by the control and the data analysis layer. This forms the second layer. The controllers in turn could be organized in a hierarchy (within this layer) based on the sophistication required. The lower-level controllers are normally regulators meaning they are mandated to maintain a control variable within reasonable bounds of a reference set point.

The higher-level controllers comprise supervisory and intelligent controllers. These controllers will be able to respond to the dynamics and actively alter the set points or cope with the nonlinearity as they present themselves using various artificial intelligence (AI) based on mathematical modeling techniques. Such controllers are very domain specific and are constructed to meet the challenges of a specific process dynamic. Also, this layer houses "Historians" for recording historical data and "Human/Machine Interfaces" for understanding the status of a process/machine visually for an operator. All of the subsystems in this layer are amenable to integration with the other layers in the architecture. Each low-level controller, historian, and supervisory controller acts as modular component in the architecture that can

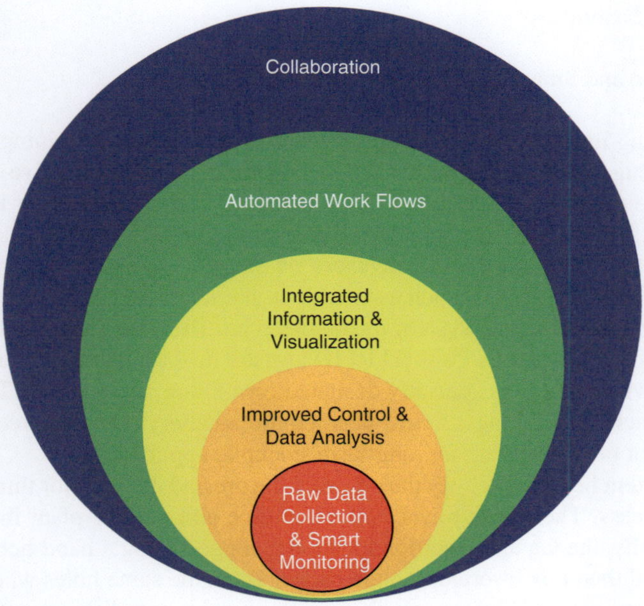

Fig. 10.2 Five layers of an Integrated Operations Enterprise (based on Sinclair 2012)

exchange data with the other elements through standard conventions across a standard network.

The third incremental layer is a plant-wide abstraction one that integrates and visualizes the data across the individual business units—this layer comprises a web and mobile platform for unified visualization. This layer houses Manufacturing Execution Systems (MES) and Manufacturing Operations Management Systems (MOM).

The fourth logical layer is the action layer one that integrates workflows between different business units across the enterprise, not restricted only to the plant or production environments. In other words, this layer comprises systems that can enable integrated and automated workflows between the different business units.

And the final layer enables "collaboration" between multiple stakeholders across different functional units within the enterprise. In other words, the collaboration layer allows for optimization across multiple departments often, one whose objective is to achieve "continuous improvement" such as minimization of costs, downtimes, variation, and maximization of the return on assets under multiple constraints spread across the different departments. The availability of the integrated enterprise architecture not only enables seamless synergy between the multiple stakeholders to effectively and actively contribute, but also allows for the collective multidisciplinary team to generate predictive actions to effect control and optimization in a future time period based on the current state.

10.3.2.1 Logical Layer 1: Raw Data Collection and Smart Monitoring (Examples)

Using Radio and Smart Tags for Ore Tracking

In the mining industry, ore tracking is becoming a necessity as mining operations are becoming immensely complex. This tracking enables effective integration between the mine and the mill. In other words, the tracking allows for information flow to be synchronized with material flow throughout the cycle. Such tracking improves efficiency and the effectiveness of the process especially when certain dynamic factors are at play, such as change in ore characteristics (Isokangas et al. 2012). Using RFID tags a system tracks ore types from the mine to the mill and through the process plants. Physical RFID tags are now available from a number of suppliers, with some tags specially designed for mining applications. They can be passive or active RFID tags. Active tags require a battery power source and therefore may not be appropriate for long-term stockpile applications.

This system helps to identify the origin of the ore and its behavior throughout the mining process. These tags travel through a mine and process plant in a series of steps. Initially, the tag and insertion location is logged using a hand-held computer or PDA, and then it is inserted into the rock mass in the same holes where blasting explosives are placed. The tag travels with the ore through digging, transport, and processing before being detected by sensors that are positioned for recording the time and the tag at various points. The RFID tag data is then loaded into a centralized database and analyzed as required.

The use of a tag system allows development of relationships between ore characteristics (available in the mine block models) and operating parameters in the mine and processing plant (such as ore dilution, fragmentation, stockpile residence times, segregation, energy consumption, ore grade) along with actual mine production and metal recovery. With this knowledge, *operating parameters can be optimized in real time* to respond rapidly to changes in ore characteristics, thus resulting in controlling operating costs and improving productivity.

Fragmentation Analysis

Optical sizing technology for measuring size distribution of fragmented rocks after blasting has been used for many years now. The WipFrag fragmentation sizing system was originally designed using a roving camera and operator-assisted analysis. Using optical sizing technology, mining blast professionals could evaluate, reassess, and redesign their blasts, while understanding the effect of their design on their final product. In addition, they could begin to quantitatively evaluate the effect of geological structure of their blasts. Although the accuracy of this method is low, it is still useful as an alternative method of screening large masses of rocks.

The new generation of the Split-Online digital image analysis has been applied for accurate, continuous and rapid measurement of rock fragmentation. Cameras are installed along key stages of crush, convey, and milling processes to determine rock fragmentation size. The critical data for mining operations, including particle size, shape, color, and texture, are calculated by the advanced split algorithms and can be reported to a centralized database to enable real-time evaluation by mine operators and management. Installation of the Split-Online camera systems in various stages of the comminution process at the Morila gold mine in West Africa resulted in a 10% mill throughput improvement. Split-Online cameras along with ore tracking system have significant potential for operations and can provide information on the less-understood interface between mine and mill, which is a key for successful implementation of an integrated operations strategy.

Using Piezo-Electric for Slurry Flow Meters and Density Gauges

A new innovative technology for noninvasive flow measurement using piezo-electric sensor array has recently achieved wide acceptability in the mining industry with some significant benefits compared to the traditional flow meter technologies such as Electromagnetic, Ultrasonic Doppler, Differential Pressure or Coriolis. These flow meters have recently been used for better quantification of recirculating loads in a grinding circuit and also allow metal balancing and reconciliation in a process plant for better Mine-to-Mill reconciliation.

Using Light and Chemometrics for Online Monitoring of Mineralogy and Assays

Online mineralogy and assays measurements provide opportunity to optimize process plant in real time. This will allow a tighter control of tailings losses, which otherwise is difficult to achieve in an offline-based mineralogy and assays measurements that are used presently. Online mineralogy measurement from BlueCube™ technology is a major breakthrough and has recently been successfully implemented in many precious and base metals industries. This technology is based on diffused reflective spectroscopy combined with propriety chemometric techniques.

10.3.2.2 Logical Layer 2: Improved Control and Data Analysis

Supervisory Control Hierarchy

The control layer is responsible for local control of the process. The sophistication of controllers depends on the complexity of the process simpler PID controllers are sufficient for linear control assumptions but more sophisticated approaches are necessary for processes that exhibit nonlinear dynamics. Control of such processes requires the development of a mathematical model of the process or plant dynamics. The parameters defined in the mathematical model are then dynamically calibrated before they are implemented as part of the controller of the process or plant. The controller is designed to cope with the dynamics of the plant/process for various set point and output conditions.

The problem arises when a critical process variable which is an input to the controller cannot be measured or the dynamic parameters of the model are imprecise or if there are external disturbances to the process that are not known or are not modeled, which are also the realities of the mining and metallurgical industry. The controllers degrade rapidly and the need for active refinement of the controller is necessary, which introduces the need for a supervisory layer. In these cases, the set point itself has to be changed to cope with the changing dynamics of a process and/or to compensate for the errors or the incompleteness in the model of the process. In such cases a hierarchical structure of controllers will be used to control and optimize the process at various levels of abstraction.

The low-level controller normally drives the system toward a local set point but does not guarantee global convergence especially in the wake of disturbances and influencing factors that are outside the scope of the local controller. In order to compensate for such factors, a supervisory control loop will be necessary. For instance, supervisory control is needed when the set point itself has to be modified actively and dynamically. Figure 10.3 shows a supervisory control hierarchy proposed by Bergh et al. (2007) for a copper solvent extraction pilot plant.

An example of the use of such a control system can be found in a copper or copper–gold extraction process. Solvent extraction is an important operation in hydrometallurgy this involves a process of transfer of soluble metal compounds occurring

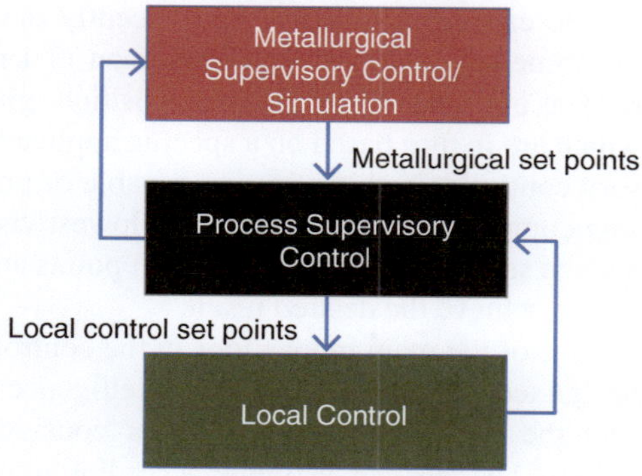

Fig. 10.3 An example of a supervisory control hierarchy (from Bergh et al. 2007)

between aqueous and organic phase. Extraction is a chemical reaction between the metal ion in the aqueous phase and the extractant from the organic phase. The control objective requires a tradeoff between the concentration and flow of the Pregnant Leach Solution, the flow and concentration in the organic phase of copper, the degree of entrainment of organic in the aqueous, and the aqueous carryover in the organic.

In such a process the control is shown to be organized in three levels by Bergh et al. (2007): (a) local control loop as specified earlier; (b) hydrodynamic supervisory control loop and (c) metallurgical supervisory control loop. In their example, field measurements included flow, level and conductivity these are provided as data inputs to a Programmable Logic Controller (PLC). The output of the PLC is the set points for the local-level controller.

The hydrodynamic supervisory control loop acts as a process supervisory controller whose mandate is to control the flow rates between different process units and the stock solution levels in a coordinated manner. This would enable every process unit to be operated in *different* internal, external or overall organic/aqueous ratios.

The metallurgical control loop can then provide the set points for local and the overall organic/aqueous ratios in order to modify the metallurgical targets of the copper concentration in different streams. The inherent problem here is the uncertainty in the measurement of the concentration of each stream. In order to achieve reasonable values for the concentration two major approaches have been reported in the literature a prediction approach that relies on the use of statistical or artificial intelligence techniques; or fitting a dynamic model by means of experimental data. In both cases, the values obtained for the concentration of copper in the various streams have been used for developing a metallurgical strategy such as the best operating conditions (set points as mentioned above) for the solvent extraction plant given the dynamic inputs.

Another example of metallurgical supervisory control driving lower-level controllers is found in iron and steel making, especially in the characterization of slags. It is well known that the composition of the slags has an impact on specific physical and thermodynamic properties of steel, such as viscosity, density, activity, and sulfide capacity. A system was recently developed based on the theoretical and experimental findings of Seetharaman (Thermoslag Ver. 2.0, 2010) to provide a reverse optimization of slag composition, given certain desired properties of steel which are in turn based on a specific application. That is, the metallurgical supervisory controller in this case was capable of providing the appropriate ranges for the slag composition for achieving the lowest viscosity, or highest sulfide capacity, for a given set of components. These set points are then usable by the low-level controllers to achieve the desired result.

One of the main innovations in the control layer is the integration of the control design technology with artificial intelligence techniques. The basic idea is to represent the equipment or plant under consideration as a *Mixed Logical Dynamic System*, i.e., systems evolving according to continuous dynamics, discrete dynamics, and logical rules. In particular, fuzzy control, genetic algorithms, and neural networks have been extensively used to provide inputs, referred to in the industry as "soft sensors," to the lower-level controllers. The values for the soft sensors are derived from process-dependent rules, which are, in turn, designed or "trained" for each specific application. The soft sensors provide a major advantage they act as "signals" of variables that are difficult or impossible to measure at high sampling rates. In other words, this technique is useful when it is impossible or difficult to physically measure a process variable that is important for the control and optimization of the process. For instance, in mineral processing, an obvious example to be used in grinding is predicting particle size average as a function of the mill state and history. Another case would be the construction of "free lime soft sensors" for cement and lime kiln control. Yet another example is assessing the temperature distribution of a kiln or furnace in a continuous manner.

Also, such soft sensors are used to provide a backup for critical process measurement devices (Fig. 10.4). In the case of a failure of critical process measurements, a soft sensor can provide the control strategy with a usable "estimate" of the missing

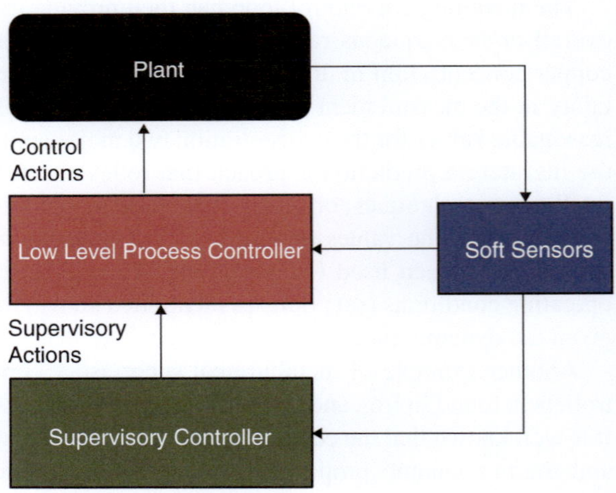

Fig. 10.4 Soft sensors serving as inputs to low-level and supervisory controllers as an alternative to unreliable or difficult to measure physical sensors

measurement. This allows the controller to continue to work to its objectives while the failed device is repaired.

The lower-level controllers are normally implemented using Programmable Logic Controllers (PLCs) or Distributed Control Systems (DCS). The tactical-level controllers are implemented using either SCADAs or PCs. The supervisory metallurgical control loop is normally executed offline using a PC. These three hierarchical levels are normally found in any sophisticated control system controlling a complex metallurgical process.

PCs and DCS systems are generally needed for implementing dynamic controllers. A commonly implemented dynamic controller in the metallurgical industry is Model Predictive Control (MPC). MPC is based on the "receding horizon" principle that allows for future optimal control actions to be computed only for a short future time horizon: $[t, t + T]$, where t is the current time and T is the prediction horizon length. The most important innovation in this type of controller is that only the first term of the sequence is implemented. Subsequently, a new sequence that replaces the previous one is computed when a fresh set of updated measurements is available. And once again the first term in this sequence is only implemented and this cycle goes on. In other words, every sequence is computed in an incremental manner.

MPC thus involves extensive mathematical modeling of the process in question, and the selection/design of a suitable objective function. MPC is often used for control and optimization of kilns, furnaces and mills.

Kiln Alternative Fuels Optimization: Case Study

With increased focus on reducing the cost of operation of cement plants, organizations have started adopting alternate fuels for kilns. This has introduced some challenges for instance, one has to cope with the different characteristics of the different alternative fuels.

ABB developed a kiln (cement) control strategy that achieved optimal kiln operation. These control strategies were based on neural networks and fuzzy control. This system incorporated an Alternative Fuels Optimization Module based on Neuro-Fuzzy controller integrated to an MPC. This advanced control application has been reported to have achieved optimized use of alternate fuels, reduction of waste and strict satisfaction of environmental, contractual and technical constraints.

The main idea here was to use the data gathered by the data acquisition layer and additional data from the market and LIMS to calculate the lowest cost fuel mix that satisfied the process and business constraints. The basic element of this algorithm as reported in the chapter was a dedicated kiln model that was used for Model Predictive Control. The mathematical model estimated cooler, flame, burning zone, back end and preheater temperatures, kiln energy requirements, emission and volatiles levels, etc. The model parameters were tuned using a combination of neural networks and Kalman filtering techniques.

Process Historian

At the turn of this century, it was envisaged that process data needed to be acquired in a real-time basis with accurate time stamps to observe the global trend of the system. This was much beyond the need of just collecting data for tracking a local controller's set point. This resulted in the development of the Process Historian. Process Historians have come a long way in acquiring plant management information about production status, performance monitoring, quality assurance, tracking, and genealogy, and product delivery with enhanced data capture, data compression, and data presentation capabilities. The historians allow for archival of time-based process data that can be used in a future period for elaborate analysis.

10.3.2.3 Logical Layer 3: Integrated Information

Mine-to-Mill Integration: Optimization of Blasting Costs Vs. Milling Costs

The use of RFID tags to synchronize information flow with material flow was discussed earlier. The emphasis therein was on the data acquisition technology. In this section a further elaboration of Mine-to-Mill strategy is provided from the point of view of illustrating how the integrated information (which is the subject matter of this chapter) can enhance the performance of mining and downstream processing activities.

One of the key objectives of the Mine-to-Mill integration is to maximize the profitability of operations through a holistic approach to the optimization of ore fragmentation. Generation of fines by blasting can have a significant impact on the following:

- SAG mill throughput increase (10–30%)
- Reduced overall energy costs (up to 30% decrease in kWh/ton)
- Higher excavator productivity
- Higher truck loading
- Better primary crusher productivity
- Better heap leach permeability

The key task for Mine-to-Mill optimization is to identify the optimum feed size distribution for the crusher or the mill. The optimum feed size distribution is typically generated through blasting (ROM size distribution), crusher settling along with stockpile and feeder management.

Mine to Mill now is a proven methodology and has been applied at many large open pit operations around the world (McKee 2013; Renner et al. 2006). This involves rock characterization, benchmarking, and process modeling. Many of the projects have achieved 10–30% increase in mill throughput. The issues facing success with this approach are not just technical, but involve cultural change and sustained implementation.

These Mine-to-Mill challenges are similar to that faced by business improvement opportunities such as in Lean and Six Sigma implementations and could be addressed through a dedicated and systematic effort. The concepts involved in Mine to Mill could be readily applied to Mine to Metal or for Ore-to-Profit applications. The key principle is to make changes upstream to improve quality of feed stream to reduce costs and improve efficiency through a holistic approach involving the entire mining value chain.

Integrated Operations Support and Quality Production Reporting

A recent development in the mining industry has been the establishment of operations centers, both on site and remote, where a substantial part of the total operation is monitored and controlled. Such centers are designed to oversee multiple aspects of an operation and this directly assists in establishing the desirable integrated thinking. While operation centers are not essential to an integrated thinking approach, there appears to be little doubt that they will help facilitate this goal as evident from recent successes from many companies. Figure 10.5 shows the various levels that an operation must achieve to obtain improved productivity.

The key target for any organization that is embarking on an integrated enterprise is to focus on obtaining quality and reliable operating data. This step is critical because some operations may not be willing to present their operating data on visible dashboards if day-to-day operating data obtained from various instrumentation, historians, and analytical tools in operations are not reliable. Most operations rely on month-end inventory to report gold production as an example. Daily accurate

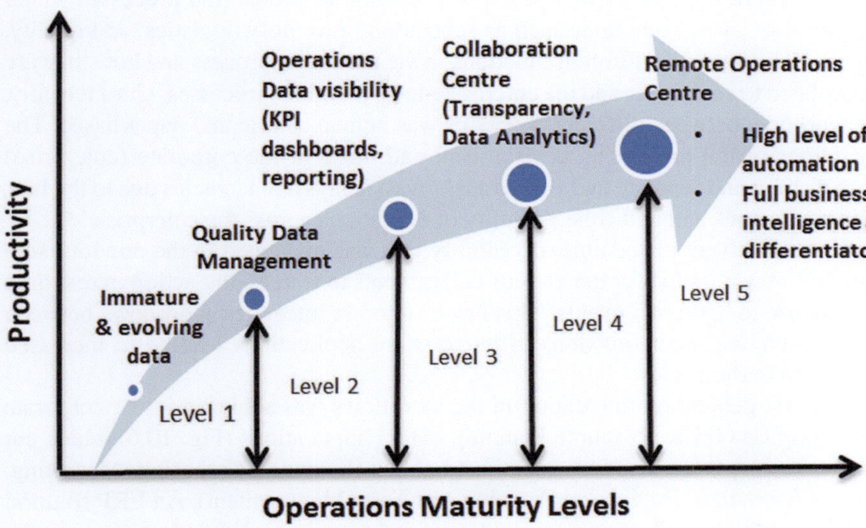

Fig. 10.5 Operations maturity levels required for improved productivity

reporting of gold production is challenging because of limitations in reliable metal balancing and accounting processes due to sometimes poor reconciliation between mine and mill and also within the different processing steps in an operation. This is not the fault of operations as embracing new ideas and relevant technologies requires a joint effort between different functions in operations and also corporates requiring an integrated thinking platform to maximize value within the mining value chain. This is an area corporate must step in to assist operations to develop this core skill-set to allow the operations to confidently report daily production and display them in integrated dashboards. This will allow operations to monitor and optimize production more regularly than possible in the present scenario.

Also, typically, a comprehensive mobile and web dashboard displaying up-to-date values for the relevant KPIs are available to all important stakeholders across the enterprise to monitor and track performance and extract useful reports from a common database. This replaces the conventional silo-based individual reports generated by disparate systems in different departments that are based on distinct and separate databases. Due to the disparity in the master data stored in the different systems, reports generated from these systems do not reflect one truth.

10.3.2.4 Layer 4: Automated Workflows

Plant to Enterprise Integrated Workflows

The corporate operations in the metallurgical industry as in the case of any manufacturing organization involve finance, purchasing, sales and marketing, human resources, asset management, and production planning actions. whereas the production operations are mainly related to the execution of production processes which will involve several functions such as scheduling, inventory, logistics, and quality. Figure 10.6 shows the different functions in the execution process and how they are interrelated to each other and the enterprise-level planning processes. Until recently, the workflow between different processes was human centric and paper based. The operations of the production environment and those of the corporate (enterprise) were distinct and separate and were fraught with delays and latencies due to the lack of instantaneous and real-time visibility of operations across the enterprise.

With the advent of the unified visibility that was explained in the previous section, it is now possible for the various departments to start taking action in real time in response to “current” events. This has resulted in integrated workflows between departments and their functions using software applications leading to increased operational efficiencies.

The first-generation integration in the workflows was achieved in the corporate side using Enterprise Resource Planning (ERP) applications (Fig. 10.6) which cut across the corporate functions mentioned above (Finance, Purchasing, Marketing, Human Resources, Production Planning and Asset Management). An ERP solution with integrated workflow management provided three immediate advantages firstly, it allowed for instantaneous and unified tracking of “operational states” of business

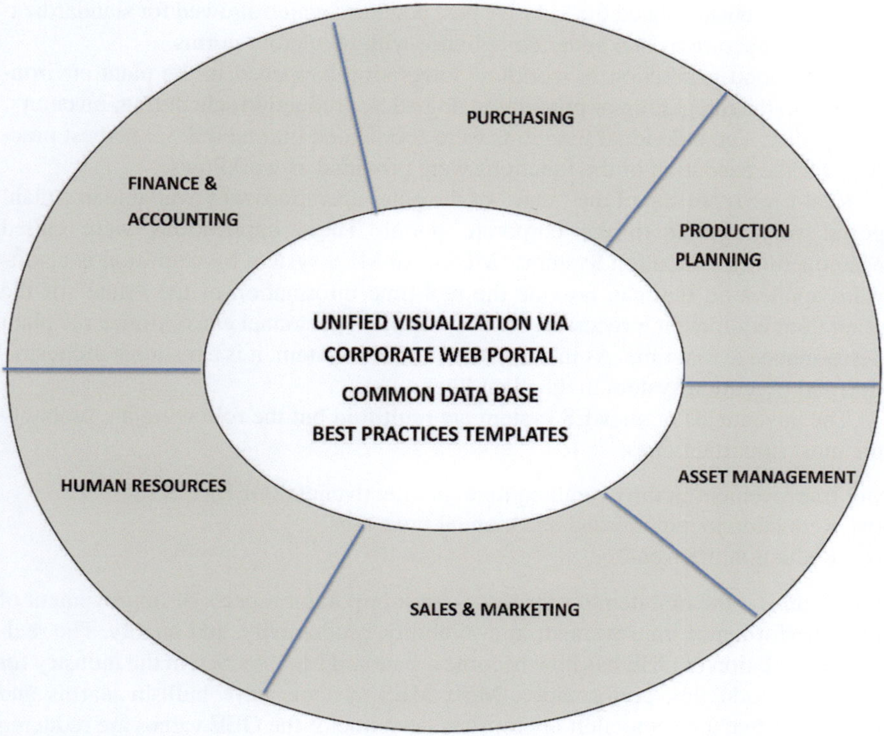

Fig. 10.6 Integrating Enterprise Operations using ERP software

processes across all related corporate functions as opposed to only getting periodic manual updates extracted from individual silos of information. Also, the dynamic occurrences of events in any division or department (within the corporate environment) were immediately notified to the other departments via alerts or information in a common corporate portal.

Secondly, ERP used a common database across all functions as opposed to maintaining individual silos of data. The common database allowed for automation applications to share data between the different departments mentioned above. As a result, inter-departmental workflows were possible. For instance, in the cement or iron and steel industries, finance could trigger (integrate) the purchasing workflows for raw materials (order processes) more effectively based on the current and actual production plans that in turn could be actively formulated based on the actual current demand. The availability of real-time or near real-time information allowed for the individual departments to take informed decisions and thereby reduce waste and costs significantly. This was not possible in manual, paper-based workflows that relied on historical or outdated information.

Thirdly, ERP allowed for the implementation of best practices across the enterprise and reduced process variance. In other words, each functional module of the

ERP system encapsulated the industry best practices which allowed for standardization of the operations and better compliance with regulatory norms.

The second generation in workflow integration happened in the plant environment with the integration of production, logistics, production scheduling, inventory, and quality. The individual functions were subdivided into modules, and best practices for the execution of the functions were provided as workflows.

Real-time reporting of the "state" of the plant operations was available in a plant portal very similar to the corporate portal. These applications were called Manufacturing Execution Systems (MES). An MES system by definition is a software application that can provide the real-time information of the "state" of the plant floor equipment, processes, etc., so that plant personnel can optimize the plant performance in real time. As in the case of an ERP system, it is a resource orchestration and execution system in the plant floor.

The advantages of an MES system are multifold but the following are probably the most important ones:

(a) Improvement in the overall equipment effectiveness (OEE)
(b) Reduction in paper based and manual processes
(c) Reduction in inventory

Of the above, the first item encapsulates three important aspects of improvement of plant performance improvement in availability, productivity, and quality. The real-time calculation of OEE has now become a standard best practice in the industry for tracking production performance. Most MES systems have built-in alarms and thresholds that are set to alert appropriate personnel if the OEE values are reducing beyond a desirable level to immediately take appropriate corrective action (Fig. 10.7).

Apart from improving production performance, MES systems help in the integration of multiple functions as stated above based on a common database very similar to the ERP applications. Since the various functions like logistics, maintenance and production can share data, it is possible to optimally schedule production taking into consideration raw material input, availability of equipment, etc. Furthermore, MES applications are used to track optimality of a process for instance, MES can track the efficiency of a process or equipment for that matter such as a boiler or a pump or a cooling tower and respond with alarms whenever the values go off limits. MES systems are also employed for tracing the complete history of the operations as they happen in the plant floor based on batches, operations within batches, equipment condition, and personnel operating the equipment. Such traceability is important in the metallurgical industry for meeting regulatory norms.

In summary, MES implementation has allowed the industry to eliminate waste, monitor, and alert suboptimal performance, integrate and automate workflows between different functions, track the execution of processes, improve standardization and quality, and because of all of the above, reduce the cost of operations.

Even though the ERP and MES applications have contributed to integrate workflows within the corporate and the plant floor environments, respectively, they were considered as two distinct islands of automation and were not integrated to each

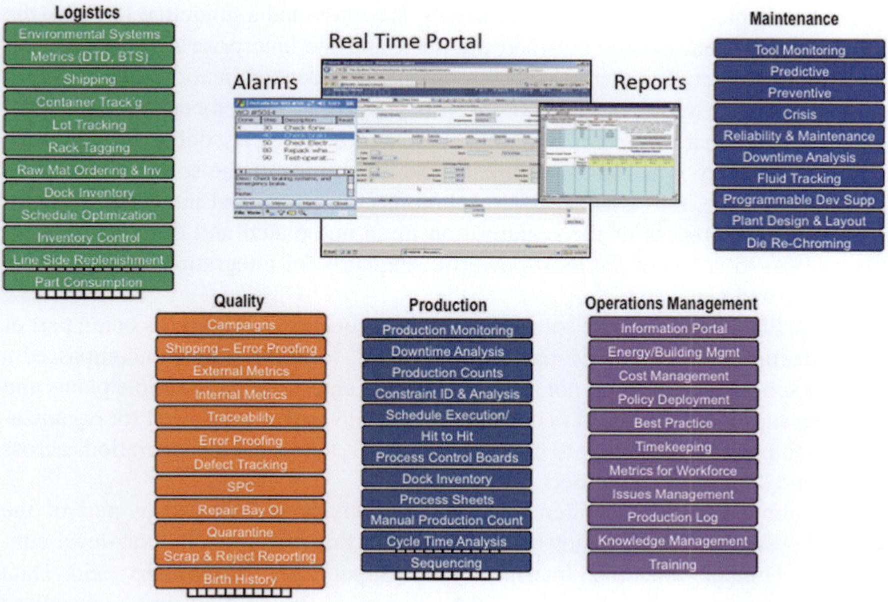

Fig. 10.7 Integrated Plant Operations using MES

other until the middle of the last decade. The need for integration of the workflows beyond the plant or the corporate was precipitated by the need for better efficiencies and to remove the latencies, in the day-to-day transactions between these two environments within a manufacturing organization. This has resulted in the birth of the third generation in workflow integration, which is widely called Plant to Enterprise (P2E) Integration. P2E integration allows for real-time integration of business processes in the plant operations with those in the corporate operations.

In parallel, the International Standards Association (ISA) introduced a standard for the integration of enterprise and control systems, ISA-95. This is not only for the metallurgical industry in particular but also for the overall process and discrete manufacturing industry in general, even though the standard has a certain bias toward the process industry. ISA-95 consists of models and terminology for defining the workflows between the control systems in the plant and other software application in the enterprise. These standards are used in the industry to determine which information has to be exchanged between systems (software applications) for sales, finance, logistics production, maintenance, and quality. This information is structured in Unified Markup Language (UML) models, which are the basis for the development of the standard interfaces between ERP and MES systems.

The ISA-95 standard can be used for several purposes, for example, as a guide for the definition of user requirements, for the selection of MES suppliers and as a basis for the development of MES systems and databases and to integrate specific departments within the enterprise regardless of whether they are part of the plant floor or the enterprise. In that sense, ISA-95 has provided a guideline to break the artificial walls that existed between the plant and the enterprise and to allow for "data sharing" between these two environments between applications that reside in each of these environments. As a result, it is possible to implement a business process workflow that transcends multiple applications today. By doing so the manufacturing industry is not held hostage by the limitations of the software packages and solutions the individual software solutions collaborate and integrate to implement business process of the organization in an automated and efficient manner. Today ISA-95 is one of the most powerful standards for integration of plant to the enterprise and vice versa.

These Plant-to-Enterprise solutions were subsequently identified as being part of "Manufacturing Operation Management (MOM)" solutions which encompassed a broader scope of connecting not just a plant to the enterprise but multiple plants and even the supply chain. MOM is different from MES in that it allowed for organizations to integrate outsourcing to internal operations, to standardize operations across plants and supply chain partners.

The above-mentioned wider scope has recently been included as part of the ISA-95 systems hierarchy shown in Fig. 10.8. In this hierarchy, the low-level controllers, Human Machine Interfaces and Supervisory Controllers and Data Acquisition occupy Levels 0–2, MOM solutions/software platforms, scheduling software, inventory control and other plant operations occupy Level 3 and enterprise operations (such as those controlled and automated by ERP, and EAM) occupy Level 4. In general, MOM platforms aggregate the vast quantities of data coming from controls, automation, and supervisory control and data acquisition (SCADA)

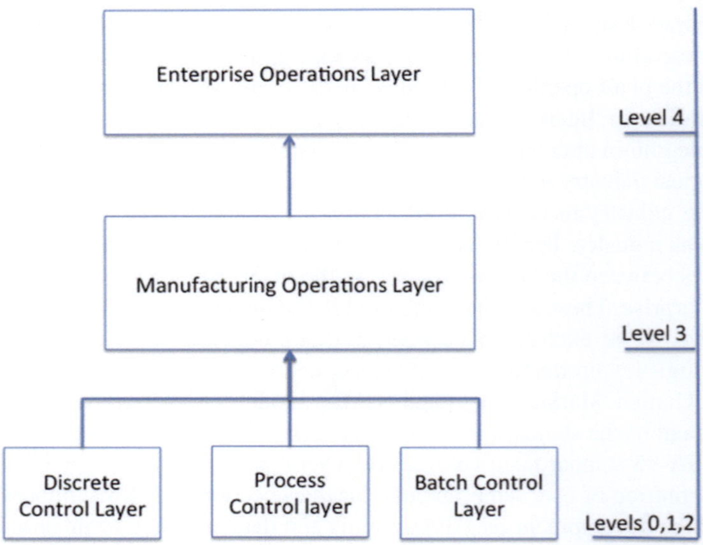

Fig. 10.8 The Manufacturing Value Chain ISA 95 Standard

systems and convert them into useful information about the production operation. In that sense, MOM platforms facilitate a more comprehensive, real-time view of all the plants and the supply chain as an integrated unit.

Typically, an MOM solution comprises one or more of the following attributes:

- A configurable solution as opposed to being hard coded to a particular application or domain
- Standard integration to Enterprise Systems such as ERP and EAM systems
- Standard integration to Industrial Automation and Plant Control Systems
- Capability to represent manufacturing data and equipment in a standard model
- Capability to perform business process modeling and integrate workflows automatically
- Capability to visualize the current status of the plant and the enterprise in multiple media
- Capability to aggregate, analyze, and respond to real-time manufacturing events

The example illustrated in Fig. 10.9 is an MOM use case that integrates production and maintenance to achieve better usage-based active Preventive Maintenance. Traditionally, Preventive Maintenance in the metallurgical industry has been time based that is based on an elapsed duration of time regardless of the usage of the machine during the elapsed time. Therefore, a machine could be either over-maintained or under-maintained based on how much the machine has been used during the said time period.

With the advent of the MOM platform, a use case for integration of actual usage of a machine to a Preventive Maintenance schedule is possible. The figure shows the MOM solution collecting the usage (run time, number of cycles, etc.) from the machines in the shop floor and automatically updating the Enterprise Asset Management (EAM) system on the corporate side in real time. The EAM system on the other hand triggers a workflow for generating a PM work order automatically once a threshold of usage is crossed for any machine. By doing so, the Enterprise Asset Management system that was an isolated system previously has now become

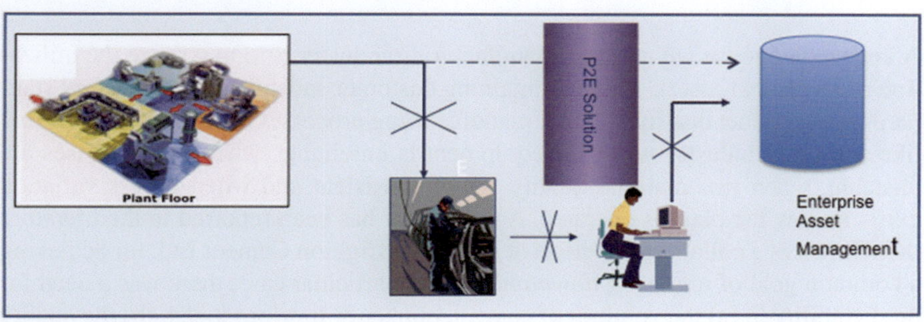

Fig. 10.9 An MOM solution Real-time production to asset management integration example showing the replacement of the human loop by an automated P2E solution (Enterprise Gateway 2010)

integrated to the plant dynamics. The wastage due to over-maintenance or under-maintenance of machines is avoided.

10.3.2.5 Logical Layer 5: Collaboration Layer

The discussion on the four layers thus far explained how automation in data access, communication, control, visualization, and workflow integration can help in building an integrated enterprise. This infrastructure can be helpful to an organization to "collaboratively" achieve optimized solutions to problems that impact multiple departments and stake holders in a holistic manner. This collaboration function constitutes the fifth and the final layer of the five-layer architecture.

There is a difference between the type of collaboration that is achieved via workflow integration (the fourth layer of the architecture) and what is discussed in this section. The workflow integration allows for a seamless and sequential execution of standard operational processes that transcend multiple departments. The dynamic in this case is more or less known and the integration is very much a routine one.

The collaboration that is referred to here is a deeper quest for achieving an optimization goal whose dynamic is understood via the synergy between the multiple stakeholders. The availability of real-time data, historical data, meta-data (information), integrated workflows (wherever possible) and unified visibility only helps in achieving this collaboration. The collaboration is enabled by automation (the integrated architecture) but involves the human loop supervisors, managers, domain experts, and analysts to collectively solve the problem in hand using the data, meta-data, and knowledge generated from the said automation. The following section outlines an example of collaboration using a case study in the cement industry for removing process variation. The first part of the discussion illustrates how the cement company collaborated to solve a problem in hand. The second part explains one way to perform collaboration to continuously *prevent* problems.

Collaboration: Cement Industry Case Study

A key challenge for any process manufacturing industry is to maximize the utilization of its critical assets, while maintaining customer satisfaction, cost, safety standards, and product quality. Cement manufacturing processes are impacted adversely like any other industry when the equipment is unreliable, when the processes are unstable, when raw material quality is not consistent and when excess variation exists in how the plant is operated. A case study has been reported in the literature that discusses a collaborative effort at Adelaide Brighton Cement Ltd. for achieving a common goal of removing *downtime*. In this particular case, there was a need for predictability in (a) the volumes of cement, clinker or lime produced, (b) the quality of the product(s) and (c) the cost of manufacturing. Previous initiatives of the company were reported to be silo based and had focused on reducing costs alone or trying to improve operations without probing into a deeper need for achieving stability in the dynamic interplay between productivity, quality, and reliability. The company's previous non-collaborative, but departmental approaches to solve the problem in parts had been more or less ineffective.

With the common goal of achieving stability in the overall process dynamics (read reduced process variation), the collaborative team started probing into the causes of variation in the process and their occurrence patterns. This required collection of data. However, the team had to collect data manually which was in itself affecting plant performance as the process engineers were spending up to three days per month simply generating reporting data and not focusing on improving plant operations. This lack of timeliness for data analysis meant that it would be days or weeks to understand the plant dynamics. The Process and Engineering teams also manually gathered and recorded downtime data using a number of sources such as Control Room log sheets, MS Access databases and MS Excel spreadsheets. As a result, there were different points of view and no single version of the truth. This motivated the company to use appropriate tools for productivity, conformance (quality), and reliability.

The company embarked on using some automated data collection and analysis tools (the first two layers of the integrated architecture). The online data immediately started providing trends that pointed to why the process is varying. A downtime reduction software was used as part of the Reliability Improvement Plan to assist the site to record and act on emerging failures in "real time," before the stoppages manifest into larger, more costly downtime events. This also improved the productivity.

The plant also introduced a production system that provided the following real-time data: Metrics such as fuel efficiency Gj/ton, power consumption kWh/ton, production data sacks/h, tph equipment reliability run hours, % utilization, % reliability, % quality, % performance factor (% of MDR [Maximum Demonstrated Rate]), wastage rejects tons, etc., were tracked (Level 3 of the integrated architecture). With the basic three-layer infrastructure in place the collaborative team was able to address the common problem of process variation.

The team identified and categorized the sources of process variation to be as follows: (a) variation in raw material attributes and quality; (b) variation in operator performance and training; (c) variations due to seasonality and demand patterns; (d) variations in instrumentation calibration and accuracy; (e) variation in maintenance checks frequency and quality; (f) variation in the work instructions, KPIs.

With the automated systems and technology, the company had access to relevant and timely information and therefore the ability to monitor many sections of the plant via centralized control rooms and effect useful actions. They identified certain trends such as differences in trip rate across shift groups. This was identified therefore as a management problem and one of standardization across the various groups.

The synergy between the groups provided for the understanding that the stabilization of the variation in the process has to be done in an incremental, evolutionary, and continuous manner taking into consideration all of the above-mentioned factors that contribute to variation. This data-driven, continuous improvement collaborative initiative has been useful in significantly improving raw mill control, kiln, and raw mill stoppages. The company concluded that the inclusive methodology adopted with all concerned employees and their collaboration and commitment with the focus on stabilizing the process continuously has delivered the improvement in kiln operation by significantly reducing stoppages and downtime by 50%.

Proactive Collaboration: Asset Performance Management Example

The previous case study demonstrated a real situation of solving an existing problem of process variation via data-driven collaboration between multi-disciplinary teams namely, production, quality, finance, and maintenance. It also demonstrated the value of real-time data and information in addressing process optimization that involve multiple stakeholders in the company. In this section, the concept of collaboration is taken to the next step to perform proactive interference in the dynamics of a plant in order to introduce desirable effects and eliminate undesirable performance in a future time period.

In asset-intensive industries (such as the metallurgical industry), one of the important strategic goals is to achieve a better return on assets as they constitute a significant portion of the investment. In order to not lose track of such a strategic goal, there is a need for synergy between multiple stakeholders of the business to foresee all possible conditions that can lead to the goal not being tracked in a future time period and avoid them beforehand. The reader must note the difference between this approach and the conventional approach of identifying the reasons for a problem after it has occurred.

Asset Performance Management (APM) is an evolving methodology that allows for *collaboratively* and proactively achieving a desired performance in the assets that will identify a good majority of failures before they happen as opposed to addressing all of the problems after the failures have occurred.

APM is not a one-time initiative but a continuous improvement process. It departs from the conventional view of looking at an operational piece of equipment as a depreciating asset, but rather as a driver of business performance and sustainability. Traditionally, there have been different primary drivers for an APM collaborative initiative most of them fundamentally aim at maximizing profit, avoiding risks, getting better return on production assets, minimizing costs, minimizing variation, etc.

The emerging trend is one that looks at APM evolving in an organization due to a synergistic combination of all of the above drivers. In other words, no longer is a mill only interested in regulatory compliance as a driver for APM, but also wants to put in place a multi-pronged strategy of maximizing productivity and quality as well to achieve better profitability. Slowly mining companies are transforming from defining multiple initiatives that address asset performance (as a part of those initiatives) to an Asset Performance Initiative that spans multiple disciplines and stakeholders within an organization to track superior performance in a future time period.

Typically, in the mining industry for example, the processing units are mandated to maximize productivity from an asset performance standpoint that means every asset (excavators, conveyors, crushers, grinding mills, etc.) must maximize their effectiveness and efficiency for a given set of dynamic conditions. The corporate governance/compliance department would like to continuously reduce the risks and exposure—from an asset performance point of view, this means the assets in question must have some redundancy built in their critical systems in addition to having a record of failures, the impact of these failures (penalties, loss) and the probability of occurrence of these failures in any given future time period. The quality department would like to reduce the Cost of Quality this means (from an asset performance point of view) the preventive and the appraisal costs of quality of a product output from the assets should be increased to reduce quality problems in a future period. The maintenance department would like to reduce its annual maintenance costs as a percentage of the overall operational costs. The corporate executive on the other hand would like to maximize the return on the assets in order to provide maximum returns to the shareholders. This would mean the increase in net profit without an increase in new capital expenditure. Therefore, an APM process by definition would have multiple, individual departmental goals that need to be simultaneously achieved in other words, the strategies employed within the APM process for the achievement of a particular departmental goal in a future time period cannot violate the achievement of the stated goal of another department.

The example in Table 10.1 shows how in an APM process (that spans multiple departments) every department strives for optimality in the achievement of the goals across multiple related departments. Let us take the first row as an example the production department's goal of maximizing profitability from its operational assets in this case is achieved without violating the goal of the maintenance department, which is to keep the maintenance costs of these assets at a minimum. The same is true with the quality department which has to decrease the cost of quality without increasing the cost of maintenance (which is the maintenance department's goal), and so on. The multi-disciplinary APM team therefore has to work out a set of "common minimum strategies" that collectively achieve the goals of every department without violating the goals of the other related departments. Typically, such strategies are defined directly for implementation at the operational levels by not only tracking certain operational metrics which are "loosely coupled" to the strategies, but also predicting the future values of these metrics and using the workflow integration layer to generate proactive actions.

Table 10.1 Optimization of multiple objectives

Department	Goal	To not violate
Production	Maximize operational profitability	Maintenance
Maintenance	Decrease maintenance costs	Production, quality
Quality	Reduce cost of quality	Maintenance
Corporate governance	Reduce risk via redundancy and prediction of failures	Production and maintenance
Executive	Return on production assets	Corporate governance production, maintenance and quality

Active Criticality Analysis for Assessment of Future Risks

Active Criticality Analysis (ACA) is a collaborative practice under an overarching APM program that is adopted to choose the "right assets" to monitor at any given time to achieve the best optimization in the future. ACA is a dynamic method that is extensively used in any risk-based asset management strategy. Simply put, ACA provides a Pareto ranking of the critical assets, at any given time, based on the consequences of the failure of these assets to the business. ACA does not result in a static selection of assets it is an active time varying dynamic. Therefore, a regular and timely assessment of criticality is required depending upon changes in the operational dynamics.

An organization can use qualitative or more detailed quantitative techniques to arrive at asset criticality both approaches have merit. For instance the calculation of the likelihood of a failure of an asset (or a component of an asset) could be considered as a probability between 0 and 1, or can be considered qualitatively as belonging to fuzzy intervals such as "frequent," "occasional," and "very rare." The likelihoods increase or decrease dynamically based on the current status of the asset ("symptoms" of failure) and the past history of failures. A real-time measurement system can monitor for the symptoms and update the criticality map dynamically (of the "current" estimation of the likelihood of failure in a future time period). A typical ACA normally identifies a relationship between such likelihoods of failure of an asset to the business consequences of such failure to the organization. Once again, the consequences can be quantitatively determined or qualitatively established to fall within certain intervals such as "catastrophic," "major," and "minor." Based on the current inputs from the real-time measurement system if there is a change in the critical ranking of the asset, then such changes are reported to the asset management system, which will, in turn, implement the appropriate maintenance strategy in response to the changes.

The outcome of ACA is normally the decider of the type of maintenance an asset would need say, preventive, corrective, and so on—in order to effectively manage the overall risks of failure in a future time period. In many instances, a previously planned maintenance regime may have to be refined in the wake of some new risks identified by ACA. The ACA therefore forces the maintenance organization to always (and actively) concentrate on the most important assets at any time (from a business point of view) and therefore renders the maintenance actions to be more effective. Figure 10.10 shows the relationship between the real-time identification of the symptoms of failure and the dynamic ranking of the criticality.

Every collaborative program needs a set of metrics and it is obviously important that the metrics are easily measurable and the measurement relates to the achievement of the goals in other words, if the goal is the reduction of maintenance costs at a strategic level in a given future time period, then there has to be one or more measurements that relate directly to the goal of reducing costs and so on. An illustration is provided in Fig. 10.10 for a typical set of measurements that can be made (and tracked) from strategic, tactical and operational points of view (this illustration provides the measurement framework and examples of some metrics, the metrics could

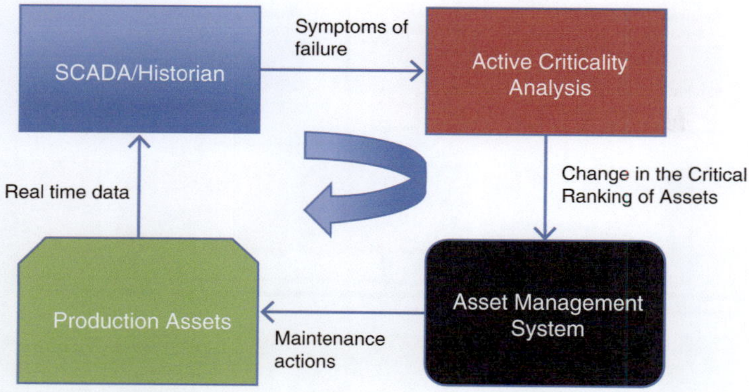

Fig. 10.10 The continuous cycle of active criticality analysis of assets metrics hierarchy

be different from the ones presented below based on the specific industry/goals) using the various layers of the integrated enterprise architecture. The figure also illustrates how the measurements are at multiple levels of abstraction and are related to each other.

The example shows three measurements at the strategic level corresponding to the strategic (and collaborative) goals of reduction of maintenance costs, reduction in risks and increase in the net return on production assets during a given future time period. The reduction of maintenance costs in this case stems from two measurements, maintenance costs as a percentage of Replacement Asset Value (RAV) and maintenance costs as a percentage of Regulatory Failures. Even though some of the cost elements that shall be accounted for in the calculation of these two metrics may overlap, the metrics independently maintain two distinct measurement points for understanding/observing the progress of achieving the collaborative goal of reducing maintenance costs. The second collaborative goal, namely, reduction of risks, is tracked (once again) by the maintenance costs as a percentage of Regulatory Failures. The third collaborative goal, achieving a better return on assets, is tracked by the NROPA measurement, which is a formula that requires net profit and total asset value as inputs. The optimization of all of these goals in a given future time period will be the objective of the "progress function." The trends (rates of change) in the progress functions (which are the gradient of the progress function) can be computed at any given time to track the velocity of progress.

The tactical-level metrics (in Fig. 10.11) act as an intermediary between the physical and strategic metrics. Continuous improvement cells/programs are normally involved in tracking the progress of the tactics to improve asset performance by measuring the tactical metrics. It is obvious that by focusing too much on the operational metrics one loses sight of the big picture (from a collaboration stand point) and on the other hand by defining the strategic metrics without a relation to their practical implementation and measurement process one loses the relevance of the strategic metric. The tactical-layer measurements bridge these two worlds.

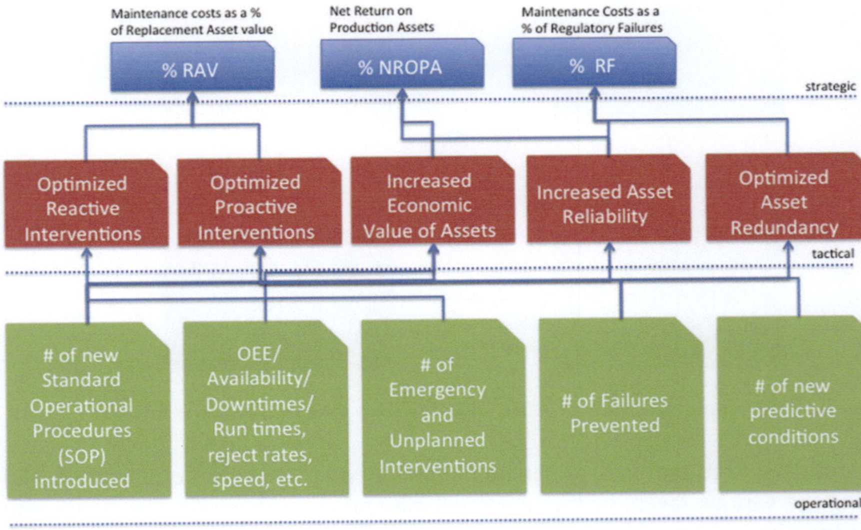

Fig. 10.11 APM metrics hierarchy example

The operational-level measurements are the fundamental layers that collect fine granular operational data that in turn feed into the tactical and strategic layers. While the tactical- and strategic-layer measurements are abstracted logical measurements, the operational measurements are physical measurements.

Optimization of Reactive and Proactive Interventions

An important outcome in the collaborative pursuit of optimizing the cross-disciplinary goals to maximize asset performance is to have a well-thought-out plan for future maintenance interventions. Too much of intervention is costly and too little is again costly. In order to implement the strategic requirements of reducing the cost of maintenance and/or increasing the net return on assets, one has to look at arriving at an optimal balance of proactive and reactive interventions.

Optimality can be achieved by understanding the elements of both proactive and reactive maintenance activities. Proactive maintenance activities include any activity that is performed for maintaining an asset before they fail. Reactive activities therefore are those that are performed after the failure. Proactive maintenance comprises two types of activities, Preventive Maintenance and Predictive Maintenance. Reactive also comprises two types: Emergency Maintenance and Corrective Maintenance. Achieving optimality would mean increasing the cost of Proactive Maintenance appropriately in order to reduce the cost of Reactive Maintenance substantially.

As stated earlier, Reactive Maintenance comprises emergency and corrective maintenance. Of these, the emergency maintenance activities are the costliest. Even if it is not possible to eliminate them, they have to be reduced substantially for the organization to reduce its maintenance costs. Emergency Maintenance results from two main reasons: (a) lack of a proper maintenance policy and (2) lack of reliability.

The lack of proper maintenance policy herein means the lack of clarity on what constitutes "emergency" within the organization. The exact meaning of emergency should be clarified in the "maintenance policy" and a common method to categorize any maintenance job as emergency or non-emergency should be available as a "standard operating procedure" (SOP). The adequacy of the maintenance policy is determined at the strategic level, and the standard operating procedures are performed at the operational level. But collaboration at the tactical level is needed to monitor the performance of the operational level as regards emergency maintenance by (a) analyzing the emergency activities in the past; using the data in (b) to predict possible emergency situations in the future (true positives); (c) predicting emergency situations that will be "hidden" until it is too late (false negatives) and (d) predicting those situations that will emerge as not emergency ones (false positives).

True positives are the right "emergency responses" as per policy that is, the responses of the maintenance organization are exactly aligned with the policy and are performed as per the standard operating procedures. False positives are emergency responses that result in a "false alarm" maintenance calls wherein the team upon reaching the site realizes that they are not actually emergency in nature. Therefore, these are wasted emergency interventions. Finally, there are false negatives emergency situations that were not identified which consequently result in a catastrophic failure.

Such real-time monitoring of the type of maintenance responses, validation of the responses against policy, adherence to the standard operating procedures and predicting false positives, false negatives and true positives will help in the understanding of the "effectiveness" of the emergency maintenance practices within the organization. Appropriate actions can be taken based on the monitored data, such as if there are emerging symptoms of a variance in the adoption of the SOP, then more training on the SOP is initiated before the variance becomes widespread. If there are a higher number of false positives (rate of change in the number of false positives in a given period), then a proactive investigation is required into why the responsible owners for making the call are suddenly unable to do their job correctly. There can also be simple hidden reasons for ordering emergency maintenance such as the rates for maintenance work being higher under emergency maintenance than in non-emergencies. It is up to the collaborative team to understand these reasons and attempt to continuously drive the emergency maintenance costs down by improving the effectiveness of emergency responses.

The second aspect that affects emergency maintenance costs (for that matter even the corrective maintenance costs) is the lack of reliability. Reliability in simple terms is inversely proportional to "failure rate." Every asset will have a history of failures and the frequency of such failures the inverse of the frequency of failures is the failure rate. If the failure rate is low, then the asset is supposed to be more reliable. Failure rates and failure histories of assets are maintained in the Enterprise Asset Management (EAM) systems. Based on the failure history maintained in the

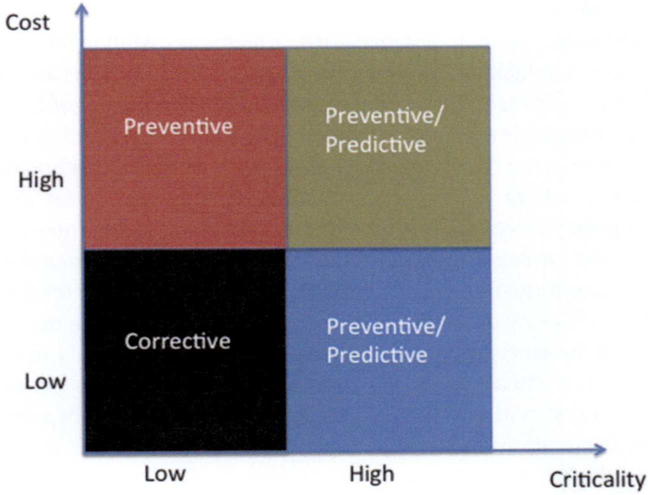

Fig. 10.12 Optimal maintenance planning based on criticality and cost

EAM systems P-F Graphs (Potential Failure Graphs) can be generated for critical assets. The P-F curves will predict the possible future failure states for a piece of equipment which is at a certain current state. This input can drive Preventive Maintenance of these assets.

While all of the above-mentioned discussed corrective and emergency responses that can be avoided, there are instances when Corrective Maintenance is the right one to adopt. Corrective Maintenance is good if the criticality of the asset and the cost of the maintenance activity are both low. Figure 10.12 illustrates the "calculus of interactions" between these two variables asset criticality and the unit cost of maintenance for the activity.

Figure 10.11 shows that if the asset criticality is low and the cost of doing the maintenance activity is low, then Corrective Maintenance is the right option. It can also be noted that three-fourths of the maintenance activities fall under the proactive maintenance (either predictive or preventive or a combination of both).

Finally, as a result of the collaboration at the operational and tactical levels the return on assets (ROA) can be monitored and continuously improved. The ROA establishes how the entire asset base of the organization returns to the shareholders, a metric that is vital to track at the strategic level of an integrated metallurgical enterprise.

10.4 The Internet of Things

The technology called internet of things (IoT) is evolving in which the industrial equipment has the capability to communicate with each other. IoT is the connection of objects such as computing machines, embedded devices, equipment, appliances, and sensors to the internet (MaRS Market Insights 2014). As a result, data from many different sources can be combined and insights can be gained using analytics that was not possible earlier. Table 10.2 shows the various components of IoT. It shows a number of industrial instruments that can be connected using IoT and the ways these instruments can be connected to analyze the data and display them for management to act upon the intelligence so derived. Table 10.3 shows the application of IoT to mining industry. It shows a number of challenges and opportunities in mining industry for tailoring solutions to improve process efficiency.

An example of improving process efficiency using IoT is the use of sensors in a scooptram. By placing sensors in an underground scooptram on the seat and the motor, the motor can be shut down remotely if it remains idle for more than a specific time and driver is not on the seat.

The IoT can also help in avoiding vehicle collisions in mining operations. For this purpose, wirelessly communicating personnel and vehicle tags can be used and vehicles can be equipped with infrared, radar and video systems, so that operators are notified of the obstacles.

Table 10.2 Internet of things (MaRS Market Insights 2014)

Components	Constituents
Human and external interface	Management Governance (e.g., regulators, safety/environment) Integrated remote operations centers markets
IoT platforms and processors	Data-processing platforms Device configuration Data analytics Risk modeling Optimization feedback loops Data security SCADA
The communicators and controllers	Networks (global system for mobile radio, satellite, ethernet, wireless area network) Programmable logic controller Signal processor
The things	Accelerometer Gyroscope Motors and pumps Pressure sensor Proximity sensor Thermometer Autonomous haulage truck Autonomous drilling rig Switch Gauge GPS RFID Video

Table 10.3 Application of IoT to mining (MaRS Market Insights 2014)

Challenges	Opportunities	IoT technology solutions
Lack of R&D Rising costs Declining labor Low commodity prices Declining ore grades Insufficient infrastructure Regulatory pressure Low levels of exploration	Plant automation Energy and process optimization System integration Consolidation of systems and protocols Increased safety Increased environmental knowledge Mineral ore predictability Ore targeting optimization	Asset utilization Energy management, e.g., ventilation, hauling and crushing Predictive maintenance Inventory and asset tracking Loss prevention Integrated remote operations centers Geostability modeling and decision-making

One example of a company that provides practical solutions to monitor equipment in harsh environments using IoT (Scanimetrics 2019) is Scanimetrics. It makes rugged sensors that can gather data from the operating site, process it, and use it to increase uptime of the equipment and productivity of the operation. Scanimetrics specializes in predictive analytics that leads to problem diagnostics and detailed analysis to further improve understanding of operational and maintenance strategies. Scanimetrics technology can be used in hazardous locations and enables predictive and condition-based maintenance strategies. The following is a list of sensors from Scanimetrics showing how companies are innovating to enable real-time measurement of various attributes that:

1. Strain, stress, fatigue, vibration, tilt, and temperature monitoring
2. Residual strain measurement
3. Wear, thickness, displacement and crack sensing
4. Bolt and stud tension
5. Flange and gasket integrity
6. Ground moisture and stability
7. Flood and scour detection
8. Slope stability and failure
9. Inspection records
10. Rugged and environmentally sealed

Another example of digitization in mining industry is real-time data acquisition. Barrick has started collecting and publishing real-time water monitoring data from the Pascua-Lama project's water monitoring stations in Chile's Estrecho River (Barrick 2017a). This is the first time that Barrick has used real-time data reporting to share water quality with local communities. The monitoring station is located in the high Andes Mountains downstream from the project and tracks the river's acidity or alkalinity (pH), its salinity, dissolved metals content, and flow rate. The monitoring station continuously collects water-quality data and publishes it every 15 min to a website setup by the Pascua-Lama project team. The data complement rigorous water sampling conducted weekly and monthly around Pascua-Lama by independent third-party laboratories on Barrick's behalf, which are reported regularly to Chile's water and environmental regulators.

10.5 Automation and Autonomous Mining

Automation is increasingly being used in the mining industry to improve safety, efficiency, and project economics. Some examples of automation in the gold and silver industry are briefly discussed in this section.

An application of automation in mining industry is sensor-based ore sorting. This is also known as ore sorting, automated sorting, electronic sorting, particle sorting or optical sorting. This is done by real-time analysis of conveyed bulk materials and diversion of undesirable material from the ore feed. Better ore grade improves process efficiency and reduces the amount of tailings. There are a number of technologies that have been developed for automated ore sorting as shown in Table 10.4.

Automatic ore sorting based on photometry and radiometry has been used since the 1970s. Prompt Gamma Neutron Activation Analysis (PGNAA) and X-ray transmission (XRT) techniques are popular for automatic ore sorting. PGNAA, also known as Thermal Activation, utilizes a source of neutrons that is positioned under the conveyor belt in a well-shielded housing. The source of neutrons is typically a radioisotope such as Californium-252 (Cf-252). The neutrons are absorbed by elemental nuclei in the ore on the conveyor belt, which generate gamma rays having an energy level related to the excited nuclei. A detector array positioned above the conveyor belt measures the energy of gamma rays and generates spectra showing peaks at energy levels from which the relative abundance of each element in the ore is calculated. Microwave moisture analysis is used to report data on a dry basis. PGNAA is suitable for steel corded conveyors, but may not be suitable for PVC conveyors. High chlorine content in PVC can obscure the spectral data from other elements. PGNAA technology has been used for bulk sorting of ore at Assmang's Khumani iron ore mining and processing operation in South Africa since 2009 and

Table 10.4 Sensor-based sorting in mineral processing (Harbeck and Kroog 2008)

Sensor/technology	Technique	Mineral application
RM (radiometric)	Natural gamma radiation	Uranium, precious metals
XRT (X-ray transmission)	Atomic density	Base/precious metals, coal, diamond
XRF (X-ray fluorescence)	Visible fluorescence under X-rays	Diamond
Color (CCD color camera)	Reflection, brightness, transparency	Base/precious metals, industrial minerals, diamond
PM (photometric)	Monochromatic reflection/absorption	Industrial minerals, diamond
NIR (near infrared spectrometry)	Reflection/absorption	Base metals, industrial minerals
IR (infrared)	Heat conductivity/heat dissipation	Base metals, industrial minerals
MW (microwave)	Differential heating	Base/precious metals
EM (electromagnetic)	Conductivity	Base metals

for plant feed analysis in real time at MMG's Sepon copper–gold operation in Laos since 2008 (Kurth 2017). Scantech's GEOSCAN system utilizing PGNAA technique has been used in copper, lead–zinc, manganese, iron ore, bauxite, and phosphate ores (Balzan et al. 2017).

XRT-based particle sorting is based on a planar projection of X-ray attenuation of ore particles. X-ray radiation is emitted on one side of the conveyor, which passes through the ore particles and an image is recorded by a line scan detector on the other side. The image is analyzed and then particle is rejected depending on the classification of the particle. XRT ore sorters are being used for diamond concentration, processing of chromite, coal, iron, limestone, phosphate, talc, tin, and tungsten ores (Robben et al. 2017).

The use of sensors in mining is increasing rapidly. Goldcorp is using smart sensors at its Éléonore Mine located in the James Bay region of Northern Quebec, Canada (Perkins 2015). The mine opened in late 2014 and extends 4000 ft below the surface. A network of sensors and monitors is used to ensure the safety and efficiency of workers and equipment. When workers are not in a particular area of the mine, lights, and electricity are turned off automatically using smart sensors and human geospatial tracking. The tracking system is used to ensure that workers are not in the area during planned blasts. Mine's air filtration system sends fresh air to the areas where workers are present, which has resulted in a 50% reduction in the amount of air required to service the mine. Savings between $1.5 million and $2.5 million per year have been realized along with a sizeable reduction in carbon emissions with the application of smart sensors and monitoring system.

Drones are being used by Freeport-McMoRan to take photographs and real-time videos for monitoring blasting operations, environmental conditions and mine security in real time (World Economic Forum 2017). Use of drones enables an objective, data-driven view of the slope angles, which may not be visible to mine workers. This enables Freeport-McMoRan to build steeper slopes, displacing less rock and expending fewer resources to access the ore body.

Other uses of automation in mining are in drilling and hauling. Rio-Tinto and BHP are at the forefront of mining companies that are adopting automation and autonomous mining in their operations. BHP has been rolling out autonomous drills to improve safety and productivity across Western Australia iron ore sites (BHP 2017). This has resulted in an increase in overall drill productivity and a reduction in wear and tear maintenance costs. BHP has also deployed autonomous hauling at Jimblebar Iron Ore mine, which has reduced costs by approximately 20%. The use of autonomous trucks and drills has shielded employees from dangerous situations. As part of rail automation, BHP has installed 4G communications system and automated track signaling to optimize logistics. BHP is using smart caps in Escondida copper mine in Chile. The caps have sensors that measure driver fatigue by analyzing brain waves. This technology is now being integrated into more than one hundred and fifty trucks.

Autonomous drills are also being used at Anglo American's Kumba Kolomela mine in South Africa's Northern Cape Province (World Economic Forum 2017). It can be operated from a remote command center away from the iron ore pit using advanced computers and monitoring screens.

10.6 Integrated Operations and Remote Operations Centre

10.6.1 Integrated Operations

The mining industry is following the lead of other industries by connecting Information Technology (IT) to Operational Technology (OT) and exchanging data throughout the mining operations and supply chain. There are several important technologies in this area that include integrated sales and operations planning, asset cyber security, IT and OT convergence, cloud-enabled backbone, smart sensors, digital monitoring, tracking and analysis of environmental health and safety indicators, integrated and agile supply chain, and advanced track-and-trace technology (World Economic Forum 2017). Embedded computing devices such as radio frequency identification (RFID) chips and sensors are being used to connect objects to internet infrastructure. IT and OT are being integrated through the internet of things (IoT).

Integration of operations using IoT has helped Dundee Precious Metals (DPM) quadruple annual ore throughput from 0.5 million to 2 million tons at its flagship Chelopech gold mine in Bulgaria (CIM Magazine 2014). DPM called its initiative "Taking the Lid Off," which started in 2009. DPM bought the Chelopech gold–copper mine in Bulgaria in 2003. The mine is located in the Panagyurishte mining district of central-western Bulgaria, which contains a number of massive sulfide and porphyry copper deposits. The mine was operated by Bulgaria state from 1954 to 1992, and later by a private owner. DPM acquired Chelopech after its previous owner went bankrupt. The mine has proven and probable reserves of 2.512 million ounces of gold at 3.26 grams per ton, 5.674 million ounces of silver at 7.37 ounces per ton and 524 million pounds of 0.99% copper. DPM applied a concept called "Short Interval Control," which uses real-time production information to update a central monitoring and control room. Activities are planned at least three months in advance using a master schedule with detailed weekly activities. Production and performance targets are checked against actual outcomes at regular intervals. Dassault Systemes Geovia InSite shift management software is used to view a shift-by-shift schedule for the next seven days. Sandvik is the main mine production equipment supplier at Chelopech. The tasks are communicated to equipment operators via a wireless network on tablet computers using software provided by Sandvik, which contains the information for each task, including the location, output, and expected time required. The control room is updated by the operator using the tablet throughout the shift. IoT is used to track the location of miners and vehicles, monitor the status of vehicles, automate building controls and utilize software that could map, model, estimate, design, schedule, simulate and manage production based on the real-time data. A wireless communications network covers nearly the entire mine and forms the backbone of the IoT system. With its "Taking the Lid Off" initiative, DPM has been able to increase their throughput along with reduced operating costs from $66 to $40 per ton.

Schneider Electric has developed Integrated Planning and Optimization Solution (IPOS) to optimize supply chain efficiency for mining companies (Schneider Electric 2013). IPOS aims to prevent delays in one area from propagating through the supply chain by providing enterprise-wide visibility across product management, procurement, energy management, and supply chain management. IPOS can boost productivity by up to 20% through optimizing the resource-to-market chain by reducing excess energy and water consumption, resolving maintenance, and production conflicts that cause delays, or minimizing excess inventory.

10.6.2 Advanced Analytics and Remote Operation Centers

Mining and metals companies are using advanced analytics to make better and faster decisions. Using advanced analytics, data from different sources can be analyzed to provide insights and identify correlations. It can be used to analyze vast amount of data for trends and identify opportunities.

Barrick has built an Analytics and Unified Operations (AuOps) Center in Nevada (Barrick 2017b). This is part of Barrick's digital transformation to get the right information to the right people at the right time. Under this initiative, the data from all areas of the operations will be collected and analyzed. The vast quantities of data from sensors, equipment, and digital tools located across Barrick's mines in Nevada will be processed at the AuOps to extract timely and relevant information to help in making data-driven decisions in real or near real time. This center will employ in-house analysts to interpret data and develop insights for site personnel to make better, faster, and safer decisions. The center will transform data into a visual representation and help the analysts understand in real time how the mines are performing against key performance indicators. This is expected to help Barrick identify potential opportunities and risks before making decisions.

Barrick is also developing Asset Health tool for maintenance work on various equipment parts (Barrick 2017c). This will make maintenance job easier because information will be automatically fed into the tool from several different data sources, including fleet management software, equipment work order information, oil analysis data, and sensors installed on equipment. The Asset Health tool will analyze this information and predict when parts are likely to fail by drawing on a library of common causes for those failures, thereby reducing the amount of time that technicians and their supervisors will have to troubleshoot and ensuring that replacement parts are ordered in advance of a failure. The Asset Health tool will help maintenance teams produce custom reports and give them advance warning when an equipment is not performing optimally. This will help technicians take actions for proactive maintenance.

BHP is using advanced sensors and real-time process control to improve quality and grade delivered to processing plants. This has improved performance and reduced energy and water usage. BHP is also using artificial intelligence to automate decision-making. At Mining Area C in Western Australia Iron Ore, BHP is introducing a system that chooses which crusher the trucks should use to minimize waiting time and idle time (BHP 2017). BHP is integrating operations that include 12 core assets, two sea ports, and more than 1300 kilometers of rail network. BHP is applying a systems engineering approach to analyze mine life cycles, identify constraints and prioritize investments, which facilitates rapid replication of best practices across its operations. BHP opened an Integrated Remote Operations Centre (IROC) in Perth in 2013 (LeeSun 2015). Now BHP can remotely control its Pilbara mine, fixed plant, and train and port operations from one central location. IROC took 22 months from the concept study phase to becoming operational and uses CCTV and radio systems to communicate with on-site personnel.

Rio Tinto has established a program called "The Mine of the Future," which is aimed at developing five different automation technologies: driverless trucks, driverless trains, autonomous tunneling and boring machines, a remote operations center, and airborne exploration drones (MaRS Market Insights 2014). Rio Tinto currently has more than 30 driverless autonomous mining trucks operating in the Pilbara mine, which are controlled from a Rio Tinto Remote Operations Centre located far away in Perth.

Remote Operations Centre can deliver approximately $65 billion in value to mining assuming an adoption rate of 50% by 2025 (World Economic Forum 2017). The need for highly skilled specialists to be on site at mines is avoided by ROCs, which can result in significant cost savings. Remote control of logistics is another source of significant cost savings. In addition, ROCs create new career options for highly skilled people and improve safety by reducing the number of personnel needed on site in potentially hazardous environments. ROCs also reduce the land footprint of mining sites, thereby reducing emissions and waste.

10.6.3 Artificial Intelligence

Artificial intelligence refers to computer systems that can perform tasks that normally require human intelligence such as visual and speech recognition, decision-making and language translation. Recent breakthroughs in artificial intelligence have been made possible by greater computing power, big data and better algorithms. Artificial intelligence is now being used to process data and provide real-time decision support and future projections. There are many important technologies in this area that include exploration, ore valuation and to build models for simulations. Also, production planning, forecasting and scheduling are increasingly being carried out using AI. Asset performance monitoring and predictive maintenance is another emerging area for asset optimization that is utilizing AI. Advanced analytics for demand forecasting along with an integrated supply chain is an area that is utilizing various combinations of AI and block-chain technologies. Other applications include digital twin simulations, digitally enabled fraud detection and anti-corruption traceability, along with operations support (World Economic Forum 2017).

Artificial intelligence and machine learning are now being utilized to better identify drilling locations with higher probability of success. It seems mining companies are now increasingly exploring the use of artificial intelligence and machine learning to enhance efficiencies in various areas of their mining business (Forbes 2018). Goldcorp and IBM Canada are collaborating to use artificial intelligence to predict the potential for gold mineralization (Goldcorp 2018). This innovative technology is called *IBM Exploration with Watson*. The aim is to review all the geological information available to find better drilling locations for gold. Their artificial intelligence platform is expected to provide a powerful search and query capability across a range of exploration datasets for this purpose. According to Todd White, Executive Vice President and Chief Operating Officer, Goldcorp, "The potential to radically accelerate exploration target identification combined with significantly improved hit rates on economic mineralization has the potential to drive a step-change in the pace of value growth in the industry."

IBM Exploration with Watson was developed using data from Goldcorp's Red Lake Gold Mines in northern Ontario. It leverages spatial analytics, machine learning and predictive models to help explorers locate key information and develop geological extrapolations in a fraction of the time and cost compared to traditional methods. According to Mark Fawcett, Partner with IBM Canada, the collaboration has following significance:

> Applying the power of IBM Watson to these unique challenges differentiates us in the natural resources industry. We are using accelerated computing power for complex geospatial queries that can harmonize geological data from an entire site on a single platform. This is the first time this solution has been ever used, which makes this project all the more significant.

At Red Lake, IBM Exploration with Watson provided new targets for drilling, which were subsequently verified.

Goldcorp was recognized by the prestigious Ingenious Award from the Information Technology Association of Canada (ITAC) for The IBM Watson initiative. The ITAC award recognizes excellence in the use of information and communications technology by organizations to solve problems, improve performance, introduce new services, and grow business.

Another company applying artificial intelligence to mining is Goldspot Discoveries Inc. This company aims to make finding gold more of a science than art by using machine learning (Goldspot 2019). It is a technology company that leverages artificial intelligence to reduce capital risk and increase efficiencies in resource exploration and investment. GoldSpot combines proprietary technology with traditional domain expertise to make full use of historically unutilized data to better comprehend resource property potential. Such applications of artificial intelligence are aimed to make the mining industry more competitive.

10.6.4 Modeling and Simulation

Modeling and Simulation tools are increasingly being used in the mining industry to carry out "what-if scenarios" with reasonable accuracy. This can result in better understanding to enable high confidence forecasting of ore properties along with mine ore and metal production to enable a high-level optimization of a mining operation. Montreal-based Mira Geoscience provides software and consulting services to mining industries, which include 3D and 4D earth modeling and data management solutions for exploration, resource evaluation and geotechnical hazard assessment (Mira Geoscience 2019). It has pioneered the "Common Earth Model" concept in the mineral industry, which brings together a wide range of geological, geophysical, geochemical, and geotechnical software tools to create integrated, multidisciplinary models with a complete geological framework. Mira Geoscience forecasts geotechnical hazards quantitatively by combining geological and geotechnical observations such as rock type, structure, rock quality, seismicity, deformation, support, geometry, production, and other data into a set of spatially modeled, normalized and weighted hazard criteria for overall hazard estimation. Mira Geoscience has applied its expertise in wide-ranging projects such as forecasting rockburst and water inflow hazards in deep Canadian mines, estimating slope stability hazards in large South African open pits, and roof fall hazards in Australian coal mines (World Economic Forum 2017).

The most advanced application of simulation modeling in mining industry is "Digital Twin." The first documented use of the phrase "digital twin" was by NASA in November 2010 in a technology roadmap (Shafto et al. 2010). The vision of digital twin by NASA was a simulation system so realistic that a mission could be flown virtually in its entirety before it was launched. NASA planned to test different parameters and flight plans, provide continuous health predictions and troubleshoot based on digital twin. After its application in manufacturing, the digital twin technology is now finding its way into mining.

The Digital Mine team at GE Transportation is using digital twins for asset and operation performance management, drill guidance, and collision awareness (CIM Magazine 2018). About a third of GE Transportation's Digital Mine customers are using digital twin concepts to digitalize an asset or their entire plant's operation. GE and South32, a Perth, Australia-based miner, announced a three-year deal in April 2017 to apply digital twin technology to its operations.

OceanaGold has engaged Andritz to apply digital twin technology at Haile gold mine in South Carolina (Schug et al. 2019). This mine became part of OceanaGold when OceanaGold acquired Romarco Minerals, the previous owner of the Haile operation, in October 2015. Inaugural gold pour at Haile operation under OceanaGold management took place in January 2017. OceanaGold had developed a digital twin for cyclone feed line over a year ago and recently added digital twins for the cyclone underflow and overflow as well. OceanaGold is planning to fully automate the plant using digital twin technology.

10.7 Paradigm Shift in Skill Development, Education, and Training

It is the age of information technology, and mining industry can use the information technology to fundamentally change the way miners and mining professionals are educated and trained. An alternative to regular classroom-based learning is E-learning, which is a learning system that utilizes computers and internet to impart education without the learner and educator being in the same location. There are many advantages to E-learning:

1. Learner can be anywhere in the world.
2. Learning material can be accessed at any time suitable to the learner.
3. The pace of learning can be up to the learner.
4. There is no limit on class size.
5. E-learning can be cost effective as it does not require a classroom facility, traveling to a specific location and finding accommodation there. Also, the learning material is delivered electronically without the need for photocopying or producing manuals or books.
6. Learning material can be delivered as soon as it has been developed.
7. Learning material can be updated easily.
8. Learners can be tested through online quizzes and tests and can get results of quizzes and tests immediately after completing them.
9. E-learning can be blended with regular exams and classroom teaching with in-classroom instructors for some parts of the curriculum.
10. Access to advanced concepts in the curriculum can be restricted till basic concepts are mastered and learner passes online quizzes or tests to show the competence to move ahead.

Gold and silver mining industry can also use virtual reality simulation-based training for skills development. Simulation-based training coupled with E-learning can offer significant benefits in the training of mining personnel (Mining Weekly 2015). This can be particularly useful for mining operations such as underground blasting. Simulation-based immersive training experience for such operations can prepare the miners for actual operations much better than classroom training. The use of virtual reality for immersive training using 3D glasses can give the miners the feel for actual underground operation, including environmental conditions, workplace temperatures and sound levels (Mining Weekly 2017). Virtual reality simulation-based training gives the opportunity to train miners in potentially risky operations in a safe and erasable environment that is difficult to accomplish with traditional methods.

In the changing nature of workplace, the opportunities for skill development, education, and training are important not only for operations, but also for employee retention. Deloitte suggests the following strategies to mining companies to attract and retain talent (Deloitte 2018).

1. *Retrain and upskill*

   As mining industry incorporates information technology in its operations, the gaps in the digital knowledge of the employee are undermining the efforts in technology transformations. The mining companies need to compare the skills of their current employees against the skillsets needed in the future operations. Accordingly, the mining companies need to hire new talent or retrain the current employees to close the identified gaps.
2. *Adopt new attraction and retention strategies*

   In order to attract and retain tech-savvy talent, the mining companies need to nurture and develop their employees. They need to create interesting and purposeful work for their employees and provide them with career flexibility.
3. *Source and integrate talent across networks*

   There is a growing need for the mining companies to partner with other organizations with needed technological expertise. The mining companies need to design and evolve their partnership networks so that they have access to the best talent for specific work. They need to cultivate talented people who may be working as freelancers or get access to digital talent by crowdsourcing.
4. *Redesign work for technology and learning*

   The future of work will involve more and more machine–human collaboration. The mining companies need to identify areas where digital technology can improve the work performance of employees and make them more productive.
5. *Create a new social contract with communities and governments*

   The mining companies need to work closely with local communities, governments and key stakeholders to prevent potential backlash. They need to work closely with vocational schools to make sure that the future labor force is trained properly.

References

Balzan, L.A., A.R. Harris, and Z. Bauk. 2017. Faster analysis: recent advances in Geoscan on-belt analysis techniques allowing for rapid real-time measurement of minerals. In *COM2017, The conference of metallurgists hosting world gold & nickel cobalt proceedings*, Vancouver, BC, Canada, August 27–30, 2017, 10 p.

Barrick. 2017a. Real-time water monitoring promotes transparency at Pascua-Lama. https://www.barrick.com/news/news-details/2017/Real-Time-Water-Monitoring-Promotes-Transparency-At-Pascua-Lama/default.aspx.

Barrick. 2017b. Using analytics to enhance decision-making. https://www.barrick.com/news/news-details/2017/Using-Analytics-To-Enhance-Decision-Making/default.aspx.

Barrick. 2017c. Barrick predicts the future… of mine equipment maintenance. https://www.barrick.com/English/news/news-details/2017/barrick-predicts-the-future-of-mine-equipment-maintenance/default.aspx.

Bergh, L.G., E.M. Lucic, and F.M. Zuleta. 2007. Supervisory control project of copper solvent extraction pilot plant. In *Process control, optimization and six sigma* (Vol.7). ed. R.L. Stephen and G.A. Eltringham. Montreal, QC, Canada: The Canadian Institute of Mining, Metallurgy and Petroleum.

BHP. 2017. Creating the future of mining integration and automation. https://www.bhp.com/media-and-insights/reports-and-presentations/2017/10/creating-the-future-of-mining-integration-and-automation.

BHP-Billiton. 2013. http://www.stepchangeglobal.com/wp-content/uploads/2013/09/IROC-Brochure-201305031.pdf.

CIM Magazine. 2014. Underground achiever. http://magazine.cim.org/en/projects/underground-achiever/.

CIM Magazine. 2018. How to simulate an entire operation before production. http://magazine.cim.org/en/technology/digital-double-en/.

Deloitte. 2018. Tracking the trends 2018. https://www2.deloitte.com/content/dam/Deloitte/global/Documents/Energy-and-Resources/gx-TTT-report-2018.PDF.

Enterprise Gateway. 2010. A MOM solution for integrating production and maintenance in real time. 5G automatika internal report 240310.

Forbes. 2018. The 4th industrial revolution: how mining companies are using AI, machine learning and robots. https://www.forbes.com/sites/bernardmarr/2018/09/07/the-4th-industrial-revolution-how-mining-companies-are-using-ai-machine-learning-and-robots/#6db359b1497e

Goldcorp. 2018. Goldcorp and IBM develop new AI technology solution to improve predictability for gold mineralization. https://www.goldcorp.com/English/investors/news-releases/news-release-details/2018/Goldcorp-and-IBM-Develop-New-AI-Technology-Solution-to-Improve-Predictability-for-Gold-Mineralization/default.aspx.

Goldspot. 2019. At the forefront of artificial intelligence in mineral exploration. https://goldspot.ca/.

Harbeck, H., and H. Kroog. 2008. New developments in sensor-based sorting. *Mineral Processing* 49: 4–11.

Kurth, H. 2017. Suitability of on-belt elemental analysis for real-time ore quality measurement and bulk sorting. In *COM2017, The conference of metallurgists hosting world gold & nickel cobalt proceedings*, Vancouver, BC, Canada, August 27–30, 9 p.

LeeSun. 2015. BHP Billiton remotely resourcing the future. Technology and Operation Management, Harvard Business School. https://rctom.hbs.org/submission/bhp-billiton-remotely-resourcing-the-future.

MaRS Market Insights. 2014. Mining & Metals + Internet of Things: Industry opportunities and innovation. https://www.marsdd.com/wp-content/uploads/2014/11/Mining-Metals-and-IoT.pdf.

McKee, D. 2013. Understanding mine to mill, a monograph published by the Co-operative Research Centre for Optimizing Resource Extraction (CRC ORE), Brisbane, Australia.

Mining Weekly. 2015. Simulation enhances preparation of mining learners. http://www.miningweekly.com/article/simulation-enhances-preparation-of-mining-learners-2015-05-01.

Mining Weekly. 2017. Virtual technology now used for blast training in SA. http://www.miningweekly.com/article/virtual-technology-now-used-for-blast-training-in-sa-2017-07-20.

Mira Geoscience. 2019. Advanced 3D modelling and data management solutions. http://www.mirageoscience.com/.

Perkins, James. 2015. Internet of things improving Goldcorp operations. 12 March 2015. http://samssa.ca/internet-of-things-improving-goldcorp-operations.

Robben, C., D. Girard, H. Cline, and A. Takala. 2017. X-ray-transmission sorting at the Kensington gold mine. In: *COM2017, The conference of metallurgists hosting world gold & nickel cobalt proceedings*, Vancouver, BC, Canada, August 27–30, 2017, 11p.

Scanimetrics. 2019. Internet connected sensors increase visibility of your assets. https://www.scanimetrics.com/index.php.

Schneider Electric. 2013. Increase your mining productivity by up to 20% Integrated Planning and Optimization Solution realizes your resource-to-market potential. http://download.schneider-electric.com/files?p_Reference=998-1223667_GMA-GB&p_EnDocType=Brochure&p_File_Id=791268470&p_File_Name=Integrated_Planning_Optimization.pdf.

Schug, B., C. Anderson, S. Nazari, and C. Cristoffanini. 2019. Real world improvement through virtual instrumentation at OceanaGold Haile. *Mining Engineering* 1: 20–25. https://www.andritz.com/resource/blob/289822/8bc1a140260ebcf9c7c3e417a73b4b78/aa-digital-twin-sme-2019-data.pdf.

Seshan, A., and B.K. Gorain. 2016. *An integrated mining and metallurgical Enterprise enabling continuous process optimization. Innovations in metallurgical processing*. Cham: Springer.

Shafto, M., M. Conroy, R. Doyle, E. Glaessgen, C. Kemp, J. LeMoigne, and L. Wang. 2010. Draft modeling, simulation, information technology & processing roadmap. *Technology Area* 11. https://www.nasa.gov/pdf/501321main_TA11-MSITP-DRAFT-Nov2010-A1.pdf.

Sherring, A. 2012. Automation and remote collaboration: will it change the future of mining? Presented at the Honeywell Users Group Symposium, extract written in the PACE magazine.

Sherring, A. 2013. Personal communication.

Sinclair, M. 2012. Remote Collaboration Centres (Honeywell). https://www.posccaesar.org/svn/pub/PCA/MemberMeeting/201210/Day2/RemoteCollaborationCenters-MikeSinclair-Oct2012.pdf.

World Economic Forum. 2017. Digital Transformation Initiative: Mining and Metals Industry. https://www.accenture.com/t20170116T084456__w__/us-en/_acnmedia/Accenture/Conversion-Assets/WEF/PDF/Accenture-Mining-And-Metals-Industry.pdf.

Zamora, C., G. Schifferli, L. Castelli and A. Bonomelli. 2010. Kairos mining: a comprehensive approach to sustained long term benefits in copper processing plant automation. Presented at the 42nd annual Canadian Mineral Processors Conference, Ottawa, June 19–21st.

Chapter 11
The Future of Gold and Silver Industry

V. I. Lakshmanan, R. Roy, A. Ojaghi, and B. Gorain

Future sustainability of the gold and silver industry is highly dependent on how mining companies, communities, governments, and nongovernment organizations work together in an ecosystem that promotes well-being of all stakeholders. Mining companies must be proactive and demonstrate their sincerity in resolving various conflicting interests to earn the "license to operate" reward so-to-speak.

This chapter highlights various multifaceted initiatives involving corporate social innovation, developing breakthroughs in mining and processing technologies along with creating innovative solutions to various environmental challenges associated with water, energy, tailings, and waste management.

11.1 Corporate Social Responsibility to Corporate Social Innovation

11.1.1 Corporate Social Responsibility (CSR)

Corporate social responsibility (CSR) has been defined by Government of Canada as "The voluntary activities undertaken by companies to operate in an economically, socially and environmentally sustainable manner beyond the minimum required by law" (Natural Resources Canada 2011).

V. I. Lakshmanan (✉) · R. Roy · A. Ojaghi
Process Research Ortech Inc., Mississauga, ON, Canada
e-mail: llakshmanan@processortech.com

B. Gorain (✉)
Hindustan Zinc Limited, Udaipur, Rajasthan, India
e-mail: Barun.Gorain@vedanta.co.in

V. I. Lakshmanan, B. Gorain (eds.), *Innovations and Breakthroughs in the Gold and Silver Industries*, https://doi.org/10.1007/978-3-030-32549-7_11

Government of Canada has advanced "Building the Canadian Advantage" initiative that is founded on four complementary pillars designed to engage multiple stakeholders and foster different aspects of CSR. The four pillars of this strategy are (Natural Resources Canada 2011):

1. Support for initiatives to enhance the capacities of developing countries to manage the development of minerals and oil and gas, and to benefit from these resources to reduce poverty.
2. Promotion of the following widely recognized international CSR performance guidelines with Canadian extractive companies operating abroad:
   - *Organisation for Economic Co-operation and Development (OECD) Guidelines for Multinational Enterprises*
   - *International Finance Corporation Performance Standards on Social & Environmental Sustainability* for extractive projects with potential adverse social or environmental impacts
   - *Voluntary Principles on Security and Human Rights* for projects involving private or public security forces
   - *Global Reporting Initiative for CSR* reporting by the extractive sector to enhance transparency and encourage market-based rewards for good CSR performance
3. The Office of the Extractive Sector CSR Counsellor to assist stakeholders in the resolution of CSR issues pertaining to the activities of Canadian extractive sector companies abroad.
4. The CSR Centre of Excellence to encourage the Canadian international extractive sector to implement these voluntary performance guidelines by developing and disseminating high-quality CSR information, training, and tools.

Government of Canada established the position of the extractive sector CSR counsellor in 2009 as part of its first CSR strategy for the Canadian extractive (mining, oil, and gas) sector called "Building the Canadian Advantage: A Corporate Social Responsibility Strategy for the Canadian International Extractive Sector" (Global Affairs Canada 2018). This strategy was designed to enhance the ability of Canadian extractive companies working outside Canada to manage social and environmental risks, to contribute to their success abroad, and to reflect Canadian values and leadership in responsible business practice.

The Government of Canada launched an updated CSR strategy in November 2014 called "Doing Business the Canadian Way: A Strategy to Advance Corporate Social Responsibility in Canada's Extractive Sector Abroad" (Global Affairs Canada 2018). This updated strategy outlined the Government of Canada's expectation that Canadian companies operating abroad respect human rights and all applicable local and international laws, and meet or exceed widely recognized international standards for responsible business conduct. This update gave the following mandate to the CSR counsellor and its office:

- Promote CSR guidelines to the extractive sector
- Advise companies on incorporating these guidelines into their operations

- Prevent, detect, and resolve disputes in their early stages
- Work closely with Canada's National Contact Point for the *OECD's Guidelines for Multinational Enterprises* on responsible business conduct
- Provide support to Canadian embassies, trade commissioners, and other Government of Canada officials promoting CSR to Canadian extractive companies operating internationally

The Government of Canada contributed to the creation of the CSR Centre for Excellence (CfE) as one of four pillars of its original 2009 federal CSR Extractive Sector Strategy. The CfE is a multistakeholder body and is currently hosted by the Canadian Institute of Mining, Metallurgy and Petroleum (CIM). The CfE is a focal point for the development and dissemination of practical tools and information for use by a broad range of extractive sector stakeholders. CfE has following accomplishments to its credit (CIM 2019):

- CfE participation in numerous extractive sector conferences/activities (e.g., multistakeholder discussions at CIM Conferences, participation at Prospectors & Developers Association of Canada (PDAC) Conferences).
- CfE identification of three priority areas where practical guidance is needed:
  - The UN Guiding Principles on Business and Human Rights
  - The OECD MNE Guidelines
  - The Voluntary Principles on Security and Human Rights
- Holding of CfE multistakeholder workshops to explore relevant issues:
  - January 16, 2015 CfE Workshop on Respect for Human Rights throughout the Supply Chain
  - March 31, 2015 CfE Workshop on Integrating Human Rights at the Site Level
  - March 28, 2014 CfE Workshop on the Remedy Aspect of the UN Guiding Principles on Business and Human Rights
- In conjunction with the federal Trade Commissioners' office, development of short handouts on relevant issues:
  - Community Stakeholder Engagement
  - Managing Requests for Community Support
  - Hiring Responsibly
- Development of video: "A Human Rights Primer for Mining Personnel."

Under the CSR Strategy of the Government of Canada, companies are expected to participate in the dispute resolution mechanisms of the CSR Counsellor's Office or Canada's National Contact Point ("NCP") for the Organisation for Economic Co-operation and Development ("OECD") Guidelines for Multinational Enterprises. Participating companies are eligible for enhanced economic diplomacy by the Government of Canada, while companies not employing CSR best practices and those refusing to participate in the dispute resolution mechanisms lose access to the Trade Commissioner services and other Government of Canada services, which include the issuance of letters of support, advocacy efforts in foreign markets, and participation in Government of Canada trade missions. A designation of noncompliance will also affect financing or other support by the Government of Canada's financing crown corporation, Export Development Canada (McMillan 2014).

The CSR strategy endorses the United Nations' Guiding Principles on Business and Human Rights (the "UN GP") and the OECD Due Diligence Guidance for Responsible Supply Chains of Minerals from conflict-affected and high-risk areas (the "OECD Guide"). The UN GP identifies three principles to guide the companies and governments regarding human rights: (1) the state's duty to protect against human rights abuses by third parties, including business; (2) the corporate responsibility to respect human rights through due diligence; and (3) ensuring greater access to effective remedies for victims (McMillan 2014). The OECD Guide provides guidelines to multinational mineral extraction companies for avoiding fueling conflict and responsibly sourcing and trading minerals in conflict-affected and high-risk areas.

Mining companies have undertaken a number of activities under the CSR initiative. These activities include investments in infrastructure (e.g., schools, roads, hospitals, health equipment, electricity, clean water, and drainage repairs), investments in building social capital (e.g., information on HIV prevention, family planning, and improving hygiene), investments in human capital (e.g., providing education, training, and skills), fostering microbusiness, aquaculture, crop cultivation, animal rearing, and textile production (E&MJ 2013).

Mining companies have realized that doing the right things in the communities where they operate is important. They are actively pursuing activities that mitigate the economic, social, and environmental impacts of mining on local communities. Endeavour Silver planted more than 40,000 trees as part of its commitment to reclaiming and restoring land disturbed during the mining process in 2014. Rio Tinto has spent over billion dollars on education and clean water projects in Ghana. IAMGOLD is running skills development programs in Burkina Faso. Barrick Gold is funding business loan programs in Peru (mining.com 2015).

Mining companies have realized that an investment in the people living where they work can have outsized returns in human impact terms, permitting time, and even the ability to continue operating (mining.com 2015). These programs need to be designed in cooperation with local communities to have a lasting impact.

11.1.2 Corporate Social Innovation

The concept of Corporate Social Responsibility (CSR) has been constantly evolving since it began in the 1990s when many companies across the globe embraced worthy community causes in areas where they operated. Most of these programs focused mainly on the company's reputation and license to operate, with little direct connection to their bottom line.

Since 2011, there has been a global movement to make CSR impacts more integral to a company's business strategy focusing on *creating shared value*. SABMiller, the world's second-largest brewer, was an early adopter. The company targeted more than 380,000 small retailers in their biggest Latin American markets: Colombia, Peru, Ecuador, Panama, Honduras, and El Salvador, with an aim to improve small retailer's business performance and therefore quality of life and leadership abilities through a combination of classroom training and in-store mentoring on business, life skills, and leadership. This involved strengthening the broader "business ecosystems" in which the retailers operate, with a special focus on improving their access to financing and technology. The effort is aimed at strengthening SABMiller's retail network and sales. The company's success hinges on the success of its retailers—a win–win situation—creating shared value (IDB Invest 2015).

The latest evolution on this CSR continuum is the advent of Corporate Social Innovation (CSI). The World Economic Forum launched the Global Agenda Council on Social Innovation in 2014, bringing together corporate leaders, impact investors, and development executives, and offered a definition for CSI that builds on shared value concepts. In this the companies proactively design and implement business models that increase incomes and better the quality of life of underserved or vulnerable communities.

One interesting feature of CSI is that a new alignment is emerging among corporate venture capitalists and impact investors. The corporate venture capitalist is seeking returns for the company and new capabilities or access to markets that are aligned with its long-term business strategy. The impact investor is interested in placing capital into companies and generating measurable social and environmental impact, together with a financial return. The impact investor also wants to expand effective development solutions and—together with development finance institutions—is beginning to understand that working with large companies may be the best route. Companies that set up corporate venture funds also have in-house expertise and distribution channels that allow them to scale up successful projects.

One example of CSI is the philanthropic arm of the oil giant Shell Foundation, who formed a strategic partnership with Husk Power Systems, a biomass electricity generator. In 5 years, Husk has installed 84 mini-power plants, providing electricity to more than 200,000 people in 300 rural villages in India. By electrifying villages, Husk is promoting economic development, as businesses are able to stay open after dark and children can study at night. Impact investors Acumen and Oasis Fund have contributed funding to the venture.

It is important to realize that business leaders have the opportunity to transform societies through driving innovation. A growing number of companies worldwide are now accepting that CSR and innovation are the foundation of business competencies. With the emergence of digital technologies and better connectivity, the ability to tackle the issues such as social justices, poverty, and climate change through innovative approaches has now enhanced. The only successful brands of the future, including that for mining and resources companies, will be the ones that see these challenges as opportunities for innovation, rather than risks to be alleviated.

According to the UK trade and industry, CSR represents "the integrity with which a company govern itself, fulfils its mission, lives by its value, engages with its stakeholders, measures its impact and reports on its activities."

It is time that mining companies should embrace CSR in the form of social entrepreneurship as change agents for the society, seizing opportunities, inventing new approaches, and developing solutions to make this world a better place to live. It is not too far away when Eco-Innovation will emerge as a new discipline with the purpose of redefining businesses, including mining, and its products and processes in terms of their contribution to sustainable development.

11.2 Breakthroughs in Mining and Metal Extraction

11.2.1 Integrated Mining and Metal Extraction

As discussed in Chaps. 2–4, an integrated approach to geology, mining, and metal extraction along with environmental and social considerations is clearly the future of gold and silver industry. This has now been possible due to various technology disruptions that have emerged or are rapidly emerging in the areas of digital, automation, artificial intelligence, robotics, augmented and virtual reality. In addition, the core mining technologies are also evolving rapidly such as the 3D seismic surveying, hyperspectral automated core scanning, continuous rock cutting technology for hard rocks, hydraulic ore pumping, ore sorting for early waste rejection, in-place or in situ leaching, bio-oxidation for refractory ores, alternative lixiviants for gold and silver, dry stacking of tailings, and ultimately these technologies are progressing toward a near zero waste mining scenario. The onus is now on the mining industry as to how quickly these technologies are embraced to make the industry safer, more efficient, sustainable, and also meet the growing expectations of communities, governments, and other stakeholders. Also, it is important to realize the future of mining industry depends on attracting the new generation who are digital savvy and expects the digital way of working as the new norm.

This chapter will discuss various other emerging technologies that have the potential to transform the gold and silver industry.

11.2.2 Bio-oxidation

Outotec Biomin, South Africa, has developed the MesoTherm process which utilizes a combination of the traditional BIOX mesophile process for the primary bio-oxidation stage followed by a thermophile bio-oxidation stage to complete the oxidation. The higher oxidation rates and more complete oxidation at the higher temperature result in lower cyanide consumption during subsequent leaching of the bio-oxidation product. Development of the process included several stages of batch and continuous pilot plant testing. The final stage in the development is the successful operation of a 21 m^3 demonstration tank at the Fairview BIOX plant which has shown a 50% reduction in cyanide consumption for similar sulfide oxidation to the commercial plant (Seaman et al. 2019). They have also introduced the OKTOP® 3105 dual impeller to give superior gas handling and oxygen mass transfer rates under typical BIOX operating conditions. It was tested in water and BIOX slurry using a 21 m^3 test reactor at the Fairview Mine in South Africa, and the results indicated overall savings for the agitator and blower of 5% in capex and 5% in OPEX compared with the benchmark dual hydrofoil (Van Niekerk et al. 2018). The second successful commercial implementation of the HiTeCC technology to combat the preg-robbing of double refractory ore has been achieved at Nordgold's Suzdal mine in Kazakhstan which treats double refractory ore with both visible and invisible gold hosted in sulfide associations of pyrite and arsenopyrite, and also carbonaceous black shale. Following extensive laboratory and demonstration scale testing of the Outotec HiTeCC technology over the period of 2012–2015, Suzdal commenced the construction of a 385 tons per day facility. The plant was designed to recover gold from both the current and historic Carbon in Leach (CIL) tailings and was successfully commissioned in 2016. In the process, the preg-robbed gold is efficiently desorbed from the carbonaceous matter by manipulating the ionic strength and temperature of the Suzdal CIL product leading to enhanced metal recovery (van Buuren et al. 2018).

The first commercially successful bio-oxidation plant was built in 2001 at Laizhou Gold Processing Plant in Shandong Province with a capacity of 100 tons/day, adopting MetBytes for BacTech technology. The largest was built in 2007 in Guizhou Jinfeng with a capacity of 750 tons/day, utilizing BIOX™ technology. The typical plant has three stages of agitation tanks in parallel for primary oxidation, and three to four stages of agitation tanks in series for secondary oxidation. Air is introduced into the tanks and dispersed with an agitator to achieve oxidation of the sulfur in pyrite concentrates. The temperature is usually controlled at 38–45 °C using cooling water in an indirect heat exchanger to provide an optimum environment for the bacteria, and the pH is controlled at 1.5–2.0 by limestone to neutralize acid, produced by oxidation of sulfur. Under suitable conditions the bacteria can be adopted to oxidize gold arsenopyrite concentrates with an arsenic content of less than 3% (van Buuren et al. 2018).

11.2.3 Pressure Oxidation

Unique operating conditions have driven designers and manufacturers to use explosion welded titanium clad construction for the world's largest pressure oxidation gold autoclave for the expansion of Polymetal International's gold processing facility in Amursk, Eastern Russia. The decision is based on titanium clad being considered to be a more economical and reliable solution than the commonly used acid resistant brick lining system (Pearson 2019). This is the first use of Ti clad for the gold POX that NobelClad is aware of. While some pilot lines use titanium for autoclaves, and have for many years, this is its first use in a full commercial production plant (Salt 2019).

11.2.4 Rapid Oxidative Leach (ROL) Process

FLSmidth, USA, is developing the application of their mechanochemical pretreatment process to oxidize refractory sulfide gold ores and concentrates under atmospheric pressure without ultrafine grinding and at temperatures much lower than traditional roasting or autoclave pretreatment (Roy et al. 2018). They are particularly targeting refractory low-grade ores with >3 g/ton gold coupled with small resources, and existing operations with low-grade stockpiles to be processed at end of mine life. Laboratory testwork indicates >70% gold extraction in 8 h. They are working with several gold producers with the goal of progressing to pilot scale testing in 2019. Projected advantages include low capex and lower environmental impact compared with existing technologies. The process is a direct application of technology developed for copper chalcopyrite ores which has been successfully piloted and is moving toward a demonstration plant in South America in 2019 (Barlow 2018; Gleeson 2019a, 2019b).

11.2.5 Heap Leaching of Refractory Gold Ores

There is an industry trend toward lower ore grades, and refractory ore types with higher arsenic level which is driving research and development into improvements to existing processes as well as the application of new technology.

Chloride leaching is being developed for heap leaching of low-grade chalcopyrite copper ores, and may have potential for low-grade refractory gold ores.

Heap leaching of refractory gold ore by integrating enhanced bio-oxidation of pyrite and chloride-based gold leaching is under research by Barrick Gold and University of British Columbia. This process involves bio-oxidation of pyrite as pretreatment to expose gold, followed by chloride-based leaching of gold.

Although refractory gold heap leaching is very cost effective, it is a possible process for recovering a number of elements from sulfide ores.

11.2.6 Mixed-Chloride Technology

Process Research Ortech (PRO), Canada, has developed an innovative atmospheric mixed-chloride leaching process for the recovery of gold from gold-bearing complex ores, concentrates, and tailings. Mixed chloride technology applied for the recovery of several products including gold from refractory ores, titanium dioxide from ilmenite ores, Rare earth element (REE) from aluminosilicate ores, base metals and PGM from sulfide ores, and base metals from laterite ores (Lakshmanan et al. 2013a; 2013b).

PRO process uses mixed chloride lixiviant (HCl + $MgCl_2$) to bring gold and silver in solution. The HCl leaching system provides the opportunity to regenerate the acid by pyrohydrolysis, while the presence of $MgCl_2$ in the lixiviant enhances the activity of the hydrogen ion by orders of magnitude, making the lixiviant very aggressive. This results in high recoveries of precious and base metals. Precious and base metals are separated from pregnant leach solution (PLS) successively using innovative solvent extraction steps (Lakshmanan et al. 2013a, 2013b).

A mixed-chloride process flowsheet, which developed by PRO for the recovery of Au from gold-bearing materials, is shown in Fig. 11.1. This process has several advantages such as (1) it is environmentally friendly as no cyanidation is involved and it can be applied in jurisdictions where use of cyanide is not permitted, (2)

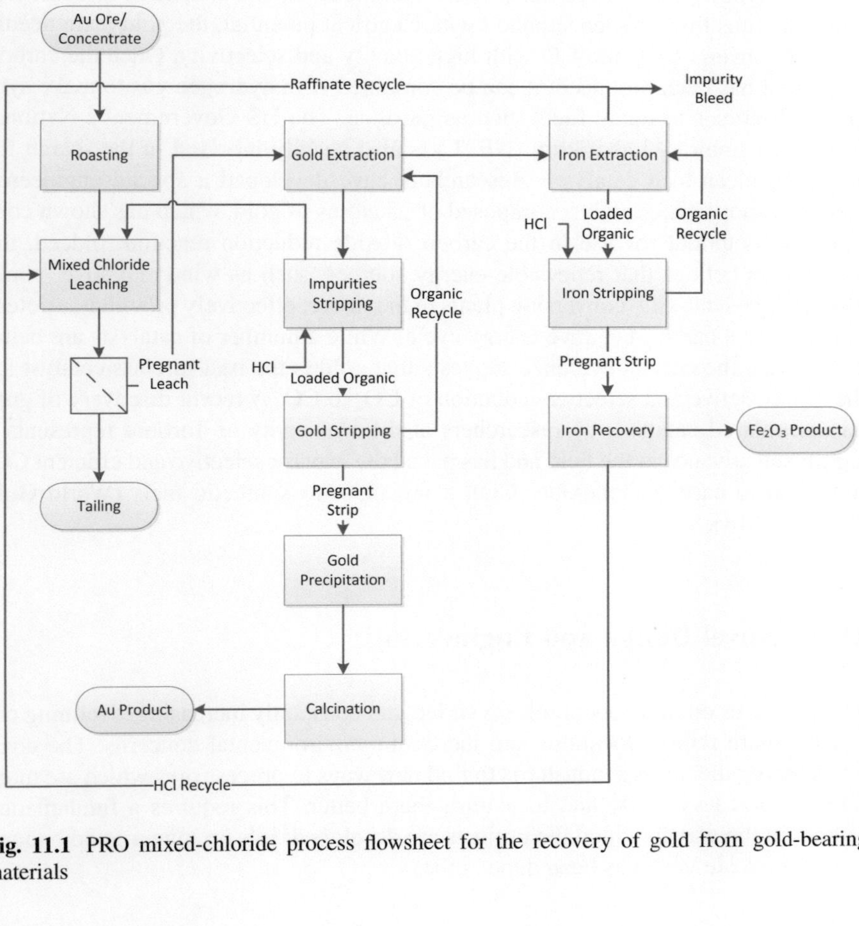

Fig. 11.1 PRO mixed-chloride process flowsheet for the recovery of gold from gold-bearing materials

selective extraction of Au from the pregnant leach solution followed by stripping, precipitation, and calcination avoid the use of cyanide, (3) this route offers economic advantage by eliminating the cyanide ion in subsequent management of process effluents and solid wastes, and (4) minimizes reagent costs by recycling HCl and iron raffinate to the leaching stage (Lakshmanan et al. 2018).

11.2.7 Nano Gold Catalyst

Recently, there has been considerable research into developing new catalysts that can selectively and efficiently convert CO_2 into fuels with steady operation for hundreds of hours. One of the most promising of these is based on gold. A recent discovery of gold nanostructured catalysts by researchers at the University of Toronto (Kauffmann et al. 2016) represents a significant advance in the field and has paved the way for selective and efficient CO_2 reduction to carbon monoxide (CO), a precursor to synthetic fuels. The nanostructured gold catalyst is shaped in a needle just a few microns thick. When supplied with electrical potential, the gold nano needle catalyst converts CO_2 into CO with high activity and selectivity. Once the carbon monoxide has been produced, it can be combined with hydrogen gas to make syngas, a precursor to many fuels such as gasoline. The US Government's National Energy Technology Laboratory (NETL) is also heavily involved in the search for advanced clean-tech catalysts. Researchers have developed a special engineered form of nanoparticle catalyst composed of 25 atoms of gold, which has shown considerable potential for use in the carbon dioxide reduction reaction. Indeed, the researchers believe that renewable energy sources, such as wind and solar, could power large-scale CO_2 conversion plants in the future, effectively providing a potential route to a carbon-negative energy cycle. While a number of catalysts are being considered, the current evidence suggests that gold is the best-in-class catalyst for the highly active and selective reduction of CO_2 to CO. A recent discovery of gold nanostructured catalysts by researchers at the University of Toronto represents a significant advance in the field and has paved the way for selective and efficient CO_2 reduction to carbon monoxide (CO), a precursor to synthetic fuels (World Gold Council 2018).

11.3 Novel Design and Engineering

The pressures on mine operators are varied and constantly increasing, declining ore grades, more remote locations and increasing environmental concerns. The challenge facing the mining industry is to find new ways to process ores which are more efficient and less costly and to control them better. This requires a fundamental rethink in the way in which the projects are developed; it is no longer good enough to just replicate what has been done before.

The future success stories will not simply be incremental improvements based on cost cutting and continuous improvement techniques. Most companies have already exhausted this route. The key to enhancing value is to do more with less, both in terms of capital and operating costs. Our focus is on processing as early as possible in the value chain and pushing the boundaries on cost, energy, water, and labor efficiencies through employing innovative processing techniques. The plant of the future will be smaller and more agile.

To make the improvements performance changes are required. First, through increased use of available and emerging technology. The second is through the identification of new design principles which address the key constraints to improvement.

The new design principles initiative seeks to facilitate step changes in performance for both greenfield and brownfield projects by addressing the key underlying constraints under which the project has to operate. It uses a divergent/convergent approach, Fig. 11.2, to identify and analyze the issues. It begins with divergent thinking in an open and collaborative workshop environment. The aim here is to answer the question what are the perceived constraints, strengths, and weaknesses. The points identified are then classified into four quadrants, strengths/weaknesses and can change/cannot change. Thinking then converges to identify what are the real problems we have to solve which are then distilled into a set of new design principles. These represent the key design principles necessary to enable the proposed project or improvement to be implemented. They may be such principles as autonomous underground ore handling, minimum tailings footprint, zero water discharge, maximum local community involvement, process underground (or as early as possible), and/or maximum use of renewable energy.

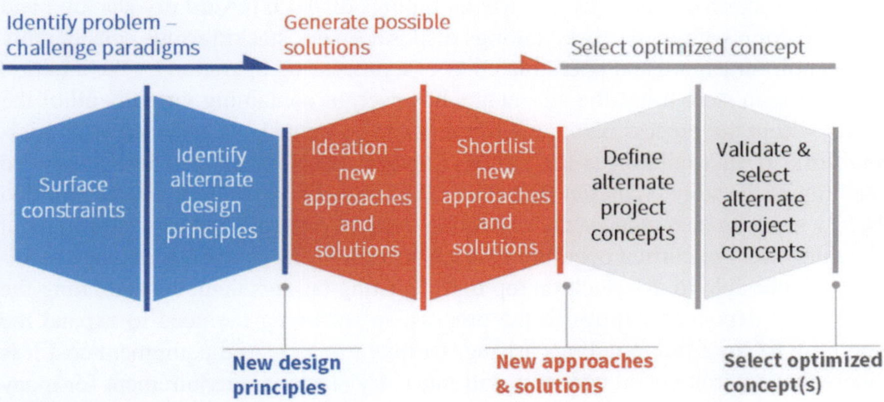

Fig. 11.2 New design principles process (Bunyard and Mullany 2017)

11.3.1 Integrated Underground System Design

A conceptual example of an integrated underground mining and surface processing system is an underground mine with ore sorting, tailings segregation, paste backfill, and tailings dewatering for water recovery and reuse. The result of such an integrated flowsheet would have multiple economic and environmental benefits, including:

- Underground mining generates minimal waste rock that would otherwise be dumped on surface.
- Underground preconcentration reduces costs for ore haulage to surface and reduces quantities of tailings produced.
- Separation and segregation of sulfide tailings reduces acid rock drainage (ARD) risks.
- Paste backfill significantly reduces size of surface tailings storage facility (TSF), reducing risk; in addition, paste backfill provides support in the mine, allowing more mining of remnant pillars, increasing mining reserves.
- Tailings dewatering and reclaim increases water recycling and reduces make-up water requirements.

11.3.2 Tailing Disposal

The long-established practice of disposing of the process tailings as slurry in an impoundment area is both an environmental risk, as several recent failure incidents attest, and a source of significant water entrainment and evaporative loss. Those operations that can use tailings as underground fill or have an exhausted open pit which can be filled are the fortunate few. The two pressures of environmental concern and water use efficiency are driving tailings disposal toward dry stacking and co-deposition with waste rock. Tailings disposal by dry stacking adds considerably to the upfront capital and operating costs of a processing operation by the addition of a filtration step. It has the advantage, however, of containing virtually all of the water within the process plant area where it can be treated and recycled with minimal loss. It also minimizes the tailings storage footprint and provides improved stability of the stored material (Fig. 11.3). The reduction in stored volume afforded by dry stacking also reduces the sustaining costs of embankment raises throughout the mine life. For current operations which have slurry impoundment storage facilities it is possible to dry stack on top of an existing tailings dam, consolidating the previously deposited tailings in the process, so removing the need to expand the footprint to store future tailings arising. Despite the increased equipment cost it is likely that environmental concerns will make dry stacking a requirement for many projects and mines, especially gold operation, in the future. The alternatives to reduce tailing contamination failure risk are explained in Table 11.1.

Fig. 11.3 Dry stacked tailings disposal La Coipa, Chile

11.3.3 In Situ Leaching

Many ore deposits are currently uneconomic due to low grades or being too deeply buried. In situ leaching (ISL) may allow economic extraction of these resources.

In-situ leaching (ISL) is receiving renewed attention as open pit mining operations face increasing pressure around decreasing ore grades, higher strip ratios, higher operating costs, safety, tailings disposal, water quality, and a range of environmental approval and community issues. This method of recovering metals directly from the underground ore body offers solutions to many of the objections to mining in the traditional manner. The surface footprint and visual pollution are significantly reduced with no headframes and smaller process buildings, noise, and dust are eliminated, heavy vehicle transport needs are reduced and energy consumption is minimized.

Electrokinetic in situ leaching (EK-ISL) is a novel in situ mining technology that uses an electric field to induce the migration of lixiviant through the subsurface to extract target commodities. The combination of electrokinetics and in situ leaching (EK-ISL) provides a unique opportunity to overcome the shortcomings of ISL in low permeability media (Martens et al. 2018).

11.3.4 Cyanide Alleviation

Factors driving new developments in gold technology include increasing environmental concern and government regulation over the use of cyanide, the trend toward refractory, complex, and lower grade resources, and the pressure to reduce operating

Table 11.1 Alternatives to reduce tailings containment failure risks

Category	Option	Opportunities	Obstacles
Generate less tons of tailings for surface disposal	Selective mining	Leads to higher mill head grade and lower processing costs	May reduce reserves and mine life
	Preconcentration	Rejecting a fraction of liberated gangue in mill feed will reduce milling costs. If mill is bottleneck, can provide relief and increase metal production	May have substantial metal losses. Some ores may not be amenable, and performance of preconcentration circuit may vary
	Flowsheets incorporating heap and dump leach	Can segregate lower grade ore and direct to heap leach operation. Lower grade ore matched with lower process OPEX	Ore may not be amenable to heap leach. May be difficult to practically segregate ore and optimize economics
	Backfill	Paste backfill can improve underground mine safety and enhance pillar recovery, increasing mining reserves and mine life	Cost: Some materials may not have suitable properties. Not all tailings can be backfilled due to expansion/swell factor.
	Beneficial reuse of fraction of tailings	May be additional source of revenue, and reduce disposal costs	Quality specifications for transport and customer delivery may be costly to overcome (e.g., impurities and size distribution)
Manage embankment design, dewatering, and deposition	Change TSF Siting (e.g., lower line of fire risks)	Improved stakeholder acceptance of project	Siting locations may be restricted by topography and land ownership position. Tailings and reclaim water transport costs may increase
	Improve drainage	Enhanced water recovery	May be limitations in effectiveness
	Increase tailings dewatering, that is, paste thicken or dry stack	Increased dewatering costs. Some materials are difficult to dewater to high solids content (e.g., clays and fines)	Significantly reduces volume of TSF, and increases water recycling
	Coarsen particle Size in tailings	Coarse tailings are cheaper and easier to dewater. Reduces comminution costs	May reduce metal recovery due to losses in coarse particle fractions

cost and increase plant performance efficiency. Alleviation of the issues around the use of cyanide, noncyanide lixiviant, processes for refractory ores and concentrates, improvements in analytical, control and monitoring systems, and upgrading of gold resources by sensor-based automatic ore sorting are reviewed in this section.

11.3.4.1 RECYN Process

Green Gold Engineering has been awarded a ReCYN™ Process design to detoxify tailings and recover cyanide and copper at its Martabe gold–silver operation in Sumatra, Indonesia (Gleeson 2019a, 2019b). Using an innovative resin-bead absorbent, known as the RECYN Process, allows for the economic recovery of cyanide from gold/silver plant tailings. RECYN Process typically reduces cyanide consumption by 50% by recovering free cyanide from the plant tailings and recycling it back into the leach circuit, while recovering dissolved base metals for sale as a by-product. The combination of recovery steps produces a fully decontaminated tailings stream from the process plant and overcomes any need for further detoxification of the tailings, also a net cost changing to a positive return. These advantages, both economic and environmental, can help change the negativity associated with cyanide.

The Mirah Gold Project in Central Kalimantan, Indonesia has, for over 2 years, successfully operated a commercial scale plant (1Mtpa ore) for the recovery and recycle of cyanide on a continuous basis. An average of 1 ton/day of NaCN has been recycled at a cost of 50% of new cyanide, with no further detoxification of the tailings required to meet environmental compliance levels (Paterson 2017).

11.3.4.2 Colorant in Cyanide

The use of a colorant in cyanide has been approved by the ICMI Board (International Cyanide Management Institute 2017). The purpose is to provide a means of visual identification of leaks or spills at mine sites or during transportation. It will require that dye be included in high-strength cyanide solutions prior to delivery at the mining operation and to solid cyanide prior to or at the time of mixing. High-strength cyanide solutions are defined as those with a minimum cyanide concentration of 15%, which would provide coverage to cyanide in both liquid and briquette form. ICIMI's decision was influenced by the use of dye addition by the Newmont Goldcorp Group, Canada, which originated at the Musselwhite Mill, Ontario, after the operations team heard about its use at the Placer Dome Mine, Ontario. After further development in cooperation with the dye supplier and cyanide supplier, Chemours, USA, the present practice of adding red dye at the manufacturing level to eliminate the need for a mixing system at site was adopted. The practice spread to other Goldcorp sites and is now mandatory company-wide (GoldCorp 2018).

The incorporation of dye to reagent-strength cyanide as a requirement of the Cyanide Code offers an opportunity to enhance protection of personnel and the environment. Use of dye by companies is believed to enhance product awareness, personnel safety, and provide a quick indication of cyanide leaks and spillages.

11.3.4.3 On-site Cyanide Generation

Synergen Met, Australia, is continuing to progress its onsite cyanide production process toward full-scale commercial application following the successful operation of a 60 tons/year pilot plant at an operating gold mine in Australia (Dunks 2018). It is targeting a modular design rated at 450–750 tons/year, with additional modules to be added for larger output. An average mine uses 500–1000 tons/year rising to over 10,000 tons/year at some large-scale operations (Leonida 2015). The process involves combining nitrogen directly from the air with methane from a hydrocarbon gas source using a plasma torch at 10,000 °C then quenching to form HCN. The HCN gas is mixed with NaOH to form NaCN, and the remaining hydrogen and nitrogen are vented to atmosphere. The NaCN solution is produced at the concentration and dosage required by the plant, and is pumped directly into a holding tank in a closed system without any handling by personnel (Synergen Met Pty Ltd 2019). Key benefits include elimination of risk of spills during transport, alleviation of environmental and community opposition, security of supply for remote sites, avoidance of worker exposure during handling and dilution, reduced cost in niche markets such as highly rugged terrain and remote mining sites, and reduced insurance cost and liability of toxic chemical storage.

11.3.4.4 Development of Alternative Lixiviant

Mixed Chloride

Process Research Ortech, Canada, has developed a Mixed Chloride lixiviant for gold leaching that has excellent stability and shown extensive applicability in the laboratory and pilot scale. The PRO's lixiviant is an alternative to cyanide and has particular application where cyanide cannot be used.

The lixiviant is containing HCl and $MgCl_2$ and acts at the atmospheric pressure leaching. In the leaching process, the leach slurry is subjected to a liquid/solid separation step to produce pregnant leach solution and subsequent solvent extraction steps to recover gold from the pregnant leach liquor. For example, a gold-bearing material containing 3.48 kg/ton of gold was subjected to mixed chloride leaching with a lixiviant of 4 N HCl, 225 g/L $MgCl_2$, at a temperature of 90 °C and pulp density of 10% solids to bring Au into solution. In addition to Au, Fe was also brought into solution. The leaching time and the oxidation reduction potential were 4 h and 1150 mV, respectively. The leaching mechanism of Au can be shown by the following reaction (Lakshmanan et al. 2018):

$$Au + 8HCl + \mathrm{NaClO_3} = \mathrm{2HAuCl_4} + \mathrm{NaCl} + \mathrm{3H_2O}$$

Thiosulfate

CSIRO, Australia, has developed a thiosulfate-based reagent system for gold leaching that has excellent stability and shown broad applicability in the laboratory compared to the thiosulfate system commercially implemented by Barrick to treat double refractory gold ore following pressure oxidation at its Goldstrike Mine in Nevada, USA. The CSIRO reagent system is an alternative to cyanide and has particular application where cyanide cannot be used and to unlock stranded high-grade deposits. CSIRO in collaboration with Eco Minerals Research Limited, Australia, commenced a project in July 2017 to undertake a demonstration at scale in the field using the CSIRO reagent system. The mobile demonstration plant setup on the Menzies Battery site in Western Australia uses a low capex vat leach process to recover gold from ores having good gold liberation at a p80 greater than 300 μm. In less than 10 months the demonstration project has taken a laboratory developed concept and transformed it into a demonstration plant involving design, build and commissioning through to successfully producing a gold ore bar. The demonstration plant processed up to 30 tons of ore per day by vat leaching and operated successfully for more than 6 months to validate the reagent performance and stability (Breuer and McCulloch 2019). The successful validation in the field at scale has paved the way for commercial application of the technology and a commercial launch is imminent.

Glycine

GlyCat™ leaching is being developed for implementation at Newcrest's Telfer Gold Mine in Western Australia to facilitate a change in circuit design that would then allow for increased concentrations of soluble copper to be tolerated in an expanded gold cyanidation circuit (Seaman et al. 2019). Reusable glycine is added to the leach to enable a fivefold reduction in cyanide usage while eliminating detox requirements. Copper is recovered by either sulfide precipitation or resin ion exchange. Gold is recovered by conventional carbon adsorption or alternatively using gold-selective resins. Over a 2–3-year period, an extensive program of batch testwork has defined the optimum leaching chemistry and proved the effectiveness of downstream processes. Three continuous piloting campaigns have shown that the process is robust and controllable, while verifying the reagent consumptions and gold recovery under steady-state conditions. Bench-scale testwork, process modeling, and engineering studies have narrowed down the circuit configurations to a preferred flowsheet involving single stage leaching and conventional downstream recovery. Implementation at Telfer will require extra leach tanks, solid–liquid separation equipment on the leach feed and discharge, and a copper recovery section. Glycine consumption is anticipated to be less than 3 kg/ton of concentrate while the resulting saving in cyanide is at least 30 kg/ton if the same concentrate was treated using cyanidation leaching alone. Glycine leaching, an environmentally benign leaching process carried out in alkaline conditions, was developed by Curtin

University in Perth. Mining and Process Solutions have acquired exclusive rights to progress the technology through to commercial application. GlyCat™ is the application of glycine with cyanide to mixed base metal and precious metal ores.

11.3.5 Novel Cyanide Analysis Methods

A few areas of improvement with regard to existing technologies include reaching lower detection limits, dealing with turbid waters, dealing with sample preservation, and developing cyanide sensors with simple technologies that can be easier to use, install, and maintain for remote use. Ideally, new technologies will also allow for the detection method to be portable and accurate for complex solutions with potentially different matrices. A key need in industry is the measurement of reliable cyanide concentrations in real time and in remote areas (Manoukian and Ahern 2017).

Quartz Crystal Mass Monitors (QCMMs) have a resonance that is sensitive to very small mass changes. In the case of cyanide, if gold is present in a crystal, and cyanide contacts this crystal, the metal dissolves, the mass decreases, and the resonant frequency increases (Ma and Dasgupta 2010).

A QCMM-based cyanide sensor with detection limits between 0.6 and 0.1 μM has been developed but has not been tested outside of a laboratory setting (Sun et al. 2005). Others have summarized various papers developing cyanide detecting biosensors. Most of the methods are dependent on measuring chemical reaction products based on enzymatic reactions (Breuer and Rumball 2007). The attractiveness of biosensors is associated to their advantages of being portable, easy to use, high selectivity, and cost effective. Despite the amount of research, there is still a lack in analytical biosensors that could be used confidently and independently in the field.

11.4 Concluding Remarks

It is becoming apparent that the future trend of gold and silver mining and processing will involve ores with decreasing head grade with a lower gold to sulfur ratios. The proportion of gold and silver production from highly refractory ore bodies involving various iron sulfides, carbonaceous matter, copper, arsenic, and mercury is expected to increase. Some of the key challenges with these deposits are suboptimal metallurgy, higher operating and capital costs, environmental issues associated with tailings and deleterious elements, increasing mining footprint, lack of resources including energy, quality water, skilled personnel, and ever-growing community issues. Major innovations with a holistic approach are essential to turn these marginal deposits into profitable mining operations in an environmentally friendly and sustainable manner. Some gold and silver mining companies are already looking at breakthroughs and step-change improvements with an integrated approach to problem solving (Bristow 2013; Dunne 2012; Kondos and Gorain 2012; Lakshmanan et al. 2012; Logan and Krishnan 2012). It is always best to address the root cause of a problem for enabling robust solutions to the present and future challenges of gold and silver mining and processing. Our industry has been very innovative, no doubt, with a strong focus on making our existing mining paradigm safer, more efficient, and automated. We have been remarkably successful so far but to address the unique challenges we now face as an industry, there appears to be no choice but to look beyond our existing mining paradigm.

The root cause of most of the problems in the existing mining paradigm is the generation of significant amount of waste and the need for handling, processing, and storing this waste with valuables representing only a very small fraction. The ideal scenario will be the case for zero waste mining, in which mining will target only the valuables without the need to remove the host rock and also all mined material will be used to create value added products with no waste dumps or tailings disposal. This is a daunting task, nevertheless, but if realized the benefits are significant. The underlying premise is that zero waste mining is economically very attractive, environmentally friendly, and fully integrated with the needs of local communities, societies, and other stakeholders. Hence pursuing "zero waste mining" is worthwhile as our ultimate long-term goal, but we must address this in small practical steps with a "horses for courses" approach.

Based on our experience and discussion with various players inside and outside the industry, some major step-change opportunities for gold and silver mining and processing being investigated or pursued by various players in the industry are presented as follows.

1. **Minimal removal of overburden to access ore body**:
   - **Borehole mining**: Use of small diameter drill holes to access the ore body along with use of novel biotechnologies to mine and recover metals (Dunbar 2014; McMullen et al. 2005). This is conceptual at this stage and will need further investigations.
   - **In situ gold leaching**: In situ recovery is presently practiced successfully in the copper, uranium, and potash industry. Companies like Rio Tinto are looking at opportunities with in situ leaching as the "mine of the future" (Batterham 2008). In situ gold leaching has been experimented before by BHP in 1989 and different lixiviants such as chlorides, thiosulfate, humic acid, and biogenic reagents have been proposed previously (McMullen et al. 2005; Zammit et al. 2013). New technologies like Discrete Fracture Networking are evolving to make this practical in an environmentally friendly manner for most ore types (Dershowitz 2011).
2. **Selective liberation of ores early in the mining process**: It is becoming important to ask ourselves "Why create waste in the first place." Comminution begins with mining and a new generation of drilling and blasting for selective mining of ores is critical to avoid or reduce waste removal early on in the mining process. This concept is referred to as "grade engineering" and is becoming a major focus of further development (CRC-ORE 2014).

3. **Minimal haulage of waste**:

   - **Preconcentration**: This allows processing to be closer to the mine site. There is a growing interest in preconcentration technologies such as sensor-based mass sorting (ROM shovels/trucks), classification using screens, stream based ore-sorting, gravity, and dense media separation, which is definitely a positive trend and important implications for the gold industry. Studies have suggested that integrated mining and waste rejection processes have high potential for deep underground mining (Bamber 2008; Batterham 2003; Dammers et al. 2013).

4. **Efficient comminution**: Technologies such as *Chemical Comminution* applied on a smaller mass after preconcentration are also being investigated (Muir 2014). Recent innovations in high intensity selective blasting to reduce footprint of comminution circuits are being tested and pursued by some operations. Use of technologies such as SelFrag to pretreat low grade ores (promoting fractures and high rock permeability) to minimize comminution energy but this technology also has important implications in maximize heap leach recoveries.
5. **Refractory ore processing**:

   - Microbial gold processing: Focus on new generation of low cost bio-heap leaching, bio-oxidation, and cyanide destruction processes using native microbial population along with phylogenetic fingerprinting is gaining momentum and has major potential (Brierley 2017; Zammit et al. 2013; Follink 2010).
   - Breakthrough flotation process: With the advent of new flotation technologies to treat ultrafine particles, and with better understanding of chemistry through state-of-the-art surface analysis capabilities, the possibility of a flotation breakthrough is likely. This has potential for a major breakthrough for processing Carlin type double refractory deposits.
   - Cost-effective sulfide oxidation: Other than innovative bio-oxidation processes, there are significant opportunities to integrate novel ultrafine grinding with new generation of pressure oxidation technologies involving lower capital and operating costs to process fine grained refractory ores that are uneconomical using existing technologies.
   - Alternatives to cyanide leaching: New generation of alternative technologies are needed such as Process Research Ortech's novel chloride leach technology for treating complex ores (Au, Cu, Ag, TCM, and As) with an ability to maximize recovery of gold along with valuable byproducts (Lakshmanan et al. 2012).

It is important to note that parallel development of many of these technologies is already happening. Also, not all these innovations will be relevant to every ore body as technologies must be tailored to suit the individual needs of an ore body. A systematic approach with a multidisciplinary collaboration involving various stakeholders is a must to realize the full benefits from a breakthrough technology.

References

Bamber, A.S. 2008. *Integrated mining, pre-concentration and waste disposal systems for the increased sustainability of hard rock metal mining*. PhD thesis, University of British Columbia.

Barlow, L. 2018. *Transformational innovation rapid oxidative leaching*. Sydney: Future of Mining.

Batterham, R.B. 2003. The mine of the future, will it be visible? In *Proceedings of the XXII international mineral processing congress,* Cape Town, South Africa.

Batterham, R.B. 2008. Current practices shaping the science and practice of mineral processing . *In Proceedings of the XXV international mineral processing congress,* Beijing, China.

Breuer, P., and J. McCulloch. 2019. Demonstration at scale in the field of a non-toxic reagent system to recover gold. In *ALTA 2019 gold-PM,* Perth.

Breuer, P.L., and J.A. Rumball. 2007. Cyanide measurement and control for complex ores and concentrates. In *AusIMM ninth mill operators' conference*, Fremantle, Australia.

Brierley, J.A. 2017. Biohydrometallurgy innovations—discovery and advances. In *Mineral processing and extractive metallurgy—100 years of innovation*, 2014. SME.

Bristow, M. 2013. Building a successful African gold mining business—managing risk and delivering long term returns. In *World Gold 2013,* Brisbane, Australia.

Bunyard, M.J., and I. Mullany. 2017. The future shape of mineral processing. In *The conference of metallurgists hosting world gold & nickel cobalt, COM 2017,* Vancouver.

van Buuren, C., et al. 2018. Application of the HITECC technology at Nordgold's suzdal operation to combat preg robbing and enhance gold recovery. In *ALTA 2018 Gold-PM,* Perth.

CIM. 2019. Centre for excellence in CSR. https://www.cim.org/en/CIMSubSites/CentreForExcellence/About-Us.

CRC-ORE. 2014. Transforming mining productivity with grade engineering. www.crcore.org.au.

Dammers, M., et al. 2013. In *6th international conference on sustainable development in the minerals industry,* Milos Island, Greece.

Dershowitz, B. 2011. Discrete fracturing network modeling in support of in-situ leach mining. *Mining Magazine*.

Dunbar, W.S. 2014. Will the mine of the future be a mine at all? In *Globe and Mail*.

Dunks, C. 2018. Technology for mine-site cyanide production. In *ALTA 2018 Gold-PM,* Perth.

Dunne, R. 2012. Refractory gold treatment—state of the art and future challenges, In *XXVI international mineral processing congress*, New Delhi, India.

E&MJ. 2013. CSR, sustainability and the mining business. https://www.e-mj.com/features/csr-sustainability-and-the-mining-business/

Follink, B. 2010. Bio-mineral processing—its potential to revolutionize the industry. In *XXV international mineral processing congress*, AusIMM. Brisbane, Australia.

Gleeson, D. 2019a. FLSmidth to move into ROL copper demonstration mode in 2019. International mining website. https://im-mining.com/2019/02/01/flsmidth-move-rol-demonstration-mode-2019/

Gleeson, D. 2019b. GreenGold's ReCYN to detoxify tailings recover cyanide and copper at Matabe International Mining website. https://im-mining.com/?s=recyn

Global Affairs Canada. 2018. About us. https://www.international.gc.ca/csr_counsellor-conseiller_rse/About-us-A-propos-du-bureau.aspx?lang=eng.

GoldCorp 2018. Seeing red: the story of cyanide dye at goldcorp [Online]. https://blog.goldcorp.com/2018/10/22/seeingred-the-story-of-cyanide-dye-at-goldcorp/.

IDB Invest. 2015. https://blogs.iadb.org/bidinvest/en/corporate-social-innovation-sustainable-business/

International Cyanide Management Institute. 2017. Use of colorant in cyanide approved by ICMI board, The Newsletter of the International Cyanide Management Institute.

Kauffmann, D.R., et al. 2016. Efficient electrochemical CO_2 conversion powered by renewable energy. *ACS Applied Materials & Interfaces* 7 (47): 26101–26107.

Kondos, P., and B.K. Gorain. 2012. Development of breakthrough technologies for complex gold and copper-gold ores. In *XXVI international mineral processing congress,* New Delhi, India.

Lakshmanan, V.I., R. Sridhar, R. Roy, and B.K. Gorain. 2012. Recovery of gold and titanium dioxide. In *Proceedings of the XXVI international mineral processing congress,* New Delhi, India.
Lakshmanan, V.I., et al. 2013a. Innovative processes for the recovery of gold and base metals. In *ALTA 2013, 4th annual nickel-cobalt-copper conference*, Perth.
Lakshmanan, V.I., R. Sridhar, and M.A. Halim. 2013b. Chloride process for the leaching of gold. US Patent 9,732,398 B2 [Patent], USA.
Lakshmanan, V.I., et al. 2018. Separation and purification of value metals from aqueous chloride solutions by solvent extraction. In *COM2018,* Columbus, BC.
Leonida, C. 2015. Gold medal. *Mining Magazine*.
Logan, A., and N. Krishnan. 2012. In *Newcrest technology—step change, XXVI international mineral processing congress*, New Delhi, India.
Ma, J., and P.K. Dasgupta. 2010. Recent developments in cyanide detection: a review. *Analytical Chemica Acta* 673 (2): 117–125.
Manoukian, L., and N. Ahern. 2017. Cyanide analysis: a review. In *The conference of metallurgists hosting world gold & nickel cobalt, COM2017*, Vancouver.
Martens Evelien, et al. 2018. Testing the Feasibility of Electrokinetic In Situ Leaching. In *ALTA 2018 ISR symposium proceedings,* Perth, Australia.
McMillan. 2014. Canada launches enhanced CSR strategy for extractive sector companies operating abroad. https://mcmillan.ca/Canada-Launches-Enhanced-CSR-Strategy-for-Extractive-Sector-Companies-Operating-Abroad.
McMullen, J., J.R. Goode, and P.D. Kondos. 2005. The recovery of gold from low grade and refractory ores what's next? In *Treatment of gold ores*, eds. G. Deschenes, D. Hodouin and L. Lorenzen. Com2005, Calgary.
mining.com. 2015. Why CSR makes sense for communities, companies and investors. http://www.mining.com/web/841426/.
Muir, A. 2014. Evolution in comminution, Keynote address in Comminution 2014. Cape Town, South Africa, organized by MEI.
Natural Resources Canada. 2011. Building the Canadian advantage: a CSR strategy for the international extractive sector. https://www.nrcan.gc.ca/sites/www.nrcan.gc.ca/files/mineralsmetals/pdf/mms-smm/pubr-pubr/pdf/bca-rac-eng.pdf.
Paterson, Malcolm Roy. 2017. The Recyn Process Changing The Negative Perception of Cyanide. In *ALTA 2017 Gold PM,* Perth, Australia.
Pearson, M. 2019. No fools gold here. Hatch Case Study. https://www.nobelclad.com/resources/NobelClad-Resources/No-Fools-Gold-Here.
Roy, G., et al. 2018. In *FL Smidth's hydrometallurgical pretreatment process for refractory gold ores/concentrates, extraction 2018,* Ottawa.
Salt, W. 2019. NobelClad website.
Seaman, B., et al. 2019. Development of a glycine-cyanide leach process for gold-copper concentrate. In *ALTA 2019 GoldPM,* Perth.
Sun, H., et al. 2005. Piezoelectric quartz crystal (PQC) with photochemically deposited nanosized Ag particles for determining cyanide at trace levels in water. *Sensors and Actuators B: Chemical* 108: 925–932.
Synergen Met Pty Ltd. 2019. Synergen Met Website. https://synergenmet.squarespace.com/news-media/2015/2/23/animation-video-of-the-worlds-first-modular-cyanide-plan.
Van Niekerk, J., et al. 2018. Continued development of the BIOX process for the treatment of refractory gold ores. In *2018 Gold-PM,* Perth.
World gold council. 2018. Gold and climate change. In *World gold council*.
Zammit, C.M., N. Cook, C.L. Ciobanu, J. Brugger and F. Reith. 2013. Possible future of applications of biotechnologies in the gold industry—a prospective approach. In *Proceedings of world gold*, Brisbane.

Chapter 12
Digital Gold and Cryptocurrency

V. Kumar Murty

12.1 Introduction

Early economic systems based on barter required coordination both to connect demand and supply, and to establish value. Together, these features created an opportunity for mutually beneficial exchange. This coordination could be facilitated in an ad hoc manner for simple economic systems. However, as the complexity of the system increased, both in terms of the number of transactions and in terms of the diversity of goods and services that were to be traded, so did the coordination requirements.

The purpose of economic markets is to facilitate this coordination, and in order to do this various instruments have evolved over time. Currency is one such instrument.

For several millennia, an almost universally accepted currency has been gold. It is a commodity that is physically mined and can be used as an asset by itself, but also to fashion articles such as jewelry. Gold has given way to many other currencies but in many cases, one speaks of "the gold standard" in establishing the value of these currencies. In general, currency represents an abstraction of value. And as there are many ways in which to abstract value, so are there many currencies. A further abstraction of currency is debt (as representing a negative balance of currency).

Simultaneous with the development of instruments which represent an abstraction of value, there has also been a development of the technology to evaluate the reliability of these instruments. The whole concept of risk, in its multifarious manifestations, is one such evaluative framework.

Moreover, there has been a development of a legislative and legal framework to establish accepted norms and resolve disputes. In particular, there is a process to assert and establish ownership or control of value.

V. K. Murty (✉)
University of Toronto, Toronto, ON, Canada
e-mail: murty@math.toronto.edu

V. I. Lakshmanan, B. Gorain (eds.), *Innovations and Breakthroughs in the Gold and Silver Industries*, https://doi.org/10.1007/978-3-030-32549-7_12

If we consider the economic history of the world, we will find this basic alignment of an abstraction of value, together with institutions to facilitate transactions based on this abstraction. In the information-based age that we live in today, new methods are being proposed of abstracting value, of creating instruments for the transaction of value (such as currency and debt), of evaluating the risk and reliability of these instruments, and of erecting the legal and legislative frameworks for dispute resolution. Moreover, the development of extensive information networks means that transactions can take place over these networks, making possible the explosion of global economic activity that we are witnessing today. The new concept of the abstraction of value is the world of cryptocurrency, and the aim of this chapter is to give an overview of this brave new world. It is a world that is still evolving and there are still problems to be solved.

Though the subject is young the literature is already quite vast and expanding quickly. An excellent exposition of Bitcoin can be found in the work of Narayanan et al. in which the reader will find many useful references (Narayanan et al. 2016). I would like to thank W. George, V. Lakshmanan, and R. Murty for helpful comments on an earlier version of this chapter.

12.2 New Issues

12.2.1 Migration to Networks

When migrating from traditional face-to-face transactions to network-based transactions, several issues appear in the digital information context. First, we need a digital method to establish identity of the parties involved in the transaction. Second, we need to know that the entities entering into the transaction are entitled to make the transaction. Third, we have to be concerned about the security of information that is communicated to enable the transaction to happen since some confidential information may have to be shared. Fourth, we need to know that personally identifiable information is revealed only through a well-configured policy engine which includes such requirements as "the need to know." Fifth, we need to have trust in the entire process. We briefly discuss how each of these aspects is cryptographically addressed in current networks.

12.2.2 Integrity

It is important to have assurance that the information that is sent is the same as the information that is received. This is the problem of data integrity that information has not been corrupted in transmission, either because of network defects (such as electromagnetic disturbances) or because of deliberate tampering.

In practice, methods of ensuring data integrity are based on the so-called hash functions. These functions extract a "fingerprint" from data and transmit the fingerprint along with the data. The recipient uses the same function to recompute the fingerprint based on the received data and compares it with the fingerprint sent. If they agree, it is assumed that data integrity has been maintained.

12.2.3 Identity

Philip Windley, author of Digital Identity states that "…managing digital identity is one of the most fundamental activities in IT and … a good identity management strategy is the key to not only protecting the enterprise from attack, but more importantly, providing flexible access for partners, customers, and employees to needed information and systems." (Windley 2005).

Many identity systems currently in use operate on the basis of digital persona known as certificates. The certificate establishes identity, as well as some other information, all of which has to be made tamper resistant. This is done by invoking several fundamental tools such as public key cryptography and data integrity algorithms. Certificates are issued by a Certificate Authority (CA) which verifies an entity's identity with the help of physical credentials. When entering into a transaction, the entities have to have their certificates verified and the mechanism by which this is done is called Public Key Infrastructure (PKI). Fundamental to this method of establishing identity is trust in the CA, which is an independent third party not involved in the transaction beyond establishing identity.

PKI continues to be a valuable tool to enable the verification of identity. However, it suffers from some drawbacks which are increasing in prominence with progress in technology. The main issues are trust in the CA, latency in execution at the time of the transaction, and a not entirely reliable revocation system. In various contexts, these issues are serious enough as to make PKI unusable. Various variants of PKI have evolved which are "lightweight" but the fundamental principle of requiring certificates to establish identity has, perhaps, to be rethought.

12.2.4 Access Control

Once identity has been established, we need to be sure that the parties to the transaction are entitled to make the transaction. For example, if payment is involved, does the entity requesting payment have the funds for the transaction? Does the entity offering the goods or services have ownership or the authority to represent the parties that have ownership?

12.2.5 Security

When information is communicated to enable a transaction, some of it may be confidential such as bank account numbers or credit card numbers. The network has to ensure that only the parties to the transaction can access this information and that an "eavesdropper" to the communication cannot. This is done through cryptography whose main function is to ensure confidentiality. The word "cryptography" literally means secret (or hidden) writing. Cryptographic algorithms turn plain text into cipher text. In order for such algorithms to work, they require as input a "key." How this key is managed is fundamental to how secure a system is. For example, we have to know whether the key is stored somewhere and if so, who has access to it? We also have to know what happens when a key is compromised.

12.2.6 Privacy

While security protects the confidentiality of information to prevent eavesdropping by unauthorized entities, privacy protects the right of an entity to control the way in which its information is used. Thus, we see that a vendor may, in the course of a transaction, ask for information that goes beyond what is needed to enable the transaction. Moreover, the system architecture may require the customer to provide that information in order to complete the transaction. This could be seen as a violation of privacy. Protocols that protect privacy and Privacy Enabling Technologies (PET) are relatively younger than other cryptographic primitives and are still evolving.

12.2.7 Trust

In a transaction, there has to be the feeling of assurance that our information is being handled in a way that is consistent with our interests. Certain aspects may require trust in a third party.

12.3 Early Attempts: Payment Gateway

Before diving into the world of cryptocurrency, it might be worthwhile to observe that at least for several decades, a large segment of our payment transactions has been managed by complex electronic systems which for the most part work in the background. These systems involve the financial and regulatory institutions that control the transfer and recording of value in its abstract form. The interaction between consumer and supplier is through a gateway which allows for the separation

of the transfer of value from the actual transfer of goods and services. Transfer of value is affected through a payment gateway.
For the most part, the architecture of such systems is based on the involvement of a Trusted Third Party (TTP). Thus, for example, there are payment systems such as PayPal, which are intermediary parties to the transaction. Perhaps the most significant difference in the cryptocurrencies to be discussed later is the attempt to eliminate the TTP.

12.4 Some Cryptographic Primitives

12.4.1 Hash Functions

A hash function is a method of producing a string of fixed length (say n) from a message of arbitrary length,

$$H : \{0,1\}^* \to \{0,1\}^n$$

We should think of it as a "fingerprint" or "message digest".

In order to really serve the role of a "fingerprint," the function H should have a property called 'collision resistance'. This means that different input strings should produce different output values. Mathematically, this is impossible to achieve as the function H maps a larger space to a smaller space. However, the requirement is a computational one in one or both of the following senses:

1. It is computationally infeasible to produce two distinct messages M and M′ with the same fingerprint: $H(\mathrm{M}) = H(\mathrm{M}')$.
2. Given a message M, it is computationally infeasible to produce a message M′ with $H(\mathrm{M}') = H(\mathrm{M})$

For a generic function that has output of n bits, it should take about O ($2^{n/2}$) steps to find a collision in the first sense (i.e., to find two messages with the same hash value). Finding a collision in the second sense should take about O (2^n) steps. Thus, if $n = 160$, it should take about 2^{80} steps to find a collision. Taking into account Moore's law (about the doubling of computing capacity every 6 months), a rough estimate of the number of operations one can perform on a single machine in a year is only 2^{64}. These numbers will be significantly reduced if quantum computation becomes a reality.

Most hash functions are based on a design strategy laid out by Damgaard and Merkle. A hash function typically has two components: compression and extender. The compression function specifies positive integers n and b and given any binary string of length b produces a binary string of length n. For example, the well-known hash function SHA-1 (which for many years was a standard prescribed by the United States government) uses a compression function that takes $b = 512$ bit blocks and produces output of $n = 160$ bits. SHA is an acronym for Secure Hash Algorithm and the SHA family has three variants of SHA-1, namely SHA-2 (or SHA-256), SHA-384, and SHA-512. The algorithm SHA-2 gives an output of $n = 256$ bits. The algorithms SHA-384 and SHA-512 take as block size $b = 1024$ and produce outputs of 384 and 512 bits, respectively. The extender gives a method of preparing a binary string of arbitrary length as iterative input for the compression function. The current standard is SHA-3.

There is another kind of hash function that can work with a stream of data rather than taking the block-based approach of Damgaard and Merkle. A family of such hash functions was developed, for example, in the work of the author and Volkovs (Murty and Volkovs 2009). These functions have some very attractive properties, including their very high speed (a simple implementation running as fast as 2Gb/s), very small footprint (of the order of a few kB), and large internal state space (of the order 2^{50} in a simple implementation) that allows them to be customizable and useful in environments such as Internet of Things (IoT) where there are a plethora of devices. The role of such algorithms in the cryptocurrency world is a subject of ongoing research.

12.4.2 Encryption

The functionality of the hash function is to ensure integrity. It is not secret in the sense that the hash function and the hash value are public. A hash value is somewhat similar to a bar code on a product. It gives assurance of integrity and not confidentiality. If we also want confidentiality, that requires a separate function of encryption.

Encryption is a function that sends a space of plain text to a space of cipher text. The inverse function decryption sends the space of cipher text back to plain text and performing encryption followed by decryption should preserve the message intact. In general, though not always, the encryption and decryption algorithm is public. The algorithms require the input of a key and that key is to be preserved as secret. If the key is compromised, the confidentiality is compromised.

12.4.3 Key Pairs

A fundamental technical tool, and one that enables most modern applications of cryptography is the concept of a key pair. In classical cryptography (as described in the previous paragraph), the encryption key and the decryption key are the same. The concept of public key cryptography is based on the idea that rather than having a single key, one has a pair of keys which are related mathematically, though it is difficult (i.e., computationally infeasible) to compute one from the other. One element of this pair K_{public} can be made public (and so it is called the public key), and the other element

$K_{private}$ is kept secret (and so it is called the private key). One can encrypt through one key and decrypt through the other (either directly or after some manipulation).To repeat, the two fundamental properties of the key pair are that one is the inverse of the other (in the sense that encryption performed with one key can be decrypted with the other key) and that we can compute one key from the other, though to do so in a practical way requires additional information. The mathematics that makes possible the generation of such key pairs is based on some elementary group theory.

Let G be a finite cyclic group, having (say) ω elements. Thus, there is an element $g \in G$ with $g^{\omega} = 1$ so that every element is a power of g:

$$G = \left\{1 = g^0, g = g^1, g^2, \ldots, g^{\omega-1}\right\}$$

We assume that in this group, if we are given an element g^t, it is difficult to compute t. Computing t from g^t is the so-called Discrete Logarithm Problem (DLP) and there are explicit instances of G in which we have no efficient algorithm for solving the DLP. Such groups G can be used to generate key pairs. Indeed, a key pair can consist of an integer a (with $1 \leq a < \omega$) and the element g^a. We can publish g^a (as well as the group G) and we have to keep a secret.

Examples of the kinds of groups that have been used for key pairs are the group $\mathbb{F}_p^{\times}$ of non-zero elements of a finite field of p elements (where p is a large odd prime), or $\mathbb{F}_{2^r}^{\times}$ (for certain values of r) of non-zero elements of a finite field of 2^r elements. Other groups are the set of rational points of an elliptic curve E which is defined over a finite field ($\mathbb{F}_p$ or $\mathbb{F}_{2^r}$). There are also algorithms involving higher dimensional versions of elliptic curves, namely Abelian varieties. The reader is referred to (Murty and Sastry 2018) for some references to the literature as well as new research toward making algorithmically possible the use of general Abelian varieties. This line of research seems to be very fruitful and there remain many problems to be solved before a practical algorithm is evolved using general Abelian varieties.

The most common use of key pairs is for key agreement to enable symmetric encryption. Another use is for signing and verifying documents. We shall describe this latter use next.

12.4.4 Digital Signature

For physical documents, we affix our signature as evidence of a completed transaction, or for asserting the validity or integrity of the information involved. The digital analogue of this is called a digital signature and it should have the property that it is tied to the document (so a forger cannot extract a signature from one document and attach it to another) and it should be verifiable (in other words, the recipient should be able to verify that this is your signature and not someone else's).

To enable digital signatures, we need a key pair (K_{public}, $K_{private}$). The public keys have to be published or available to a third party which wants to verify signatures.

We then encrypt the document using $K_{private}$ and send it to the verifier. The verifier looks up our public key K_{public} and uses it to decrypt the document. Since we are the only entity to have access to $K_{private}$, this verifies that we had sent the document. For efficiency considerations, we do not sign the entire document, but the message digest (the hash value of the document). Implicit in this kind of scheme is that identity can be linked to our public key K_{public}.

12.4.5 Blockchains

A block is a piece of data. Together with this block B, we compute and store its hash value $H(B)$. This value can be stored in another location, and when we retrieve the contents of B, if we want to ensure that its integrity has been maintained (in other words, if we want to know that it has not been altered, knowingly or unknowingly), we can compute its hash value $H(B)$ and compare with the stored value. If the two values match, we accept the integrity or reliability of B.

A blockchain is formed when we have a number of blocks $B_1, B_2, \ldots, B_t$ (say). If in B_2 we incorporate the value $H(B_1)$, then we can use B_2 to verify the integrity of B_1. More generally, if each B_i consists of data D_i and the hash value $H(B_{i-1})$ $(i \geq 2)$, then each successive block can be used to verify the integrity of the preceding blocks. In order to verify the integrity of the last block in the chain, namely B_t, we also need to store $H(B_t)$.

The blockchain is the underlying platform technology on which are built a variety of applications. The most popular of these is cryptocurrency, and perhaps the most famous cryptocurrency is Bitcoin. The blockchain is used to maintain a tamper-resistant ledger of transactions, where the tamper-resistant property is derived from the ability of the integrity algorithm (hash function) to avoid collisions and thus detect if any change has been made. Value is abstracted as cryptocurrency and is related to the maintenance of the blockchain. The security of the currency depends on the management of private keys and this is where wallet technology comes in. It runs at each node but off-chain to manage the currency and keys of the node.

12.5 Decentralization and Consensus

12.5.1 Decentralization

The concept of a decentralized architecture which either eliminates or minimizes the role of a trusted intermediary is an important one in the innovation of cryptocurrency. Narayanan et al. (2016) give the example of email using the Simple Mail Transfer Protocol (SMTP). Anyone can establish an email service using the open SMTP standard and one does not need a large central administrator. However, the market has created dominant email providers who have more influence on the entire market than smaller players. Moreover, there is an ecosystem that supports the use of email in which there are dominant players. Similarly in cryptocurrencies, there are exchanges (that facilitate the conversion of one coin to another), wallet technology providers that allow users to manage their cryptocurrency, and so on. Moreover, one aspect of cryptocurrency, namely their mining or the creation of new coins, is capital intensive and so introduces a bias in terms of accessibility.

12.5.2 Consensus

A problem that has to be solved in a decentralized context is how the collection of users can agree on the state of affairs. This is called the problem of consensus. If there is a central authority, the state of affairs is what that authority declares it to be. In the absence of such an authority, there has to be some mechanism to bring about consensus.

This is not a new problem as it already occurs in distributed systems. If multiple users are editing a common file, such as a database, how do we maintain coherence? The same problem arises when a service is operated using mirror sites. In the cryptocurrency context, we need consensus on which transactions have occurred and in which order. Moreover, in the cryptocurrency context (or more generally, the blockchain context), any consensus protocol has to take into account rogue sites or dishonest sites that might broadcast false information. Since cryptocurrencies such as Bitcoin have a loose definition of identity, consensus protocols cannot largely depend on the concept of identity.

In the world of cryptocurrency, we need three forms of consensus: the first is about the rules under which the system will operate, the second is about which transactions have occurred and the balances of each node, and the third is about the inherent value of the currency.

In very simplified terms, the first kind of consensus about the rules is established by imitating existing mechanisms for the development and preservation of open-source software. In the case of Bitcoin, the rule consensus protocol is the Bitcoin Improvement Proposal framework. Ultimately, there is a core group of real people which decides on the rules, though the community has the freedom to decide whether to agree with them, or to create a "fork" which is a new blockchain with a new "genesis block." In practice, changes to the protocol require a social consensus which is complex because of the diverse parties that make up the community. Thus, there has to be agreement between those who develop the basic client and the nodes (which consists of the users of the client, merchants who accept the cryptocurrency, wallets that manage the currency and exchanges). This is why changes to the protocol are so difficult, and this difficulty gives a certain stability to the system.

For the second and third kind of consensus we need to understand what we mean by a transaction in this context.

12.6 Transactions

Typically, the basic transactions consist of instructions to pay a certain amount of currency to a particular user. A user possesses currency that they can spend if they have previously received transactions that they have not yet spent. One needs a way to refer to a user. In Bitcoin, this is done by using the public key of the user or the hash of the public key (the "Bitcoin address"). The instruction has also to be signed by the private key of the user or node initiating the transaction. This total string: the instruction, together with the signature, is the transaction and also proof of payment.

To reach consensus about which transactions have occurred, a typical protocol involves the following steps:

1. At each unit of time t, each transaction is broadcast to all nodes; each node will verify the validity of the transaction by verifying the signature and that there are sufficient funds to enable the transaction.
2. Each node collects the transactions broadcast at time t into a block. These are represented using a data structure known as a Merkle tree that allows one to verify each transaction in a block.
3. The node adds to its own account a certain amount of currency as a reward. This is the last transaction of the block. All new currency is created as such rewards.
4. The header for the block contains a field for a nonce which is a digital string. The node searches for a value of the nonce so that the hash of the entire block has a sufficiently large number of zeros. More generally, by some mechanism (understood at the outset), a random node is selected to broadcast its block to all other nodes.
5. Nodes accept this block as valid if the transactions that are recorded in it are valid.
6. Once accepted as valid, each node computes the hash of the block and uses it in the next block.

12.7 Creation of Currency

In the above description of a transaction, we stated that the currency itself is a digital string that was created through the processing of transactions. If we want to associate real value to the currency—and this is necessary for it to be usable for actual goods and services—then we will need to have more structure in the creation of currency. This is done through the concept of mining. The process of finding a nonce that has a hash value with a prescribed number of zeros is the mining process. It tends to be very computationally intensive. Note that the miners are the nodes on the network that are engaged in the processes of verifying transactions and maintaining the blockchain, in particular of forming blocks by computing nonces appropriately.

This brings us to the third aspect of consensus, namely the value of the currency. In the case of traditional currencies, value is determined by fiat, that is by the decision of some central authority. In the case of cryptocurrency, the value is partially determined by miners in terms of the reward for mining a new block as opposed to the costs of doing so. These costs, which involve computing power as well as electricity to run the servers, are denominated in some fiat currency such as dollars. Thus, there has to be a consensus that the reward for mining a block offsets the real costs of doing so.

This is only one component in the way a currency is valued. The value of a cryptocurrency is also determined in the economic marketplace based on economic fundamentals such as supply and demand and utility. For example, the value of gold is determined by its availability (or lack thereof), and by its usefulness, both by itself and as a hedge against uncertainty in other currencies. Similarly, the value of a currency such as Bitcoin is determined by these two factors. There is a cap on the amount of Bitcoin that can be created (21 million) and this is a well-known limit on availability. On the other hand, its usefulness is that it enables global, decentralized transactions, thus eliminating costs arising from intermediate infrastructure. When these aspects are quantified in the marketplace, Bitcoin is assigned a value in terms of fiat currencies or in terms of gold. We have mostly mentioned the cryptocurrency Bitcoin, but there are now a plethora of so-called altcoins, or alternative cryptocurrencies. One particularly interesting currency is ethereum. The distinctive feature of ethereum is that it incorporates smart contracts which execute payment under certain circumstances. More generally, a smart contract is a script that executes certain functionality under certain conditions. This feature enables new uses of the blockchain, including possibly Digital Rights Management.

12.8 Identity, Security, Privacy, and Trust

Earlier in this chapter, we indicated various important aspects of traditional transactions that network-based transactions need to adopt and incorporate. While cryptocurrencies have addressed issues of security, privacy, and trust, the question of identity seems to be more elusive. Some in the community will claim that this is by design. We have already stated that Bitcoin, for example, does not depend on identity beyond a public key. On the other hand, we know (Meiklejohn et al. 2013) that being able to read the activities of a public key along a series of transactions may indirectly reveal the actual identity of the individual or individuals to whom the public key belongs. More research is needed to understand how one might design a blockchain-based identity system that more completely protects privacy. Zcash and Monero are two currencies that have made some progress in this direction.

References

Murty, V. Kumar, and P. Sastry. 2018. Explicit arithmetic on Abelian varieties. In *Geometry, algebra, Number Theory and their information technology applications*, eds. A. Akbary and S. Gun, 317–374. New York: Springer.

Murty, V. Kumar, and N. Volkovs. 2009. ERINDALE: A polynomial based hashing algorithm. In *International workshop on coding and cryptography*, eds. C. Xing et.al., 157–170, Lecture Notes in Computer Science 5557, Berlin: Springer.

Meiklejohn, S., M. Pomarole, G. Jordan, K. Levchenko, D. McCoy, G. Voelker, and S. Savage. 2013. A fistful of bitcoins: characterizing payments among men with no names. In *IMC 2013*. http://dx.doi.org/10.1145/2504730.2504747.

Nakamoto, N. 2008. Bitcoin: a peer-to-peer electronic cash system, https://bitcoin.org/en/bitcoin-paper.

Narayanan, A., J. Bonneau, E. Felten, A. Miller, and S. Goldfeder. 2016. *Bitcoin and Cryptocurrency technologies*. Princeton: Princeton University Press.

Windley, P. 2005. *Digital Identity*. Sebostopol: O'Reilly press.

Index

V. I. Lakshmanan, B. Gorain (eds.), *Innovations and Breakthroughs in the Gold and Silver Industries*, https://doi.org/10.1007/978-3-030-32549-7

I

K

L

M

MIX
Papier aus verantwortungsvollen Quellen
Paper from responsible sources
FSC® C105338

If you have any concerns about our products,
you can contact us on
ProductSafety@springernature.com

In case Publisher is established outside the EU,
the EU authorized representative is:
Springer Nature Customer Service Center GmbH
Europaplatz 3, 69115 Heidelberg, Germany

Printed by Libri Plureos GmbH
in Hamburg, Germany